INGENIOUS MECHANISMS
FOR DESIGNERS AND INVENTORS

VOLUME III

INGENIOUS MECHANISMS

FOR DESIGNERS AND INVENTORS
VOLUME III

Mechanisms and Mechanical Movements Selected
from Automatic Machines and Various Other Forms
of Mechanical Apparatus as Outstanding Examples
of Ingenious Design Embodying Ideas or Principles
Applicable in Designing Machines or Devices Requiring Automatic Features or Mechanical Control

Edited by
HOLBROOK L. HORTON

INDUSTRIAL PRESS INC. NEW YORK, N. Y.

Industrial Press Inc.
200 Madison Avenue
New York, New York 10016-4078

INGENIOUS MECHANISMS
FOR DESIGNERS AND INVENTORS—VOLUME III

25 26 27 28 29 30

THIRD VOLUME OF INGENIOUS MECHANISMS

IN this third volume of INGENIOUS MECHANISMS FOR DESIGNERS AND INVENTORS a large number of mechanisms and mechanical movements not previously described in Volumes I and II have been brought together for convenient study and reference. The steady demand for Volumes I and II indicates the continuing need on the part of machine designers, engineers and students for detailed information about unusual, yet practical mechanical movements.

As in the previous volumes, the mechanisms described are the work of numerous contributors to MACHINERY and represent successful applications of a wide variety of types and designs. Of particular interest to many will be the chapter on Hoppers and Hopper Selector Mechanisms for Automatic Machines which appeared as two articles in MACHINERY by J. R. Paquin.

While it is not feasible in any work of this kind to include mechanisms that are directly applicable to every type of machine and operating condition, it is believed that the numerous designs found in Volumes I, II and III embody mechanical principles which may be utilized in the solution of practically any mechanism designing problem likely to be encountered. Although this volume is an independent treatise, the same general classification and chapter headings used in Volumes I and II have been retained, as far as possible, in Volume III to facilitate the use of all three volumes as a correlated reference library on the subject of mechanism.

CONTENTS

CHAPTER   PAGE

1. Cam Applications and Special Cam Designs.............. 1

2. Intermittent Motions from Gears and Cams.............. 20

3. Intermittent Motions from Ratchet and Geneva
Mechanisms 63

4. Overload, Tripping, and Stop Mechanisms.............. 86

5. Locking, Clamping, and Locating Devices.............. 109

6. Reversing Mechanisms of Special Design.............. 137

7. Reciprocating Motions Derived from Cams, Gears,
and Levers 162

8. Crank Actuated Reciprocating Mechanisms.............. 198

9. Variable Stroke Reciprocating Mechanisms.............. 214

10. Mechanisms Which Provide Oscillating Motion.......... 246

11. Mechanisms Providing Combined Rotary and Linear
Motions 282

12. Speed Changing Mechanisms......................... 301

13. Speed Regulating Mechanisms....................... 329

14. Feed Regulating, Shifting, and Stopping Mechanisms 350

15. Automatic Work Feeding and Transfer Mechanisms.. 379

16. Feeding and Ejecting Mechanisms for Power Presses 420

17. Hoppers and Hopper Selector Mechanisms for Automatic Machines 446

18. Miscellaneous Mechanisms 471

CHAPTER 1

Cam Applications and Special Cam Designs

In the design of mechanisms to obtain irregular move-ments of various kinds, cams are frequently employed. Those which are described or illustrated in connection with the mechanisms covered by this chapter are notable for some ingenious arrangement or design. Other applications of cams and cam-operated mechanisms will be found in Chapter 1, Volume I, and Chapter 1, Volume II, of "Ingenious Mechanisms for Designers and Inventors."

Cam Designed to Provide Longer Stroke without Enlarging Operating Space.—The cam shown at B in Figs. 1, 2 and 3 serves to impart a reciprocating motion to the machine slide G. The slide is required to have a longer stroke than could be produced by a cam of conventional design and of a size which could be assembled in the recess W. This unusual and interesting design of the cam was therefore necessary to provide the long stroke required without increasing the diameter of the cam.

Plan views of the mechanism are shown in Figs. 1 and 3, Fig. 2 being an end view. The complete assembly is shown by Figs. 2 and 3, whereas Fig. 1 shows the mech-anism with the slide G removed. This view, however, shows the cam follower rollers $C, D, E,$ and F in the respective positions they occupy when assembled on the under side of slide G.

1

Referring to Fig. 1, the shaft *A*, to which cam *B* is keyed, rotates in the direction indicated by the arrow. Cam *B* is provided with a series of projecting surfaces or cam tracks indicated by reference letters *H* to *N*, inclusive. The four roller followers *C*, *D*, *E*, and *F*, assembled on the under side of slide *G*, are equally spaced on the center line.

Rollers *C* and *D*, Fig. 1, act as cam followers on opposite sides of the track *H*. Rotation of cam *B* in the direction indicated causes these rollers to move toward the periphery of cam *B* at a rate of speed governed by the shape of track *H*. As roller *C* reaches the periphery of cam *B*, roller *D* reaches the position previously occupied by roller *C*, Fig. 1,

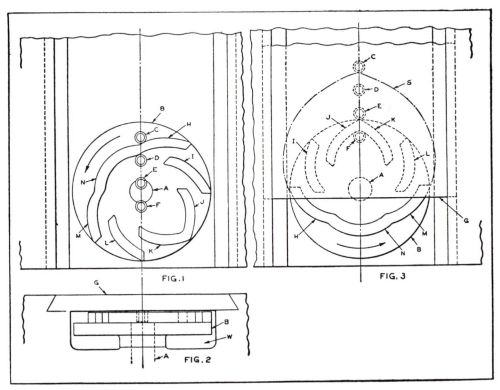

Fig. 1. View of Cam for Operating Slide Shown with Slide Removed but with Slide Follower Rolls in Positions They Occupy with Slide at Lowest Point of Travel. Fig. 2. End View of Cam B and Slide G. Fig. 3. Plan View of Assembled Cam B and Slide G with Latter Member at Highest Point of Travel.

since the center distance between rollers C and D is equal to the rise of cam track H.

When roller C reaches the periphery of cam B it ceases to function as a cam follower, but the motion of the slide G is continued by rollers D and E, acting as followers on track I, which at this point is in the position previously occupied by track H, as shown in Fig. 1. Track I then functions as a cam between rollers D and E until track J reaches the same position, when rollers E and F act as followers to continue the movement of slide G.

In Fig. 3, the rollers E and F are shown at the highest point to which they are carried by the track J, slide G having reached the extreme end of its upward travel. As cam B continues rotation, tracks K, L, and M pass in succession between rollers F, E, D, and C, returning the slide G to its original position. As track N is concentric with the center of rotation of cam B, the rollers E and F and slide G remain stationary until track H is again positioned, as shown in Fig. 1, to complete the cycle.

In the design illustrated, the upward stroke or travel of slide G at a uniform speed is produced by cam B during 135 degrees of its rotation. The reverse or downward stroke of slide G at uniform speed is also produced by cam B while rotating through an angle of 135 degrees. Following this complete reciprocating movement, the slide G remains stationary at its lowest point of travel while the cam B rotates 90 degrees to finish one complete revolution. With this type of cam, an irregular motion of the slide can be produced by varying the shape of the cam tracks, provided that each track is so designed that its leading end will continue the motion produced by the cam which preceded it. In order to compare the size of cam B with a cam of conventional design which would be necessary to produce the same length of stroke or movement of slide G, an outline of the track for such a cam is indicated by the dot and dash lines at S, Fig. 3.

Cam Mechanism with Variable Quick-Drop Adjustment.—The cam mechanism illustrated in Fig. 4 was designed to raise follower-roll A at the end of lever B at a uniform rate until the highest point of its travel is reached, and then to permit it to drop quickly a predetermined adjustable distance before resuming its downward

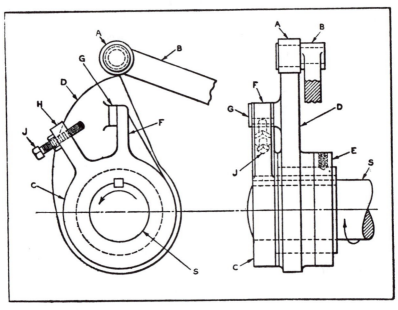

Fig. 4. Cam Mechanism Designed to Impart Upward Movement to Follower Roll A at Uniform Rate, Then Permit Quick Drop Followed by a Slow Descent to Lowest Point.

movement at a slower rate. This adjustable quick-drop cam mechanism is used on a wire-fabricating machine to transmit the particular motion required on one of the machine parts through lever B.

Driving shaft S revolves in a horizontal position in the direction indicated by the arrow, and carries with it flange C, to which it is keyed. Cam D is a free running fit on the hub of flange C, on which it is retained by collar E. With the members of the mechanism in the position shown, shaft S transmits motion to cam D through contact of arm F with the projection G on cam D. The arm F forms an integral part of flange C.

The profile of cam D is designed to transmit a slow uniform upward vertical movement to roll A, followed by a rapid drop. Arm H, which is a part of flange C, carries stop-screw J. Roll A is kept in contact with cam D by a spring (not shown), which is attached to lever B.

As shown in the illustration, roll A is nearly at the top of its vertical movement. As soon as the peak of cam D passes under the center of roll A, the downward pressure of the spring attached to lever B reacts on the angular face of cam D, causing its rotation to be rapidly accelerated in the direction in which it is turning until projection G comes in contact with stop-screw J. Since this movement takes place rapidly, as controlled by the tension of the spring, there is a rapid drop of roll A, which is limited by the contact of projection G with adjusting stop-screw J.

Continued rotation of shaft S permits cam D to rotate at a slow rate of speed, the drop of roll A at this point being at the same rate as though cam D were keyed directly to shaft S. As the heavier side of cam D reaches a position opposite that shown in Fig. 4, it remains at rest until arm F again comes in contact with projection G.

Multiple Cam and Lever Mechanism.—The multiple cam and lever mechanism shown in Fig. 5 was designed to impart a movement to rods D and E, in the direction indicated by the arrows, through rotating shaft B. Interfering members and the considerable distance between shaft B and the rods made a simple drum cam and lever mechanism impractical for this particular application.

Rods D and E are prevented from rotating by slots in their ends, which slide along keys in the frame of the machine. Collars G and H, which are pinned to the rods, are machined on one side to form face cams. These cam surfaces slide on the cam surfaces of the two lever type collars J and K. The collars are joined by connecting-rod L. The roller P

which is mounted on one end of this connecting-rod, is free to rotate on the pin Q and rolls along the pad at the upper end of lever R.

With cam A, roller C, lever Z, adjustable connecting-rod F, and lever R in the positions shown, the face cams have advanced rods D and E to their extreme forward positions. As cam A continues to rotate with shaft B in the direction

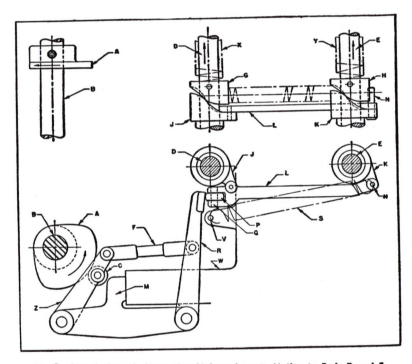

Fig. 5. Cam A through Connecting Linkage Imparts Motion to Rods D and E at Right Angles to Its Plane of Rotation.

indicated by the arrow, and roller C leaves the lobe of the cam, spring S returns the parts of the mechanism to their original positions. This spring is hooked over pin V in the frame of the machine and stud N in collar K. Springs X and Y, which are compressed during the advance stroke of rods D and E, are then free to return the rods and collars G and H to their starting positions.

Compound Cam Mechanism.– A mechanism employed on a French net-making machine that embodies an interesting application of a compound cam movement is shown in Fig. 6. The object of the mechanism is to give to the needle point n, Fig. 6, a closed path of the form shown in Fig. 7, and, additionally, an endwise movement normal to the plane of the paper. To effect this, there are two large gears, A_1 and A_2, meshing together, each of which carries a cam. In

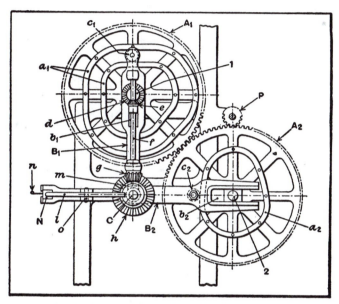

Fig. 6. Compound Cam Mechanism Employed on a
Net-making Machine.

the case of gear A_1, the cam is formed as a closed track by attaching to the wheel the inner and outer members a_1, and in the case of gear A_2 by providing a rim a_2. A lever B_1 is slotted at b_1 so as to embrace the shaft 1, and is provided with a slipper block (not shown) for this purpose. At its upper end this lever carries a roller c_1, fitting the cam track previously described. A second lever B_2 is slotted or forked at b_2 to embrace the shaft 2, being provided with a slipper block as shown. A roller c_2 is employed to engage the cam

a_2, being kept in contact with it by means of a strong spring, not shown.

At C levers B_1 and B_2 are connected by a pin, so as to have a free turning movement. As shown, lever B_2 is continued to the left of C and carries a needle bar N.

So far, then, it will be seen that if gear A_1 were held stationary and A_2 were turned (not practically possible, of course), then the cam a_2 would give to the needle a substantially horizontal to and fro motion. Similarly, if we imagine gear A_2 held and A_1 turned, then the needle would receive a substantially vertical reciprocatory motion. Therefore, when both wheels are rotating, being driven by pinion P, a combination of these movements is effected, giving rise to an enclosed locus, as shown at the upper right in Fig. 7.

It is instructive to consider the method of finding the cam profiles for a given locus. When the needle point n is at position 1 of the locus, the levers are at right angles, as shown by heavy lines in Fig. 7. First, it is necessary to mark off on the locus, positions which will be occupied at successively equal time intervals, this being done from a knowledge of the required rapidity of movement between each such position. Such a series of points is shown as 1 to 12, at the upper right of Fig. 7. Next, circles are described on centers 1 and 2, Fig. 7, and are divided into twelve parts, as shown; the commencing position being noted, also the direction of numbering, thereon.

Consider a point such as 8. It can be seen that the center lines of the levers B_1 and B_2 must always pass through centers 1 and 2, respectively, and their lengths being fixed, it is a simple matter to determine the lever positions, and therefore the roller centers c_1 and c_2. Now, for this disposition of the linkage, the cam wheels will have turned through an angle such that division lines 8_1 and 8_2 will be at the positions first occupied by division lines 1_1 and 1_2, respectively, as shown in the diagram. The offsets x and y must also be noted. Now transfer the positions of points c_1, c_2 around to divisions 8_1, 8_2, respectively, and thereby locate points P_1 and P_2, which will be points on the required cams. Repeating this process for the other positions of the levers will enable the whole of the cam profiles to be determined.

A further motion is incorporated in this mechanism, and this serves to move the needle point sidewise, transversely to the plane of the gears in Fig. 6. This is accomplished as

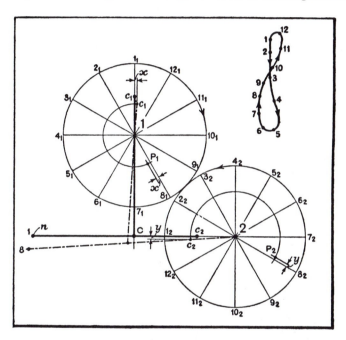

Fig. 7. At Upper Right is Shown Path of Needle Point Controlled by Mechanism Shown in Fig. 6. The Central Diagram Shows Method of Determining the Cam Profiles.

follows, and during the time the points are moving from 2 to 3 in the locus diagram.

Attached to shaft 1 so as to rotate with it is a bevel gear d, and meshing with this is a bevel pinion e, keyed to shaft f which is carried in bearings on lever B_1. At the lower end of this lever is another pinion g, and at the junction C of the levers is a bevel gear h driven from pinion g, gear h carrying a face cam m. A boss is formed on lever B_2, and a transverse pin o in the boss provides an axle for a lever l, one end of which is engaged by cam m, the other end coming in contact with the end of the needle-point bar at N.

Thus, as the mechanism turns, the bevel gear train causes rotation of the face cam m, and accordingly produces the endwise movement desired.

The whole mechanism provides an unusual example of the compounding of movements, and is suggestive of other applications where complex paths have to be traced out by a moving element.

High-Lift Cam with Low Pressure Angle.— Cams acting on straight-line followers are generally limited to moderate pressure angles because of the side thrust developed against the follower bars. In order to produce a high lift without excessive side pressure, the cam shown in Fig. 8 was designed. It was applied in a wire fabricating machine.

The drive-shaft A, rotating in the direction indicated by the arrow, has keyed to it a cam body B that carries a cam-shaped bar C. This bar is shaped at each end to produce one-half of the total cam lift, and is a slide fit in the gibs at each side of it. A slot in the bar provides clearance for the driveshaft, and the bar is normally held in the position shown by the spring D.

The follower bar G, mounted on the machine member E, carries a block F in which is contained a roller H. Another roller J rotates on a stud which is held in a fixed position in its support I. The rollers H and J are offset relative to each other, so that each one contacts only one end of the cam bar C, which is relieved accordingly at both ends. This arrangement prevents the occurrence of two movements of the follower bar G in one complete revolution of the cam.

In the position shown, the follower bar is about to be raised by the upper end of the cam bar which contacts roller H, and the lower end of the cam bar is about to engage the fixed roller J, so that the cam bar will be raised at the same time that its upper end is raising the follower bar. As a result, the movement of the cam bar, added to the normal rise produced by the upper end, moves the roller H to the position indicated by phantom lines at X. To produce this lift, with a conventional cam having the total rise on one lobe, a form such as that shown at Y would be required.

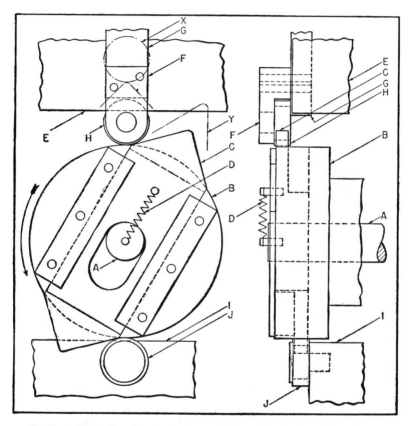

Fig. 8.   A Sliding Cam Bar Having Each End Formed to Produce One-Half the Total Rise Provides a High Lift for the Follower Without Excessive Side Thrusts.

Cam Designed to Operate on Alternate Revolutions.—

The design of a cam that transmits an irregular oscillating motion to a shaft on each alternate revolution of the cam-shaft is shown in the accompanying illustrations. This is accomplished by guiding the follower roll C along two dif-

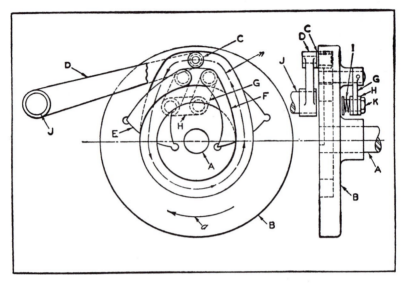

Fig. 9. Cam Mechanism Designed to Raise and Lower Lever D on Each Alternate Revolution of Shaft A.

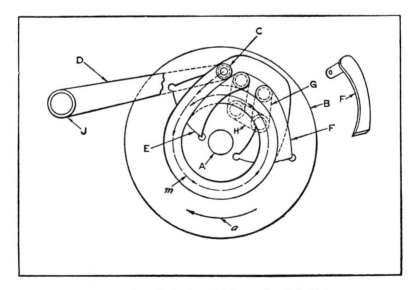

Fig. 10. Cam Mechanism with Levers E and F Set to Guide Roller C into Concentric Groove.

ferent tracks, the concentric track or path indicated by dotted lines at m, Fig. 10, which imparts no movement to lever D, and the cam path n, Fig. 9, which actuates the lever D, producing a rise and fall of this lever with a short rest period at the peak.

Referring to Fig. 9, which shows two views of the mechanism, shaft A carries the cam B, which rotates in the direction indicated by the arrow a. Shaft J receives its motion from cam B through the follower roll C carried on lever D. Cam B is machined to receive the levers E and F, which are recessed into it to less than half the depth of the cam groove.

A better idea of the shape of lever F can be obtained from the view in the upper right-hand corner of Fig. 10. The inner edges of levers E and F are shaped to form a part of the outer edge of the circular follower groove. The outer edges of these levers are shaped to form a part of the inner edges of the follower groove that produces the rise and fall of the roll C. Two levers G on the back of cam B are mounted on the shafts that carry levers E and F and that extend through the body of cam B. Two links H, one on each side of levers G, connect the lower ends of the levers by means of two screws K. Screws K carry the springs I, which act on links H to produce a light friction between levers G and links H.

As shown in Fig. 9, cam B has been rotated in the direction indicated by arrow a, causing roll C to follow the rising side of the cam groove, as indicated by the dotted arrow path n, to the point where the roll is at the peak of the rise. It will be noted that lever E is nested in the recess in the outer wall of the cam groove, while lever F is nested in the inner wall on the opposite side.

Continued rotation of cam B guides roll C to the falling side of the cam track, as shown in Fig. 10. However, this side of the cam track is obstructed by the lever E, as seen in Fig. 9. When roll C comes in contact with lever E, the latter is caused to swing to the opposite side of the groove, thus opening the falling side of the cam groove, as in Fig. 10.

Since lever E is connected to lever F by levers G and links H, lever F is caused to swing an equal amount in the same direction, thus obstructing the rising side of the cam groove.

Further rotation of cam B, Fig. 10, causes roll C to come in contact with lever F, but as lever F is locked against outward movement, roll C is guided in a circular path, as indicated by the dotted arrow path m. Continued rotation of cam B causes roll C to return lever E to its original position, so that roll C may again be guided up the rising side of the cam groove, as in Fig. 9.

The friction caused by the pressure of the springs I serves to hold the levers E and F in position until moved by the roll C. In this manner, shaft J is given an oscillating motion during one revolution of cam B while roll C is following the irregular groove, but is allowed to remain at rest during the next rotation of cam B while roll C is following the concentric path.

Cam Actuated Toggle and Lever Mechanism for Operating Pressure Pad.—The pressure pad A of the mechanism shown in Fig. 11 is moved up and down through a distance B along arc Z by the rotating cam N. The movement of the stud H of the pad along the path indicated by arc Z gives the pad a horizontal movement. This horizontal movement is so slight, however, that it can be disregarded in the case of the mechanism shown.

The action of the toggle mechanism operated by the cam N, roller P, lever M, and rod L enables the pad A to exert considerable downward pressure. The illustration shows the lever M and the pad A at the top of their strokes. When the rotating cam moves the lever M to its lowest position, shown by the dotted lines at Q, rod L will have pivoted lever J about rod F in the direction indicated by arrow K until the center line coincides with the line V.

This action results in bringing link R downward until it is almost in line with lever J, and causes lever S to pivot about the pin D fixed in the housing E until it is in a vertical position on line *W*. The link *T* of the toggle arrangement is thus forced into a vertical position, causing pad *A* to move downward in the direction indicated by arrow *C* along the path of arc *Z* for a distance *B*. The function of lever *G* is to restrict the downward movement of the stud *H* to the path indicated by arc *Z*.

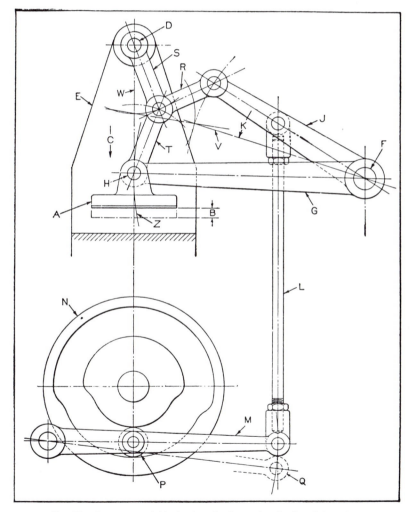

Fig. 11. Cam-actuated Mechanism for Imparting Stroke of Length B to Pressure Pad A.

Cam Mechanism Designed to Control Shaft Speed.—

A shaft on a wire-forming machine must occasionally be rotated by a hand-crank. Owing to the nature of the product, it is necessary at one point in the cycle that the shaft be rotated at a predetermined speed. A speed that is either too fast or too slow results in defective work, although a fairly wide range is permissible between the maximum and minimum speeds. Fig. 12 shows how a cam is used to maintain the speed within the required range.

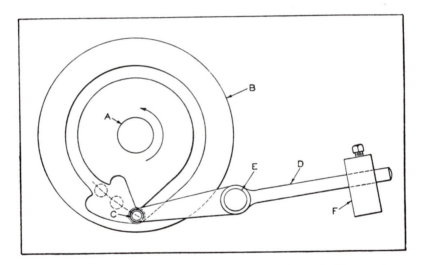

Fig. 12. Cam Arrangement for Controlled Speed of Shaft Driven by Hand-crank.

Shaft A, which is rotated by a hand-crank in the direction indicated by the arrow, carries the cam B. The cam roller C is carried on the end of the lever D which fulcrums on the stud E on a stationary part of the machine. The adjustable weight F holds roller C in contact with the inner cam surfaces.

In operation, if the speed of rotation of shaft A is within the required range, the roller C will follow approximately the path shown by the dotted circles. If the speed of rotation is too high, roller C will have insufficient time to follow its normal path and will lock in the pocket of the outer cam surface. If the speed of rotation is too low, roller C will roll down the sharp incline on the inner cam surface and become locked in the inner pocket. In either case, the cam roller must be returned to its normal path before further rotation is possible. The position of weight F is determined by experiment.

Cam and Planetary Gear Mechanism for Indexing Work-Table and Feeding Drill.—The diagram in Fig. 13 shows a mechanism for automatically feeding and withdrawing a drill and then holding the drill in the "out" position while the work-holding fixture is indexed to the next operating position. After the indexing movement has been completed, the succeeding drilling cycle begins automatically and the cycle of movements is repeated.

The drilling fixture is mounted on the indexing plate L, and is indexed by planetary gearing, which also drives a cam that controls the feeding and withdrawing movements of the drill. The sun gear E, which is driven continuously by the worm K and worm-wheel J, furnishes the drive for both the indexing and drilling movements. An interlocking device is provided which locks the indexing plate L while the drill is being fed, and then locks the drill feeding cam while the table is being indexed. Thus the indexing and drilling movements follow each other automatically and continuously.

Immediately below the indexing plate or table is the indexing ring, which is provided with slots P. These slots may be located at any desired position in the ring to give the angular indexing movements desired. Plate L and the indexing ring are connected to the spider H which carries the planetary gears F, only one of which is shown. Rod A is moved in and out by cam C and is connected to the drill feeding mechanism which it operates. The interlock or stop-bar N is supported by spring M and is moved vertically by cam D. Cams C, D and internal gear G are fastened together.

When the mechanism is in operation, with the members in the position shown in the diagram, cams C and D revolve while the interlock holds the drilling plate or fixture stationary. Cam C provides a short dwell at the end of the withdrawing stroke for the drill, at which time cam D pulls the interlock N downward, thus releasing the fixture plate for indexing, while locking the feed-cam C. When the next slot

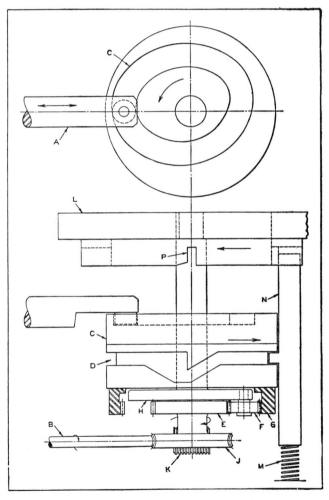

Fig. 13. Mechanism for Indexing Work-table and Feeding Drill.

P in the indexing ring reaches the position above the inter-lock N, the latter member is pushed up by the spring M, thus locking the indexing plate L and unlocking cams D and C.

In some cases, it may be desirable to provide means for preventing the interlock N from being forced back into the same slot P. In other cases, a light drag or brake on the plate L may be needed to insure having the cam D move around far enough to permit the interlock N to move upward freely into the indexing slot. The latter condition was found to exist in the case of the particular mechanism illustrated.

CHAPTER 2

Intermittent Motions from Gears and Cams

When the designer is called upon to provide for the intermittent motion of some part or parts of a machine, he may need to obtain only a stopping of movement for one or more periods in the operational cycle, such as would be required for loading or unloading a work-piece. On the other hand, the designer may also need to provide for the advance of the work or tool for a given distance at specific intervals with periods of rest between. Here, regular intermittent motion of an indexing type is called for.

Either type of intermittent motion may be obtained in various ways. The use of cams which are designed to produce periods of "dwell" is quite common. The employment of gears which are modified to produce intermittent motion is not as general. The mechanisms described in this chapter employ these two methods separately and in combination to produce one or more rest periods in the motion cycle. They supplement those presented in Volumes I and II of "Ingenious Mechanisms."

Intermittent Gear Mechanism.—A change required in a wire fabricating machine presented an interesting problem which was solved in an ingenious manner. On this machine, a uniformly rotating driving shaft transmitted its motion, in direct ratio, to the driven shaft through a pair of spur gears. Owing to a change in the manufactured product, it was required that the driven shaft be given alternately a half revolution and a period of rest corresponding to a full revolution of the driving shaft, and that the speed of the driven shaft and the center distance between the driving and the driven shafts remain unchanged. Figs. 1 and 2 show how these conditions were met without resorting to complicated declutching mechanisms.

Referring to Fig. 1, which shows three views of the assembly, gear A is carried on the driving shaft B. The driven shaft H carries the two gears C and D, the faces of which are half as wide as the face of gear A, so that both can mesh with gear A from the same center of rotation. Gear D is free on shaft H and meshes with gear A. Gear C is pinned to shaft H and has the teeth removed at two diametrically opposite points, the number of teeth removed being sufficient to prevent the rotation of gear A from being transmitted to C at these points.

Gear E rotates freely on stud F and has a definite number of teeth removed from the upper half of its face, so that during a portion of its rotation there is no contact with gear C. Gear E is idle during the greater portion of the cycle, its main function being to assist in the control of the timing of the rotation of driven shaft H. The diameter of gear E and the number of teeth removed from it are governed, of course, by the motion required. In this case, gears A, C and D each have 24 teeth.

The number of teeth removed from gear C must be sufficient to prevent contact with gear A at the required points. The removal of the teeth, however, does not affect the operation of the rotative cycle, the only requirement being that the number of teeth remaining in each section be sufficient to maintain contact between gears A, C, and E during part of the cycle. The number of teeth in the full gear E equals the number of teeth in the driving gear A which will pass any given point during a complete cycle. As the complete cycle includes one rest period corresponding to a full revolution, and a rotating period corresponding to a half revolution of the driving gear A, the ratio between gears A and E will be 1 to 1.5; in other words, gear E in the lower or full-toothed side contains $24 \times 1.5 = 36$ teeth. The number of

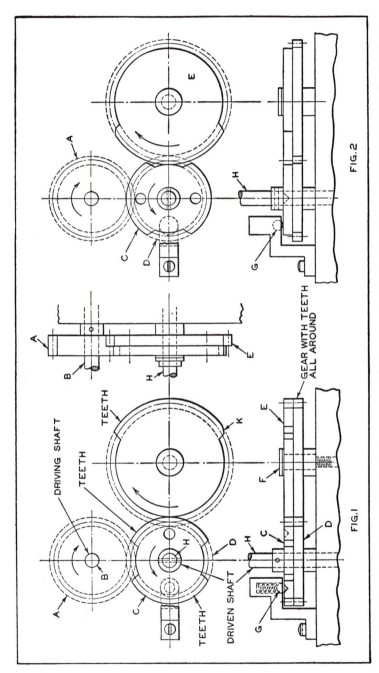

Fig. 1. Mechanism by Means of which Shaft B Transmits a Half Revolution to Shaft H, Followed by a Rest Period Equal to One Revolution of Shaft B. Fig. 2. Same Mechanism with Shaft B Driving Shaft H while Gear E Remains Idle.

teeth removed from the upper half of gear E corresponds to one rest period, or 24 teeth.

Fig. 1 shows the assembly at the middle of the rest period. Gear A, rotating in the direction indicated by the arrow, transmits its motion in the reverse direction to gear D, rotating idly on shaft H. As there is no connection between gears A and C, the latter remains stationary. The rotation of gear D is transmitted to gear E, which, at this point of the cycle, forms no connection with gear C. The leading tooth in the upper half of gear E must be cut away somewhat at K, for free entry when contact is made with gear C.

When the first tooth of gear E engages gear C, the latter begins its rotation, transmitting it to the shaft H. It will be noted that, at this point, although gear E is responsible for the rotation of gear C, it is not acting as a driver; the teeth of gear E merely act as keys to lock gears C and D together so that they rotate in unison. As the teeth of gear C mesh with those of A, the drive is direct from A to C, gears D and E operating idly.

Fig. 2 shows the mechanism with half of the rotative portion of the cycle completed. At this point, gear E serves no useful function, its work being completed as soon as gear C comes under the driving action of gear A. Although the toothed portion of the upper half of gear E is passing through the toothless portion of gear C, this is merely incidental and would not occur if the driven shaft H were to receive a full revolution between rest periods. As the last tooth preceding the toothless portion of gear E passes out of contact with its mating tooth in gear C, the latter ceases rotation and the rest period begins, as at this time the toothless section of gear C is again in position to clear the teeth of gear A, as shown in Fig. 1. The spring ball-stop G was added, to prevent accidental rotation of shaft H.

Intermittent Spur Gear Drive Mechanism.—The intermittent drive shown in Fig. 3 is incorporated in the design of a machine used in manufacturing a wire product. The

two shafts M and N are required to rotate in opposite directions. Shaft N must make a complete revolution while shaft
M makes one-half of a revolution, shaft N remaining stationary while the other shaft completes its revolution.

Gear A on the driving shaft M meshes with the gear B on
the driven shaft N, the latter revolving at twice the speed
of the former. Gear B has three teeth removed at one point

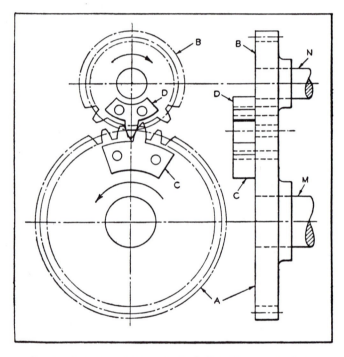

Fig. 3. Mechanism that Permits Shaft N to Remain Idle during
One-half Revolution of Driving Shaft M.

where it receives no motion from gear A. The single teeth
C and D are fastened to the outer sides of gears A and B,
respectively, so that they make contact with each other and
transmit motion the same as the regular teeth.

When teeth C and D are in contact, the rotation of gear A
in the direction indicated by the arrow produces rotation
of gear B in the opposite direction. Gear B receives its mo-

tion from gear A through teeth C and D until gear B is rotated sufficiently to bring the standard teeth into mesh. The single teeth are made slightly longer than the standard teeth to insure contact without danger of clashing. When the standard teeth have made contact, rotation of gear B continues until the space from which the teeth have been removed permits the teeth of gear A to pass, at which time the rotation of gear B is discontinued. At this time, tooth C is 180 degrees from the contact point with tooth D, so that gear B remains stationary until gear A has completed its revolution and teeth C and D are in contact again.

Adjustable Intermittent Rotary Movement.—A conveyor belt used on an assembling table required a dwell or rest period at specified intervals in order to permit loading. As the table was used for a variety of assembling operations,

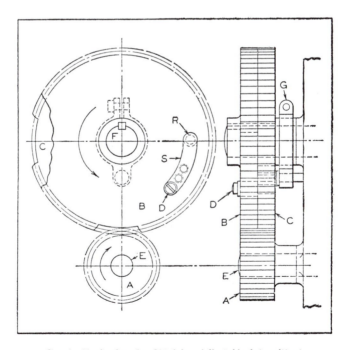

Fig. 4. Mechanism for Obtaining Adjustable Intermittent Rotary Motion.

some means for adjusting the rest period was necessary. The diagram in Fig. 4 shows the adjustable intermittent gear mechanism designed to meet these requirements.

The shaft E carries the driving pinion A, which rotates in the direction indicated by the arrow. Shaft F, which carries the conveyor belt, also supports the mutilated gear B which is keyed to it, as well as the mutilated gear C, which is free to rotate on shaft F. Both gears B and C mesh with the pinion A. Gear B has a slot S in it, while gear C carries the stop D, which is located in one of the tapped holes provided for it and is free to move in slot S.

The tapped holes for stop D in gear C are so located that when stop D is in any one of them, and in contact with either end of slot S, the teeth in gears B and C will be in exact alignment. The brake G, acting on the hub of gear C, provides sufficient resistance to prevent accidental rotation of gear C due to friction.

In the diagram, pinion A is shown in mesh with gear C, which it is driving in the direction indicated by the arrow. Gear B, however, remains stationary, as the mutilated portion is in line with pinion A. This arrangement provides the rest period required for the conveyor belt. Rotation of gear C causes stop D to be moved to the opposite end of slot S, as shown by the dotted circles at R. When it reaches this position, the movement of gear C is again transmitted to gear B through stop D, both gears then being rotated in unison and transmitting motion to the belt.

As the mutilated portion of gear C reaches pinion A, its movement ceases while gear B continues to rotate. Gear C remains stationary until stop D is again located in its original position in the slot in gear B, when both gears again rotate in unison. During this period the conveyor belt has continued its movement, the only purpose of the gap in gear C being to permit the two gears to change their relative positions, so that the stop D will again be in position for the next rest period. The relative positions of the mutilated sections of gears B and C are immaterial, provided that no part of the two gaps overlap.

When a shorter rest period is required, an additional stop D is placed in one of the other tapped holes, thus reducing the effective length of the slot in gear B. As the amount of movement of the conveyor belt between stations is controlled by the number of teeth in gear B, any change in the period of rest does not affect the length of travel between stations.

Modified Helical-Gear Indexing Mechanism.— Helical gears modified as shown in Figs. 5, 6 and 7 can be used directly for indexing. With the arrangement shown in Fig. 5, shaft A will be indexed one revolution for every four revolutions of the driving gear B. In other words, shaft A will be

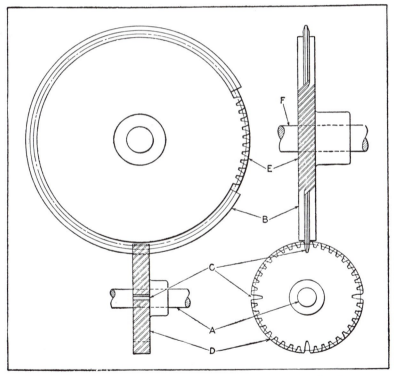

Fig. 5. Modified Helical Gears Index Shaft A One-fourth of a Revolution During One Revolution of Driving Shaft F.

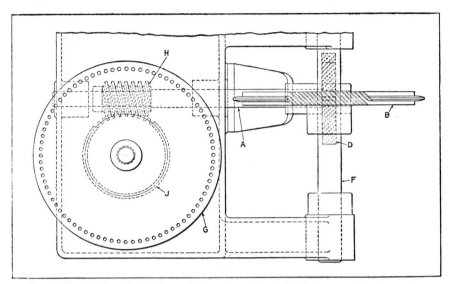

Fig. 6. The Number of Indexings of Plate G per Revolution is Increased by Interposing Worm Gearing Between It and Modified Helical Gears B and D.

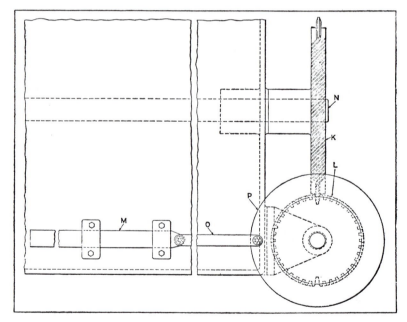

Fig. 7. Modified Helical Driving Gears N and L Arranged to Produce a Reciprocating Movement of Slide M with a Dwell at Each End of Its Travel.

indexed through an angle of 90 degrees during one revolution of the driving gear. The frequency of the indexing can be increased, with a corresponding decrease in the amount of rotation of shaft A per indexing, by increasing the number of slots C in gear D and reducing the number of helical teeth E in gear B. Similarly, shaft A can be indexed through a greater arc by decreasing the number of slots C. The number of helical teeth E in driving gear B, must equal the number of helical teeth in gear D, between any two adjacent slots C.

The modified helical gears B and D shown in Fig. 5 can be used as illustrated in Fig. 6 to increase the number of indexing stations without changing the number of revolutions per minute of the driving shaft F. This is accomplished by interposing worm-gearing between shaft A and index-plate G. The pitch of worm H may be so selected that worm-wheel J will index plate G the required number of divisions. In the case illustrated, a 1 1/2-inch pitch, double - thread worm was used to produce the required indexing.

In the mechanism shown in Fig. 7, modified helical gears K and L are used to reciprocate a slide M. This slide is moved to the right in one revolution of driving shaft N, and to the left (to the position shown) in the next revolution of the shaft. The motion of the slide in either direction is accomplished during one-half of a revolution of shaft N, the slide dwelling at the end of its travel during the other half revolution of the shaft. Link O connects the slide to plate P, which is keyed to the shaft on which gear L is mounted.

Cam-Actuated Intermittent Worm-Drive Mechanism.—
A worm drive of rather unique design developed for use on a wire-forming machine is shown in Fig. 8. The mechanism comprising this drive converts a continuous rotary motion into an intermittent rotary motion at a reduced rotative speed. The object of employing the worm and worm-wheel is to give a compact high-ratio speed reduction in combina‧ tion with the intermittent motion.

Referring to Fig. 8, shaft *A*, mounted in bearing *C*, carries the single-thread worm *B* and rotates in the direction indicated by the arrow. Shaft *A* receives its motion from the driving shaft *E* through the splined sleeve *D*, which per-

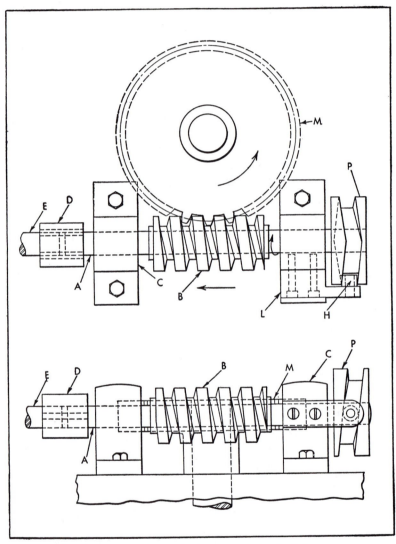

Fig. 8. Worm and Worm-wheel Drive with Cam that Provides Intermittent Rotary Movement.

mits axial movement of shaft *A*. Worm *B* meshes with the worm-gear *M*, to which it transmits motion in the direction indicated. Shaft *A* carries cam *P*, which rotates with it. Bracket *L*, attached to bearing *C*, carries roller *H*, which operates in the groove of cam *P*. It is obvious that, owing to the fixed position of roller *H*, the rotation of cam *P* will cause shaft *A* to be moved axially.

The groove in cam *P* is shaped to produce a uniform axial motion in one direction during one half revolution, and in the reverse direction during the other half revolution. The lead of cam *P* is equal to the lead of worm *B*.

If worm *B* were fixed against axial movement, one revolution of worm *B* would produce a movement of gear *M* equivalent to the lead of the worm. However, in addition to rotative motion, worm *B* is also given an axial motion by cam *P* acting against roller *H*, as mentioned; thus the rotation of gear *M* is effected by both the rotative and axial movements of worm *B*. As the lead of worm *B* and cam *P* are equal, the motion of gear *M* equals that which would be produced by an axially fixed worm of double the lead of worm *B*.

As shaft *A* continues to rotate, roller *H* is passed by the high point of cam *P*, which reverses the axial movement of shaft *A*. When this occurs, there is no movement of gear *M*, as the axial movement of shaft *A* produced by cam *P* is equal to the lead of worm *B*, but is in the reverse direction. In this manner, the axial movement of shaft *A* neutralizes the lead of worm *B*, the worm merely turning or threading itself back to its original position without imparting any motion to gear *M*. The effect is to produce a series of partial revolutions of gear *M* with equal rest periods between the movements.

Intermittent Mechanism with Hourglass or Cylindrical Cam.—The mechanism shown at the left of Fig. 9 consists of an hourglass cam with a helical cam surface that extends part way around the hourglass surface and then forms one side of an annular groove for the remainder of the circumference. This cam engages a toothed wheel as shown at the left of Fig. 9. As a tooth of the wheel is engaged by the helical cam surface, the wheel is revolved until the tooth reaches the annular groove. The wheel then ceases to rotate

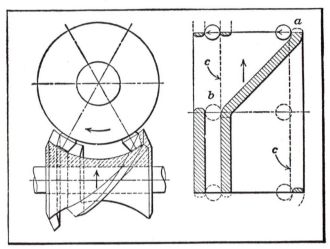

Fig. 9. Intermittent Worm-gear Drive in View at Left, with Development of Circumference of Similar Parallel Worm in View at Right.

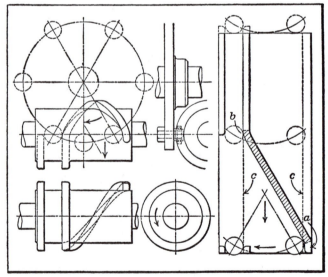

Fig. 10. Alternative to Intermittent Worm-gear Drive Design Shown in Fig. 9.

until the next tooth is engaged by the cam helical surface. The teeth on the wheel are spaced so that at the moment when one tooth has passed out of the annular groove the helical cam surface comes into contact with the next tooth, thus providing intermittent rotation of the wheel.

The helical cam surface is modified somewhat at the beginning and end to insure a gradual start and stop of the wheel. The wheel shown has six teeth so that the cam makes six turns for each turn of the wheel.

The view to the right of Fig. 9 shows a development of the pitch surface of a cylindrical cam with a similar helical cam surface and annular groove.

In Fig. 10, at the left, is shown a cylindrical cam with a cam surface and annular groove engaging a roller toothed disk. The view at the right of Fig. 10 shows a development of the pitch surface of this cam.

Obtaining an Intermittent Motion from a Uniformly Reciprocating Slide.—In tooling a wire-forming machine for a certain job, it became necessary to have a uniformly reciprocating vertical slide intermittently actuate a horizontal slide. The horizontal slide was to finish one-half of its cycle in each complete cycle of the vertical slide. This was accomplished by means of the mechanism shown in Fig. 11.

Slide A reciprocates uniformly in a vertical plane, carrying lever B, which is free to swing about stud K. At the lower end of the lever is a pin C, which enters a recess milled in fixed member D.

Horizontal slide E, which consists of a right- and a left-hand section, is free to reciprocate in ways provided in member D. The two sections of this slide are connected by plate F, which serves as a cam to produce the required motion. Rocker cam G can be swung about stud L, within member D, against light frictional resistance created between the under side of the rocker cam and member D by spring H.

As shown, vertical slide A is in its uppermost position and horizontal slide E is in its extreme right-hand position. At this point, the portion of pin C within the recess in member D is in contact with rocker cam G on one side and the cam surface of plate F on the other. When the vertical slide begins its downward stroke, the pin will be lowered in the groove formed between the rocker cam and member D until it has passed the lower end of the cam. At this point, the vertical slide completes its downward movement, and lever B will swing to a central position, as indicated by the broken outline Y.

On the upward stroke of the vertical slide, the pin will rise along the left-hand side of the rocker cam until it reaches the position indicated by broken outline Z. Here

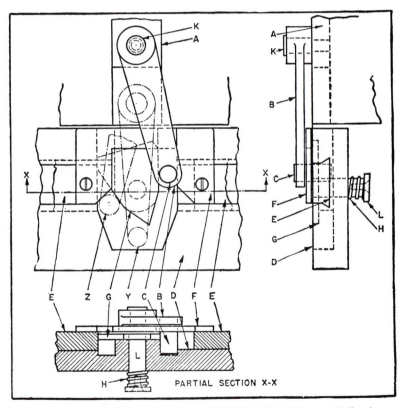

Fig. 11. Intermittent Motion of Horizontal Slide E is Obtained from a Uniformly Reciprocating Vertical Slide A by Means of Rocker Cam G and Cam Plate F.

the pin will contact the angular cam surface of plate F, causing the plate and the vertical slide to move to the left.

Simultaneously, the rocker cam will be pivoted so that its lower pointed end will lie on the left-hand side of the vertical slide center line. This completes half the cycle of the horizontal slide. On the next cycle of the vertical slide, the reverse action takes place, completing the cycle of the horizontal slide as it is returned to the position shown.

Thus the horizontal slide E is intermittently traversed first to the right and then to the left continuously by the reciprocating vertical slide A.

Mechanism for Converting Uniform into Intermittent Reciprocating Motion.—A machine for fabricating a wire product is required to move the work at various stages of the operating cycle by means of reciprocating push-rods. Because of a change in the product, it became necessary to reduce the length of the reciprocating movement and provide a period of rest without any major alteration in the actuating mechanism, which is required to operate other units of the machine. The changed mechanism by means of which the desired motion was accomplished is shown in Fig. 12.

Referring to the two upper views, a channel was machined in the original reciprocating push-rod B to carry the auxiliary rod C, which was made to slide within the channel. The block A serves as a guide for the assembly and, by means of the irregular slots D in its outer walls, aids in converting the uniform motion of rod B into the intermittent motion required for rod C. The irregular slot D is machined in both outer walls of part A so that the slots are in alignment and pin G will slide freely therein. A slot E likewise is machined in both walls of rod B so that pin G will slide freely. Rod C is also provided with a slot for pin G, as shown at F. This completes the assembly, except for the two plates P which serve as retainers for pin G. These plates, however, are not shown in the two

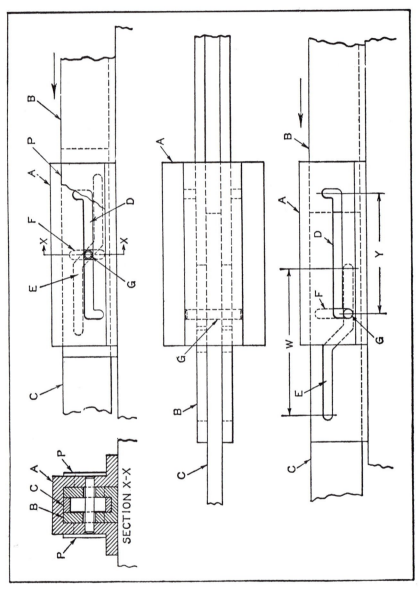

Fig. 12. Mechanism to Convert Uniform Reciprocating Motion of Bar B into Intermittent Reciprocating Motion of Bar C.

lower views, which illustrate the operation of the mechanism.

Referring to the top view, assume that bar B is moving in the direction indicated by the arrow and that pin G, at this point, lies in the angular section of slots E in the bar B, and in the center of the slot F in bar C. Any horizontal motion given to pin G, which passes through the vertical slot in bar C, must produce a corresponding movement of bar C. In the position shown, pin G is free to move horizontally in the slots D in part A, but is restricted from any vertical movement. As bar B is the actuating member and as pin G is locked in the angular section of slots E in bar B by the restricting influence of slots D, bar C is carried along with bar B, which transmits its motion to bar C through the slot F.

The rest or dwell portions of the cycle are accomplished in the following manner: Continued movement of bar B in the direction indicated by the arrow moves pin G to the ends of slots D, thereby preventing further horizontal movement of pin G. When this occurs, the angular portion of slots E in bar B forces pin G into the pockets at the ends of slots D, as shown in the bottom view. In this position, bar C is incapable of further horizontal movement, because it is locked to part A by the pin G. As pin G is now in the horizontal portions of slots E, bar B continues its motion in the direction of the arrow without transmitting any motion to pin G or bar C.

Reversal of bar B takes place before the ends of slots E strike pin G. The horizontal movement of bar C is thus controlled by the length of slots D, or the distance indicated by Y in the bottom view. The maximum movement of bar B equals the distance Y plus twice the distance indicated by W. The mechanism operates in a similar manner on the reverse stroke of the reciprocating driving bar B, pin G moving upward, however, when it reaches the ends of slots D, instead of downward.

Indexing Mechanisms for Small Film Projector.—The film indexing mechanism shown in Fig. 13 was designed for a small motion picture projector, the object being to obtain the desired indexing motion with the minimum number of moving parts. The indexing movement imparted to the shoe H by the mechanism carries the film from the position indicated at M to that indicated at N.

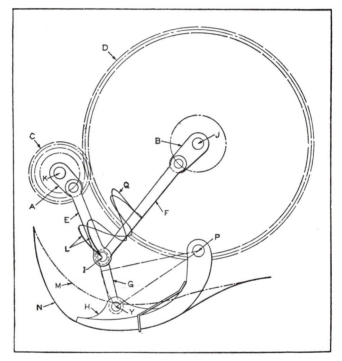

Fig. 13. Double-crank Link and Gear Mechanism for Indexing
Motion Picture Film.

The cranks A and B are mounted on shafts carrying the two spur gears C and D, which are always in mesh. Gear D has four times as many teeth as gear C, so that crank A makes four revolutions to one revolution of crank B. The ends of cranks A and B are connected to rods E and F, the free ends of which are united by a third rod G.

This rod actuates the shoe or lever *H* which swings about the pivot pin *P* when indexing or moving the film from *M* to *N*.

During one revolution of crank *B*, which corresponds to four revolutions of crank *A*, a very complicated curve *Q*, having eight single strokes or four double strokes, is traced by the point *I*. The range of the curve is limited by four circles, the radii of which are equal, respectively, to the lengths of the following members: $F + B$ and $F - B$ drawn around point *J* as a center and $E + A$ and $E - A$ drawn around point *K* as a center. For the movement of the film, only the last two single strokes, indicated by the heavy lines at *L*, are utilized, whereas the other three double strokes are not used. During the time in which the

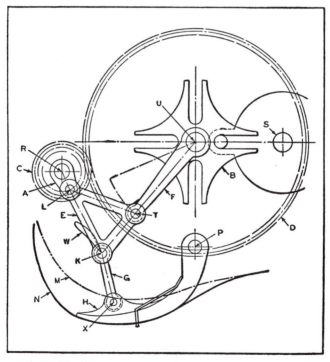

Fig. 14. Geneva Stop Motion Link and Gear Mechanism
for Indexing Motion Picture Film.

point *I* traces the lines representing these three double strokes, gear *D* makes three-fourths of a revolution without transmitting any motion to the film.

Fig. 14 shows a similar driving mechanism that has been developed, which gives an equivalent indexing movement, but avoids the idle time period. This mechanism rests at first for three-fourths of a revolution of the driving shaft *S*. For this purpose, a Geneva stop motion with four stops and one roller is used. During one-fourth of the rotation of the driven member *D*, the point *K* of the mechanism is required to make a double stroke. Therefore, between the driving member *B* of the Geneva stop motion and crank *A* of the mechanism there must be a ratio of 4 to 1. Also, the index dial of the Geneva stop drive should be provided with a large gear *D*, whereas the crank *R* should have a small gear *C* which is one-fourth the size of gear *D*.

During the indexing motion, the crank *A* makes one full turn or rotation. This rotational movement is then changed by a four-bar-link motion involving the members *A, E, F,* and a fixed member, causing them to give a swinging motion to the point *K*. Point *K* follows the path indicated by lines *W* if the connecting lines from the point *K* through the free links *L* and *T* pass through the respective fixed axles *R* and *U* of the mechanism.

For the sake of simplicity, the fixed axle *U* of the four-bar-link was selected as the fixed axle of the index dial. With the position of the curve point thus fixed, the mechanism can be laid out or drawn. If the lengths of crank *A* and the swinging lever *F* are so proportioned that crank *A* can make a full rotation, the point *X* will follow a path similar to that of point *Y*, Fig. 13.

Intermittent Rotary-Motion Gear and Cam Mechanism.— Shaft *A* of the mechanism shown in Fig. 15, rotating in a clockwise direction at a constant speed, is required to transmit intermittent rotating motion in a counter-clockwise direction to the driven gear *J*. Shaft *A* is keyed to a

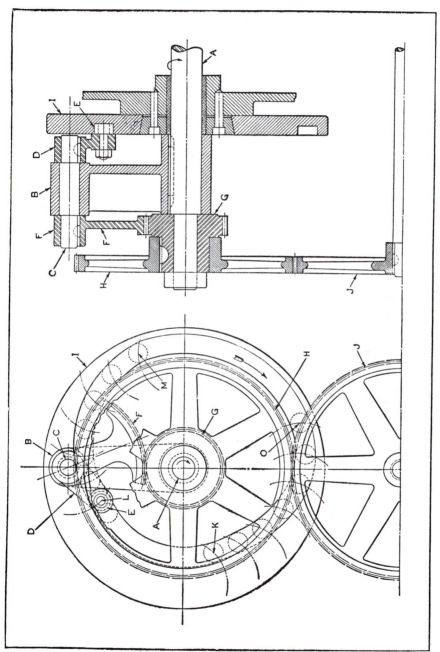

Fig. 15. Rotation of Shaft A at Constant Speed Imparts Intermittent Motion to Gear J.

driving arm *B*, which carries a short shaft *C*. Keyed to one end of shaft *C* is a crank-arm *D* with a cam-follower roller *E* which, traveling in a groove in stationary base cam *I*, transmits an oscillating rotary motion to shaft *C*.

To the front end of shaft *C* is keyed a segment gear *F*, which is in constant mesh with gear *G*. Gear *G* is a running fit on shaft *A*, and has a hub to which gear *H* is keyed. Gear *H*, in turn, is in mesh with gear *J*, keyed to the shaft that is to be given the intermittent rotating motion. Since gear *J* has the same number of teeth as gear *H*, it will have the same intermittent motion, but rotation will be in the opposite direction, or counter-clockwise.

Referring to the view to the left in Fig. 15, it will be clear that so long as the cam-follower roller *E* is traveling in the concentric portion of the cam groove from *K* to *L*, there will be no rotary motion of shaft *C* in arm *B*. During this period, arm *B* and segment gear *F* simply transmit rotary motion to gear *H* in a clockwise direction at the same speed as that of shaft *A*. Thus, so long as there is no rotary motion of shaft *C* in its bearing in arm *B*, the latter member, together with segment gear *F* and the gears *G* and *H*, are effectively locked together and rotate as a single member.

When shaft *A* and arm *B*, rotating in a clockwise direction, cause the follower-cam roller *E* to travel from *L* to *M*, the rise in the groove of cam *I* causes segment gear *F* and its shaft *C* to rotate in a clockwise direction in arm *B*. Since segment gear *F* is in mesh with gear *G*, a rotating motion is transmitted to gears *G* and *H* in a counter-clockwise direction, which is opposite to the clockwise rotation imparted to them by arm *B* alone.

Now, since the counter-clockwise movement imparted to gears *G* and *H* by segment gear *F* is equivalent to the clockwise movement imparted by arm *B* alone in making the quarter revolution required to carry roller *E* from *L* to *M*, gears *G* and *H* will remain idle during this period.

To obtain this intermittent or idle period, the rise of the cam groove from L to M must be just sufficient to cause segment gear F to transmit counter-clockwise motion at the same rate as clockwise motion transmitted by arm B

Continued clockwise movement of arm B during the next quarter revolution carries roller E from M to O. As this portion of the groove in plate I is concentric with shaft A, there will be no rotation of the segment gear in arm B, and shaft A, arm B, segment gear F, and gears G and H will revolve together for one quarter revolution.

During the next quarter revolution of shaft A, in which roller E travels from O to K, the fall in the cam groove will cause segment gear F to rotate in a counter-clockwise direction, and, consequently, transmit motion to gears G and H in a clockwise direction, which is in the same direction in which arm B is driving these gears. Therefore, in this case, the driving motion imparted to gears G and H by segment gear F is added to that imparted by arm B alone, thus doubling the rotating speed of gears G and H during this quarter revolution of shaft A. Gears G and H are, therefore, rotated through an angle of 180 degrees, while shaft A rotates through an angle of 90 degrees.

Summarizing the operation of the intermittent action of the mechanism, rotation of driving shaft A at constant speed has the following results: There is no movement of gears G, H, and J while arm B carries roller J through one-fourth revolution from L to M; the driven gears rotate at the same speed as shaft A and in the same clockwise direction while E travels from M to O; the driven gear J rotates at twice the speed of shaft A in a counter-clockwise direction while E travels from O to K; the driven gears rotate at the same speed as the driving shaft A while roller E is carried from K to L. The angular position of shaft A and arm B at which the dwell period of gear J begins can be varied by adjusting cam I and clamping it in place when the proper adjustment has been made.

Simple Gear and Star-Wheel Indexing Device.—The indexing device shown in Fig. 16 constitutes the main mechanism of an automatic stamping device. It has proved almost as efficient as the well-known Geneva stop motion, and is much easier to produce. The main drive-shaft *A* carries the disk *B*, which has a number of teeth cut on its periphery. This disk acts in the same way as the check disk of the Geneva motion.

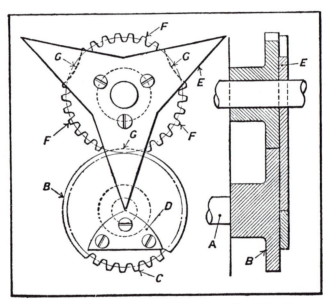

Fig. 16. Indexing Mechanism Developed for Automatic Stamping Device.

Fastened to the face of the disk *B* immediately above the toothed sector *C* is an operating cam or lug *D*. The driven member *E*, which is shaped like a three-pointed star, has three gear sectors *F* secured between the points of the star. As the main drive-shaft *A* is revolved, one of the sides of the operating piece *D* pushes against the side of the star-wheel and causes it to revolve. The toothed sectors on disk *B* and on star-wheel *E* will then engage each other, and in the particular case shown in the illustration, the star-wheel will be rotated through 120 degrees until its motion is arrested by means of the check portions *G* coming into contact with the untoothed periphery of the disk *B*, similar to that in the Geneva stop mechanism

High-Speed Intermittent Gearing.—One difficulty experienced with all types of intermittent gearing, including Geneva movements, is their inability to function properly at high speed. To permit ready engagement and disengagement at relatively high speed, it is necessary to have a certain degree of freedom of motion between the mating parts. This, in turn, causes clashing, incorrect timing, excessive operating noise and wear, and jamming of the mechanism in a short time.

To overcome these difficulties, designers resort to various expedients, such as spring- or inertia-operated prestarting elements which are intended to lessen the initial shock of engagement. These devices, when properly designed and applied, enable high speeds to be employed, but do not extend the range to the high speed sometimes required.

Fig. 17 shows an intermittent gear mechanism designed to meet the requirements of nearly noiseless operation at extremely high speed, and positive locking during the rest period. The mechanism consists of driving shaft *A* which carries a cylinder *B*, a driven shaft *C*, and an indexing shaft *D*. The driven and indexing shafts *C* and *D* are at right angles to the drive-shaft *A*. The cylinder *B* actually consists of two gears, a spiral gear *E* and a circular rack *F*. From each of these two gears certain numbers of teeth are cut away, so that there will be no engagement between the spiral teeth *E* and the spiral gear *G* on shaft *D* when the rack teeth *F* are engaged with the spur gear *H* on shaft *C*. The two shafts *D* and *C* are interlocked by gears *J* and *K*.

When the gearing is in operation, shaft D is locked in position during the time that circular rack F is engaged with spur gear H. The spiral teeth E engage spiral gear G at the moment when circular rack F becomes disengaged from gear H. While spiral teeth E and gear G are engaged, the shaft D is rotated a predetermined portion of a revolution, as determined by the part of the circumference occupied by teeth E. This movement is transmitted to shaft C through gears K and J.

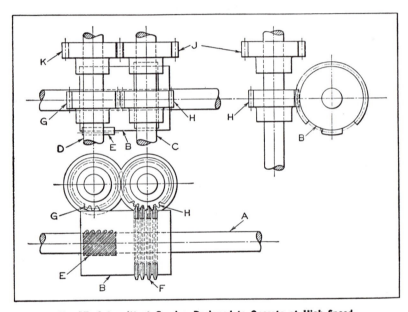

Fig. 17. Intermittent Gearing Designed to Operate at High Speed.

As the teeth E become disengaged from gear G, the rack teeth F enter gear H and lock shaft C against further rotation. The accuracy of the mechanism is not affected by the amount of backlash existing between gears E, G, K, and J, because the effective locking action is between F and H. To facilitate engagement, the entering ends of teeth F are pointed, the same as teeth E.

The mechanism described is positive, accurate and quiet

in operation at extremely high speeds. These desirable features are obtained by having all contacts between the driving and the locking members made by sliding surfaces.

Irregular, Intermittent, Rotary-Motion Mechanism.—The design of an irregular, intermittent, rotary-motion mechanism which is used on a machine for producing a twisted wire product is shown in the accompanying illustration. With this mechanism, the driven shaft C is given a complete revolution during a half revolution of the driving shaft A; and during the second half revolution of the driving shaft, the driven shaft remains stationary, except for a slight change of position.

Referring to Fig. 18, left-hand view, driving shaft A carries gear B, which transmits its motion through gear D to shaft C. Gear D has a full complement of teeth, while gear B, which is of a pitch diameter equal to twice the pitch diameter of gear D, has only a sufficient number of teeth to produce a full revolution of gear D. Gear B carries an internal cam E, the groove of which receives roller F mounted on gear D.

In this view, gear D is shown beginning its rest period. The last tooth in the toothed section of gear B is just leaving its mating tooth in gear D, and roller F has entered the groove in cam E, the leading end of which is shaped to suit the path in which roller F travels while gear D is still being driven by gear B. From this point, the groove in cam E is formed to a true radius with the center of shaft A, which permits cam E to continue its motion without imparting any motion to gear D. It will be noted that, during the rest period of gear D, roller F is off the center line between shafts A and C, which locks gear D in position, preventing accidental rotation.

As the rise in cam E reaches roller F, the effect is to give a slight rotative motion to gear D in its original direction. Gear D then remains stationary in its changed position for a short time, returning to its original rest

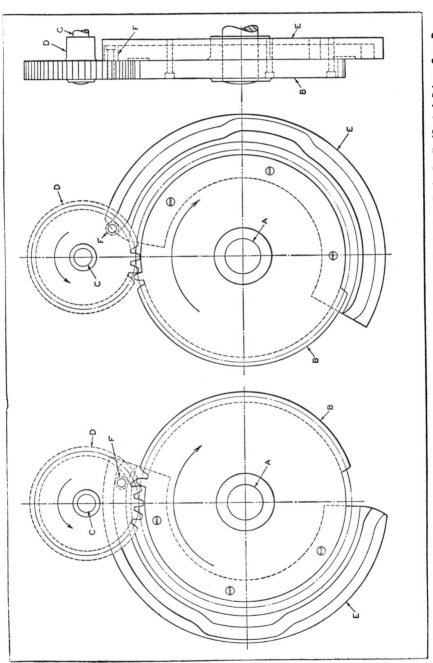

Fig. 18. (Left) Irregular, Intermittent, Rotary-motion Mechanism with Driven Gear D about to Begin Rotation; (Center) Driven Gear D about to Begin Dwell; (Right) End View of Mechanism.

position as the fall of cam E passes roller F. Gear D now remains stationary for the remainder of the rest period.

In Fig. 18, center view, gear D is shown just about to start its rotation, the first tooth of the toothed section in gear B engaging its mating tooth in gear D. Before these teeth become engaged, the rise at the end of cam E produces a gradual partial rotation of gear D corresponding to that which would be given it by one tooth. In this manner, gear D is already in motion when the teeth engage, thus eliminating shock on the first tooth. Gear D then makes a complete revolution before roller F again enters the groove in cam E. An end view of the assembled mechanism is shown in Fig. 18, at the right.

Intermittent Rotating Mechanism Designed for Smooth Operation.—The twisting spindle of a machine for fabricating a twisted wire product was required to finish its cycle in approximately half the time needed for the complete cycle of operations performed by the machine, and then to rest while the remaining portion of the cycle took place. Owing to space limitations, a mutilated gear was selected as the simplest means for producing the required movement. As the driven spindle was required to be positively locked during its rest period, a locking arrangement was attached to the gears; but, on trial, it was found that although the driven spindle rotated at comparatively low speed, the momentum, due to the weight of the rotating parts, was sufficient to produce a severe hammering effect at the end of the rotating period. The design that finally proved satisfactory is shown by the left and center diagrams of Fig. 19.

The plan view at the left shows the mechanism shortly before the termination of the rotating period of the driven spindle D. Driving shaft A carries the mutilated gear B, which rotates in the direction indicated by the arrow. Spindle shaft D, rotating in the opposite direction, carries the full gear C which meshes with gear B. Gear B carries

slower rate. As roller H reaches the end of the groove, the leading end of plate E comes in contact with the right-hand foot of plate F, locking gear C against accidental rotation. The reduced speed of gear C toward the end of its period of rotation serves to eliminate the objectionable hammering effect.

As shown in the center diagram of Fig. 19, the leading end of plate F is riding on the periphery of plate E, thus locking gear C and spindle D against rotation.

A side view of the assembled mechanism is shown at the right of Fig. 19.

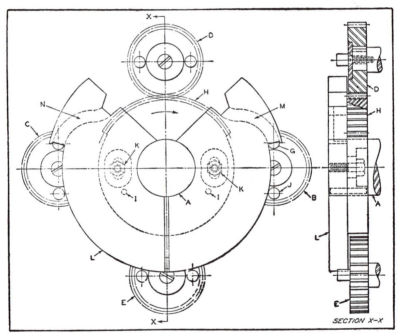

Fig. 20. Mechanism by Means of which Shaft A Intermittently Drives and Locks the Four Gears B, E, C, and D.

Mechanism for Intermittently Rotating and Locking Four Equally Spaced Shafts.—The shaft A of the mechanism shown in Fig. 20 transmits 1/3 horsepower at a speed of 180 revolutions per minute, and can be run at

this speed in either a clockwise or a counter-clockwise direction. In turning in a clockwise direction, as indicated by the arrow, shaft A rotates the gears B, E, C, and D intermittently and successively in a counter-clockwise direction. The drive to these gears is obtained through the combined or successive action of cam groove M, segment gear H, and cam groove N. The segment gear is keyed to shaft A and the cam members are adjustably fastened to the segment gear by studs K, dowel-pins I being used to maintain the permanent settings of the cams after they have been adjusted to the required position.

Starting with the various members of the mechanism in the position shown, one-half turn of shaft A will rotate gear B one revolution. Gear B is then locked in a fixed position until shaft A makes one complete revolution. As the four gears B, E, C, and D are spaced 90 degrees apart, the driving movement imparted successively to each of these gears begins when the preceding gear has made only one-half of a revolution. For example, when gear B has been given one-half turn, the gear E begins to turn. Similarly, when E has made one-half revolution, gear C begins to revolve.

A feature of the mechanism is its silent operation, which is obtained by employing the cam grooves M and N for respectively starting and stopping the driving movements imparted to the four gears, the segment gear H providing the intermediate driving movement.

In operation, the cam groove M engages the pin G and thus drives the gear B, accelerating its speed until the speeds of the two members are the same. This engagement of the members is accomplished without shock. The segment gear and the pinion B are thus engaged without shock when their pitch line velocities are the same. The segment gear continues to drive gear B until pin J enters the second cam groove N, the gear remaining in contact with the teeth of the gear segment until the cam groove has fully assumed its driving function. The cam groove *N* is designed to impart a decelerating movement to gear *B* until it has stopped. The concentric edge or rim *L* of the cam member now being in contact with the two pins *G* and *J* serves to lock gear *B* in a fixed position and hold it thus until cam groove *M* again engages the pin *G*. Each of the gears *E*, *C*, and *D* is successively rotated one revolution and then locked in a fixed position, the same as gear *B*.

This mechanism can, of course, be changed so that the intermittent drive will be imparted to one, two or three of the gears as desired. Also, it can be so modified that the gear or gears will be driven only by the cam groove. The contour of the cam groove *M*, for example, can be changed so that it will impart either accelerating or decelerating movements to the gear, and can also be made to include a dwell. The speed of shaft *A* can also be increased, and its rotation can be reversed.

Irregular Intermittent Motion Using Friction Drive.—The mechanism shown in Fig. 21 is designed to transmit an irregular intermittent rotating motion to shaft *J* from the driving shaft *A*, which rotates continuously at a constant speed. It is used on a machine that fabricates a wire screening material having a mesh of alternately increasing and decreasing size. The variations in the size of the mesh are controlled by the mechanism that varies the number of revolutions or fraction of a revolution that is made by the driven shaft *J* between the periods of dwell.

Referring to the illustration, the driving shaft *A* carries the disk *B*, which is keyed to it, and rotates at a uniform speed in the direction indicated by the arrow. Gear *I* is supported freely on shaft *A*, and carries a series of pins *G* which control the various steps in the operating cycle. Gear *I* also carries two studs *E*, which serve to connect it to the two-piece band *D* which encircles the disk *B*.

Band *D* is provided with a lining of frictional material *C*, and is held in place by two clamping bolts which can be adjusted to regulate the frictional driving force transmitted by disk *B*.

Gear *L* meshes with gear *I* and serves to rotate the shaft *J* which operates the feed mechanism of the machine. The

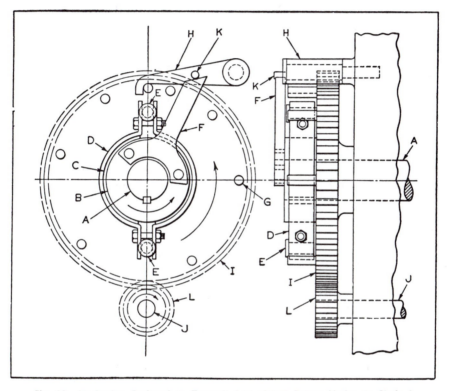

Fig. 21. Mechanism Designed to Transmit Intermittent Rotary Motion to Shaft J from Shaft A with Varying Length of Rotating Periods between Eight Successive Dwell Periods.

latch *H*, which is supported freely on a stationary part of the machine, successively makes contact with the pins *G* on gear *I* as they arrive at the top position. Disk *B* carries the arm *F*, which actuates latch *H* through contact with pin *K*.

In operation, shaft *A*, disk *B*, and arm *F* rotate uniformly in the direction indicated by the arrow. The frictional driving force exerted on band D by disk B tends to rotate gear I through the connecting studs E. When the various members of the mechanism are in the position shown in the illustration, with the hook end of latch H in contact with one of the pins G, the gear I is restrained from rotating, and therefore no rotating motion is transmitted to shaft J, the disk B rotating against the frictional resistance of the band D.

Referring to the illustration, which shows arm F in contact with pin K on latch H, it will be evident that continued rotation of arm F will cause latch H to be lifted out of contact with pin G, permitting the friction drive from disk B to band D to transmit motion to gear I and shaft J through the medium of gear L.

When arm F has passed under pin K, latch H drops ahead of the next pin G, again stopping the rotation of gear I. Gear I remains stationary until arm F has completed a revolution, when it again lifts latch H through pin K. As the pins G are unequally spaced, shaft J will be rotated through a varying number of revolutions or through different fractions of a complete revolution between each of the dwell periods, as determined by the location of these pins.

Quick-Acting Intermittent Feeding Mechanism.—The feeding mechanism shown in Figs. 22 and 23 was designed for winding paper intermittently on an automatic machine. In this mechanism, advantage is taken of the toggle joint locking principle for quickly applying and securely holding a split nut in contact with the lead-screw.

The design of the machine necessitated placing the cam-shaft A, Fig 22, at a considerable distance from the lead-screw B. The two halves C and D of the split nut are shown in Fig. 22 disengaged from the lead-screw, so that the mechanism to which they are attached will not be moved transversely by the lead-screw. In Fig. 23, the

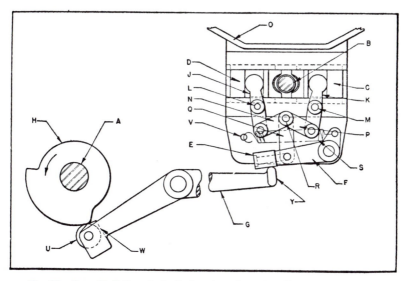

Fig. 22. Cam H, Bellcrank G, Rocker Arm F, and a Toggle Joint Provide Intermittent Feed by Periodical'y Engaging and Disengaging Lead-screw B and Split Nut C and D.

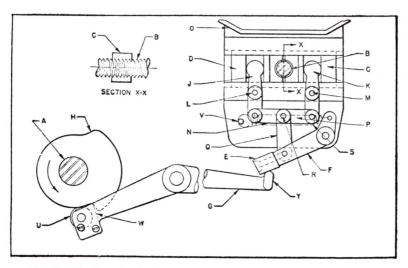

Fig. 23. Closed or Feeding Position of Mechanism Shown in Fig. 22. In this View, the Split Nut is in Engagement with the Lead-screw to Feed the Mechanism.

split nut is engaged with the lead-screw, and the mechanism carried on bracket O has been fed about 2 inches. Intermittent feed is accomplished by bringing together and separating the two halves of the nut at predetermined intervals in the operating cycle of the machine.

The upper ends of arms J and K engage slots in the sides of the split-nut halves. These arms pivot about studs L and M, and their lower ends are joined to a stud R by links N and P. Link Q connects this toggle joint to rocker arm F.

Roller E, at the lower end of the rocker arm, engages extension pad Y on the longer arm of bellcrank G. The bellcrank is pivoted by roller U or block W coming in contact with cam H, which rotates with shaft A in the direction indicated by the arrow.

The function of block W is to prolong the period during which the lever holds the split nut in its open position after the roller leaves the rise on the cam. The shape of the block provides a fast drop-off at the end of the "open" travel, as indicated in Fig. 23, and permits the split nut to snap quickly into engagement with the lead-screw. Spring S, which is attached to the upper end of rocker arm F and is hooked over pin V, exerts sufficient tension to hold the roller and block in contact with the cam, and also locks the toggle joint in either the open or closed position.

Rack and Gear Assembly for Intermittent Rotary Motion.—In the design of a wire-forming machine, a shaft was required with an intermittent rotary motion that exceeded the radial travel obtainable with ordinary ratchet and pawl mechanisms. The rack and gear assembly shown in Fig. 24 provided the desired motion efficiently. The reciprocating rack A meshes with gear B, to which it transmits an alternating rotative movement. Gear B is in mesh with gear C, which is smaller in diameter than gear B and does not mesh with rack A. Gear C is keyed on shaft D,

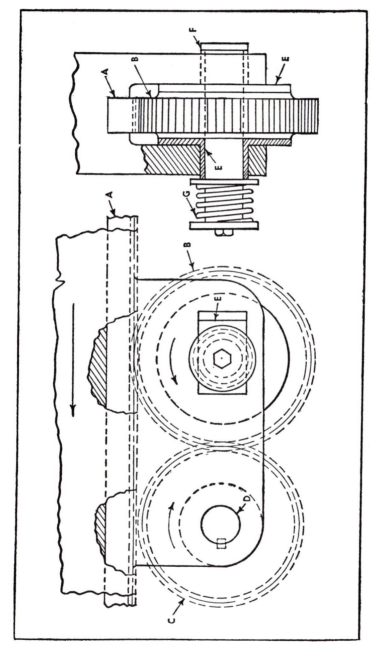

Fig. 24. Rack and Gear Assembly that Provides Intermittent Rotary Motion in One Direction.

which is given an intermittent, rotary motion in one direction.

Driving gear B is splined to shaft F, which is supported by two flanged bearings E. Bearings E are rectangular in section where they pass through the supporting member. The rectangular sections of bearings E are mounted in rectangular slots, which are somewhat longer than these sections to permit a horizontal sliding movement of the bearing. The hubs of gear B are of large diameter and are in contact with the flanges of bearings E. Shaft F is flanged on one end and is provided with a spring G on the opposite end, the pressure of which draws bearings E together so as to apply frictional resistance to the rotative movement of gear B.

As illustrated, rack A is moving in the direction indicated by the arrow, causing gear B to rotate in the same direction. Gear B, meshing with gear C, causes it and shaft D to rotate in the reverse direction. When the movement of rack A is reversed, the tendency for gear B to rotate in the reverse direction also is resisted by the friction applied through spring G. Inasmuch as there is no resistance to the horizontal movement of bearings E, the latter will slide in the rectangular slots, thus disengaging gear B from gear C. When bearings E come in contact with the ends of the slots, further sliding movement is prevented, and continued movement of rack A causes gear B to rotate; however, as gears B and C are out of mesh, no rotary motion is transmitted to gear C. When the movement of rack A is once more reversed (being then in the direction indicated by the arrow), bearings E immediately slide gear B into mesh with gear C and shaft D is rotated.

An idler between gears B and C will permit rotation of shaft D in the opposite direction to that illustrated.

Intermittent Motion for Changing Timing Interval for Air-Valve Functioning.—Fig. 25 shows how a shaft D, which originally served to actuate an air valve once for each revolution of the chain-driven sprocket A, was equipped with an intermittent indexing mechanism designed to actuate the valve once every sixth revolution of the driven sprocket. Since the slow constant speed at which sprocket A rotated could not be changed, it was necessary to provide some means of driving shaft D from sprocket A at a reduced speed in the ratio of 6 to 1 to accomplish this change in the valve operating cycle.

Fortunately, the complete operation of opening and closing the air valve (not shown) could be accomplished in one-eighth of a revolution of shaft D. This made it possible to employ the cam-actuated intermittent mechanism shown, which indexes shaft D one-sixth of a revolution at each

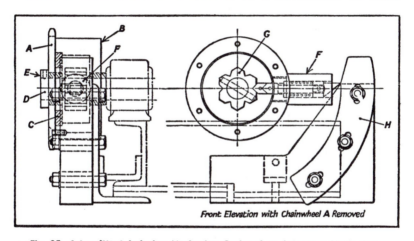

Front Elevation with Chainwheel A Removed

Fig. 25. Intermittent Indexing Mechanism Designed to Actuate a Valve Once Every Sixth Revolution of the Driven Sprocket.

revolution of sprocket A. With this mechanism, every sixth indexing movement of shaft D through one-sixth of a revolution served to open and close the air valve as required.

The necessary modifications in the drive included the securing of sprocket A to the idling drum B, which is a free running fit on shaft D, and the provision of a collar E for retaining the drum and sprocket assembly on shaft

D. The sprocket wheel is located on drum *B* by the machined ring *C*, and is held in place by countersunk-head screws. A steel housing *F*, attached to drum *B*, contains a spring-loaded pawl-ended follower. The follower, as shown by dotted lines in the view to the right, is made in two pieces, the pawl end being a square section bar with a V-shaped end. The rear end of this pawl is turned down to form a shank which is a sliding clearance fit inside the loading spring. The end of this shank is threaded to fit the threaded hole in the follower shown in contact with the cam-plate *H*.

Secured to shaft *D* is a hardened tool-steel hexagonal member *G*, which has a V-slot centrally located in each of its six flat faces. Normally, the loading spring holds the pawl out of contact with member *G*, but when the follower end comes in contact with cam *H*, the pawl end is forced into one of the slots in member *G*, which causes shaft *D* to rotate with sprocket *A*. The length of the indexing movement is determined by cam *H*, which keeps the pawl in engagement with member *G* through one-sixth of a revolution and then permits the pawl to withdraw, so that member *G* remains idle for the remaining five-sixths of a revolution.

Operation of the air valve is, of course, accomplished during one of the six indexing movements of member *G*, the other five indexing movements performing no valve-operating function. Cam *H* is provided with radial slots for the clamping screws to permit adjustment of the cam so as to insure correct timing of the engagement of the pawl and member *G*.

CHAPTER 3

Intermittent Motions from Ratchet and Geneva Mechanisms

Two methods of producing intermittent motion in which the periods of rest are evenly spaced and of equal length are by means of ratchet gearing and by using some modification of the Geneva motion. In its basic form this motion is obtained by means of a Geneva wheel, acting as a driven member, which has four radial slots located 90 degrees apart that successively engage a roller or pin on the driving member. The Geneva wheel thus turns with the driving member through one-quarter of a revolution and is idle for the remainder of the revolution of the driving member.

A number of ingenious mechanisms in which a ratchet arrangement or a Geneva motion play a prominent part are described in this chapter. For other mechanisms of a similar type, the reader is referred to Volumes I and II of "Ingenious Mechanisms."

Ratchet Mechanism with Device for Controlling Engagement of Pawl.—A rather novel method of controlling the action of a pawl on a ratchet is incorporated in the ratchet mechanism shown in Figs. 1 to 6. In the particular application for which this mechanism was developed, the ratchet is required to operate at a slow and uniform rate, and at periodic intervals to skip one or more movements.

The ratchet wheel A, Fig. 1, is mounted on a shaft B, which, in turn, rests in a bearing C. The pawl D forms a part of the operating unit, which consists of the actuating rod E, connected to the upper mechanism, and the bar F, which serves to keep the ratchet and pawl in the same relative positions throughout their movements. A stud G holds the three members of the operating unit to gether. The pawl control unit consists of a displacing collar H, Fig. 3, which has a sliding fit on the pin J. The key K slides in a keyway in the pin J and serves to keep the member H in a fixed position relative to ratchet A. A cam follower L makes contact with the surface of the cam

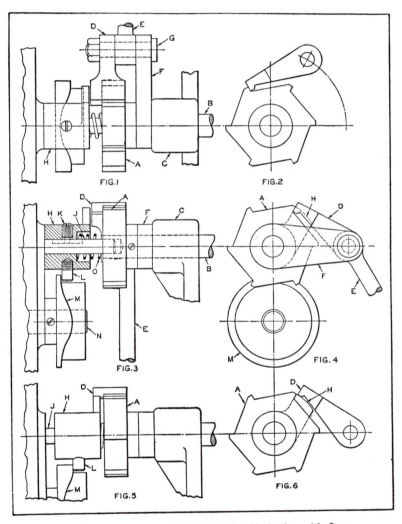

Figs. 1 to 6. Diagrams Showing Ratchet Mechanism with Cam Arrangement for Disengaging Pawl.

M, which is fastened to a shaft *N*. Spring *O* tends to keep the displacing collar *H* away from ratchet *A* and at all times under the action of cam *M*. To properly align the ratchet mechanism and displacing collar, the pin *J* is extended into a hole in the end of shaft *B*.

In Fig. 4, the pawl is shown in position ready to rotate the ratchet through a distance equivalent to one tooth space. Fig. 2 shows how the pawl travels through an arc determined by the the length of bar *F*. In Figs. 1 and 2, the position of member *H* is as shown in Fig. 3, where it will be noted that the cam follower *L* is at the lowest point of the cam surface and the displacing collar is away from the ratchet. Fig. 5 shows the cam rotated to the position where follower *L* has caused member *H* to be moved forward toward the ratchet. The result is shown in Fig. 6, where it will be seen that the pawl has been raised, to prevent it from coming in contact with the next tooth, thus interrupting the ratchet movement.

The cam *M* can be arranged to provide any form of interrupted ratchet motion desired. It can be arranged to rotate continuously or intermittently, depending upon the nature of the application. The cam action is so timed that member *H* is moved forward into position to prevent the pawl from engaging the ratchet wheel at the moment the pawl is in the position shown in Fig. 2.

Ratchet Movement with Idle Period.—A ratchet movement operated by an oscillating lever in the conventional manner, except that the pawl is rendered inactive at a predetermined period, is shown in Fig. 7. This movement is used to operate a conveyor belt on a wire-forming machine, the purpose of the idle period being to increase the loading time at a certain point in the cycle.

Lever *B* is free to oscillate on shaft *A*. Ratchet wheel *C* is keyed to shaft *A* and carries on its hub a similar but narrower ratchet wheel *D*. The latter wheel is free to turn on the hub of wheel *C*. Pawl *F*, which transmits the

The ratchet wheel *A* is directly connected with the rope pulley or drum. The operating lever *B* supports a ratchet pawl *C* which produces only a forward motion of the ratchet wheel. Any reverse motion is prevented by another pawl *D*, fixed in the base of the mechanism and pressed against the ratchet teeth by a spring. Operating lever *B* causes rotation of the wheel and the rope drum.

When the winding work is completed, the hand-lever is brought into the position shown in the middle diagram, where it is stopped by a pin which comes in contact with lever *E*. Pawl *D* remains in mesh with the ratchet teeth, so that any unintentional motion of the wheel is prevented should pawl *C* be accidentally disengaged. If it is desired to withdraw the rope, the hand-lever is placed in such a position that the fixed pin lies in front of lever *E* and causes it to tilt, as shown in the right-hand diagram. In this position, the cam-like surfaces of the pawls disengage the pawls from contact with the ratchet teeth, permitting the rope drum to rotate freely.

Ratchet Movement with Remote Control.—A reversing ratchet movement in which the operating pawl is tripped from a distant point is shown in Fig. 9. This movement is used to control the work-table of a metal polishing machine. Referring to the illustration, *M* is the work-table carrying the rack *B* which meshes with gear *A*. Gear *A* is free to turn on shaft *J*, which is supported on bearings (not shown). Shaft *J* carries the levers *F* and *C* at opposite ends, both levers being keyed to the shaft. Rod *D* transmits an oscillating motion to lever *C*.

Lever *F* carries pawl *G* and bar *H*, both of which are keyed to shaft *P* which passes through lever *F*. Shaft *P* is a free turning fit in lever *F*. Shaft *K* passes through shaft *J* carrying levers *E* and *L* at opposite ends. Lever *E* carries a plunger *N*, backed by a spring, which makes contact with bar *H*, thus engaging pawl *G* with gear *A*.

Referring to the view at the left, rod *D* is assumed to

be moving in the direction indicated by the arrow, the motion being transmitted through levers C and F, pawl G, and gear A to rack B, so that table M moves in the direction shown by the arrow. As lever E rests against the pin in lever F, motion is transmitted to lever L through shaft K. On the return stroke of rod D, gear A remains stationary, the pawl G riding back over the teeth.

As the movement of table M continues, pin O eventually strikes lever L, giving shaft K a partial revolution within shaft J, so that lever E is brought against the upper pin on lever F. This causes plunger N to act on the opposite end of bar H, swinging pawl G so that its lower end engages

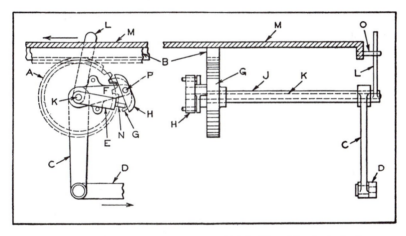

Fig. 9. Reversing Ratchet Movement which is Tripped by Contact of Pin O on Table M with Lever L.

gear A. In this manner, gear A is given a partial revolution on the forward stroke of rod D instead of on the pulling stroke as shown, so that the motion of table M is in the reverse direction. This continues until a pin at the opposite end of table M strikes lever L, again tripping pawl G and repeating the cycle.

Noiseless Ratchet Mechanism for Preventing Reversal of Shaft.—The ratchet mechanism shown in Fig. 10 was

designed for noiseless or silent operation, and was originally intended for use on the head-shaft drive of belt
conveyors, where a peripheral speed of 100 or more feet
per minute is attained. However, it could easily be adapted
to other uses where it is desirable to prevent a shaft from
turning backward.

The mechanism is very simple, centrifugal force being
utilized to keep pawls A from contact with the ratchet
teeth B while the rotating member C is in motion. Member
C is keyed to the shaft K. The instant C stops rotating,
one of the three pawls A engages teeth B, preventing shaft
K from rotating backward. When forward motion of the
shaft is resumed, the pawl is instantly thrown out of contact with the ratchet teeth, the outward motion being
restricted by stop-pins S.

The rotating member is composed of two indentical plates

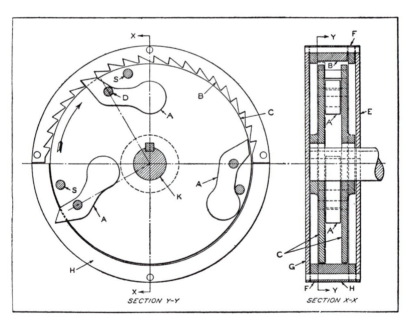

**Fig. 10. Silent-operating Ratchet Designed to Prevent Shaft from Rotating
in Reverse Direction.**

C, as shown in the cross-sectional view to the right. The three pawls *A* are suspended between plates *C* in such a way that they are free to swing on their pivot pins *D* while the member *C* rotates in a forward direction. The ratchet teeth *B* are cut in a 180-degree segment mounted in the upper half of a dust-tight housing, the segment being held fixed between plates *E, F,* and *G.* The inner plate *E* is attached to the frame of the mechanism. The part *H* occupies the same relative position between the plates in the lower half of the housing as that occupied by the ratchet-tooth segment in the upper half.

Ratchet Mechanism that Converts Reciprocating Movement to Continuous Rotary Motion.—In designing a certain mechanism, the problem arose of providing a rotary drive for a shaft when the only available motion was reciprocation in a plane at right angles to the axis of the shaft. It was required that the rotation be continuous in one direction, but it did not need to be absolutely uniform. The problem was solved by the ratchet mechanism shown in Fig. 11.

In the illustration, the shaft to be rotated is shown at *A.* It is supported in suitable bearings (not shown). Ratchet *B* is keyed to the shaft *A.* On each side of the ratchet and turning freely on the shaft are pawl arms *C* and *D.* These arms are held in place by collars *E* which are pinned to shaft *A.* At the outer end of arms *C* and *D* are pins *F* and *G,* about which pawls *H* and *J* are free to swivel. These pawls are held in contact with the teeth of ratchet *B* by springs *K.* The latter are attached to the hubs of the pawl arms and bear against spring pins *L* mounted on pawls *H* and *J.*

The reciprocating member *M* is connected by links *N* and *P* to the outer ends of the pawl arm pins *F* and *G.* One link is above ratchet *B,* and the other link below the ratchet.

As reciprocating member *M* moves toward the right,

the ratchet is rotated counter-clockwise by the pawl J engaging a ratchet tooth. During this movement, pawl H rides over the ratchet teeth. When member M moves to the left, pawl H engages a ratchet tooth and continues the rotation of both the ratchet and shaft A in a counter-clockwise direction. During this movement, pawl J slips over the ratchet teeth.

The pawls are beveled at their outer ends on the side adjacent to each other, as shown by pawl H in the plan view, so that the two pawls can pass each other without interference when they are at the extreme right-hand end

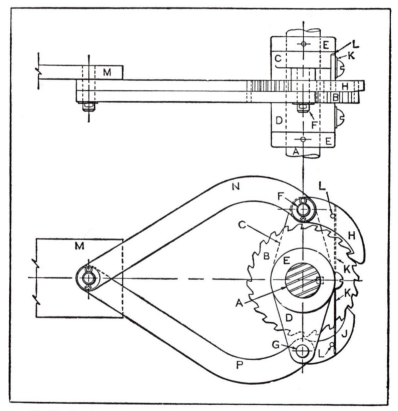

Fig. 11. Ratchet Mechanism which Provides a Continuous Rotary Movement that is Derived from a Reciprocating Motion.

of their travel. Likewise, links N and P are curved, so that they will readily clear ratchet B when member M is at the extreme right-hand end of its movement. A flywheel (not shown) promotes uniformity of motion of the driven shaft.

Lever-Operated Self-Locking Indexing Mechanism.— One continuous motion of lever C of the mechanism shown in Fig. 12 serves to index shaft A through angle T and lock it in position. This mechanism is employed on a hand miller in making two saw cuts in a steel arm. There are various sizes of arms and the angle between the cuts varies with the size of the arm. Provision is made for adjusting the indexing plates S and R to suit any required indexing angle T.

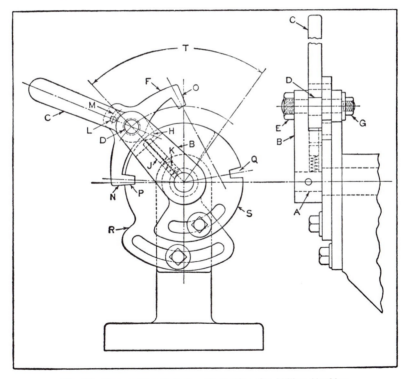

Fig. 12. Two-position Indexing Mechanism for Milling Machine.

Shaft A is part of the work-holding cradle of the fixture, which is to be indexed and locked in either of two positions. Arm B is pinned to shaft A. The stud D is held in arm B by nut E. The operating lever C and pawl F are free to swing on stud D, being retained on the stud by a washer and nut G.

The end H of lever C terminates in a point formed to a small radius. This pointed end engages a pointed plunger J which slides in a hole in the arm B and is forced against the pointed end of lever C by helical spring K. The action of the spring plunger against the operating lever C causes the lever to turn away from the plunger when the lever is swung to either side.

A pin L driven into a hole in lever C slides in a slot M in the pawl F. The length of the slot is such that the lever can be swung until its point snaps over the plunger before the pin reaches the end of the slot and turns the pawl F. The pawl projections N and O engage the slots P and Q, respectively, in the index-plates R and S.

With the members of the mechanism in the position shown, the pressure of the plunger J against point H of lever C causes the lever to turn counter-clockwise. The pin L is thus brought against the end of slot M in pawl F, turning the pawl and forcing projection N into slot P in the index-plate.

To be indexed to the other position, the lever C is turned clockwise until its point H snaps over the point of the plunger J and the pin L has moved to the other end of slot M in pawl F. The pressure of the spring plunger J on lever C now holds pin L against the end of slot M. The projection N of the pawl will be withdrawn from slot P and the pawl will be rotated clockwise until its projection O comes in contact with the edge of the index-plate S. Continued rotation of lever C causes arm B and shaft A, with its cradle, to be rotated through angle T, pawl projection O sliding on edge of index-plate S.

When projection O reaches slot Q, the spring plunger forces the pawl projection into the slot, thus completing the indexing operation. The reverse indexing is effected in the same manner by moving the lever counter-clockwise.

Double-Action Reversing Ratchet Movement.—A machine used for polishing a wire product has a traveling table which is given an intermittent motion by means of a rack and pinion actuated by an oscillating lever through a ratchet and pawl. In the original design, the pawl actuated the ratchet during one-half of the oscillating cycle of the lever, the pawl riding over the ratchet teeth on the return stroke in the conventional manner.

In the improved design, shown in Fig. 13, two pawls G and H are employed to rotate the gear A in a clockwise direction on both the forward and reverse strokes of the oscillating lever E. Referring to the illustration, rod T is given a reciprocating motion by means of a crank, thus transmitting an oscillating motion to lever E, which is free on shaft C. Gears A and R are keyed to shaft C, gear R meshing with the rack P, which is carried on the work-table S.

Referring to the view at the left, lever E transmits motion to lever F through the link I. Lever F and gear B are free to rotate on shaft D and gear B meshes with gear A on shaft C. Levers E and F carry the pawls G and H, respectively. Pawls G and H are slotted to receive pins on the ends of rods U and V, which slide in dovetailed grooves in levers K and L. Rod U is drawn upward by a spring, while rod V is drawn downward by a similar spring. Levers K and L are connected by the link J, which carries pin W at its center. Any horizontal movement of pin W causes levers K and L to move in unison. Stops M and N in work-table S serve to trip the pawls at both ends of the work-table travel.

Referring to the view at the left, the rod T, moving in the direction indicated by the arrow, transmits motion

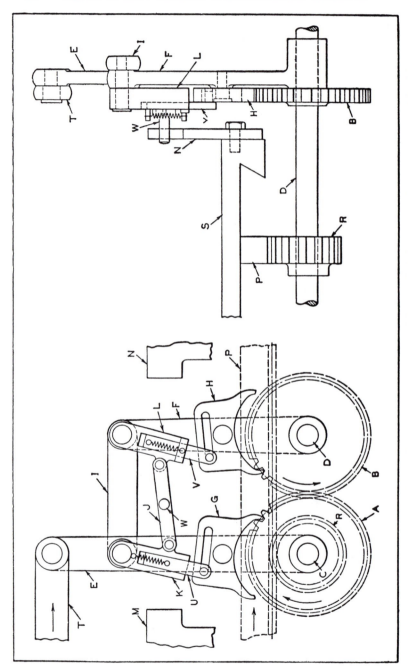

Fig. 13. Reversing Ratchet Mechanism Designed to Rotate Ratchet Wheel on Both Forward and Return Movements of Pawl Lever.

through lever E and pawl G to gear A, giving the latter a partial rotation in the direction indicated by the arrow. As both gears A and R are keyed to shaft C, the motion of gear A is transmitted through gear R to the rack P in the direction indicated. Gear A transmits its motion to gear B in the reverse direction, gear B serving no useful purpose at this time, merely turning under the pawl H.

As rod T reaches the end of its travel and reverses its direction, levers E and F also reverse their direction of travel. At this point, the reverse motion of lever F is transmitted to gear B through pawl H, the motion being continued through gears A and R to the rack P. In this way, both forward and return strokes of rod T are utilized to transmit motion to rack P in the same direction.

As the stops M and N are attached to work-table S, they move with the rack P. In the position shown in the view to the left, each movement of rack P brings the stop M closer to pin W. As stop M strikes pin W, the levers K and L are made to swing to the right, which, in turn, causes the outer ends of pawls G and H to engage the gears A and B. As this causes gears A and B to rotate in the reverse direction, the movement of rack P is reversed. This motion continues until stop N strikes pin W, when the direction of movement is again reversed.

Geneva Wheel Designed for Precise Intermittent Indexing Movements.—A modified type of Geneva wheel that was developed to obtain precise intermittent indexing movements is shown in Fig. 14. This special mechanism, shown diagrammatically at J, was designed to eliminate any backlash or over-running action of the driven member F at the end or beginning of each indexing movement, when the pin A of the driving member E enters or leaves the slot in one of the arms of the driven member F. The relative positions of the driving and driven members at the start and end of the indexing movement are shown in the diagrams J and L, respectively.

With the mechanism designed as shown, the driven shaft H, keyed to wheel F, is indexed one-fourth revolution for each revolution of the driving shaft G, which is keyed to wheel E. Each indexing movement starts smoothly as the long driving pin A enters one of the slots in wheel F and pin C passes out of contact with the flange on the rim of wheel F.

The driving pin A, as shown in section $X—X$, is long enough to make contact with the sides of the slots in wheel F for the full depth or thickness of the slotted

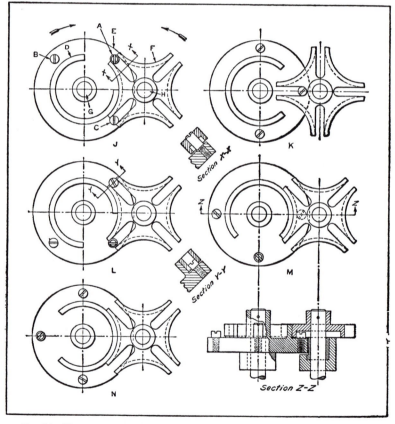

Fig. 14. Diagrams Illustrating the Operation of a Geneva Wheel Designed for Precise Intermittent Indexing Movements.

part of the wheel. The two shorter pins *B* and *C*, one of which is shown in section *Y—Y*, are made to clear the bottom of the recess machined in the under side of wheel *F*, as shown in section *Z—Z*, but are not long enough to contact the straight sides of the slots contacted by pin *A*.

The pins *B* and *C* are so positioned, however, that they make close running contact with the flange formed by the recess on the under side of wheel *F* (see the broken section *Z—Z*). The flange *D*, machined integral with wheel *E* to a close running fit with each of the four segments of wheel *F*, has a 90-degree section cut away opposite the driving pin *A* to provide clearance space for the projecting arms of the driven wheel during the indexing movement.

Referring to the diagram at *J*, pin *A* is just entering a slot in wheel *F*, while pin *C* is passing out of contact with the rim on the under side of wheel *F*. The pin *C*, being a close running fit on the inside of the flange on wheel *F*, and flange *D*, being a close running fit on the outside of the flange on wheel *F*, serve to hold wheel *F* stationary until the indexing actually begins, and also prevent it from further movement the instant the indexing is terminated. During part of the revolution of driving wheel *E*, the flange *D* alone serves to hold the driven wheel stationary in the dwell position, as indicated in diagram *N*.

The accurate positioning of pin *C* also prevents any movement of wheel *F* before pin *A* engages a slot. Thus the indexing movement of wheel *F* is started with a smooth, accelerating motion and stopped as smoothly with a decelerating motion without any over-run or backlash.

Diagram *K* shows the mechanism with the driven wheel *F* rotated through one-half of the first indexing movement, at which point it has reached its maximum speed of rotation.

Diagram *L* shows the driven wheel at the end of the indexing movement, with the short pin *B* of the driving wheel making contact with the flange on the under side of the driven wheel, so that it holds the outer side of the flange in contact with flange D of driving wheel E. Thus pin B prevents any rotational movement of the driven wheel as driving pin A leaves the slot in the driven wheel at the end of the indexing movement.

The diagram at M shows the driven wheel in the dwell position with pin B still in contact with the flange on the under side of the driven wheel. The broken section Z—Z shows the short pin B clearing the recess in the driven wheel, which is held stationary in the dwell position. Section Z—Z is broken, part of the view to the left being shown in full lines, in order to indicate the difference in the heights of the driving pin A and the two pins B and C.

The diagram at N shows the driven wheel still in the dwell position, where it remains stationary until rotation of the driving wheel E brings pin C and driving pin A into the positions indicated in diagram J. The indexing movement and dwell period described are then repeated, driven wheel F being indexed one-fourth turn for each complete revolution of driving wheel E.

Geneva Motion Mechanism of Unique Design.—In designing a machine for the automatic stamping of consecutive numbers on the corners of envelopes, it became necessary to develop some interesting mechanisms, among which was the Geneva dial motion shown in Figs. 15 and 16. This unique mechanism performs its intermittent indexing movements at a uniform rotational speed instead of at the accelerating and decelerating speed of the harmonic motion characteristic of the Geneva mechanism of conventional design ordinarily employed for such purposes.

Referring to Fig. 15, the mechanism is driven by a pinion at A which meshes with gear B. Indexing of turret C, as required to bring the correct numbers into their respective stamping positions, is controlled by a separate mechanism (not shown) which actuates trip-rod D. When rod

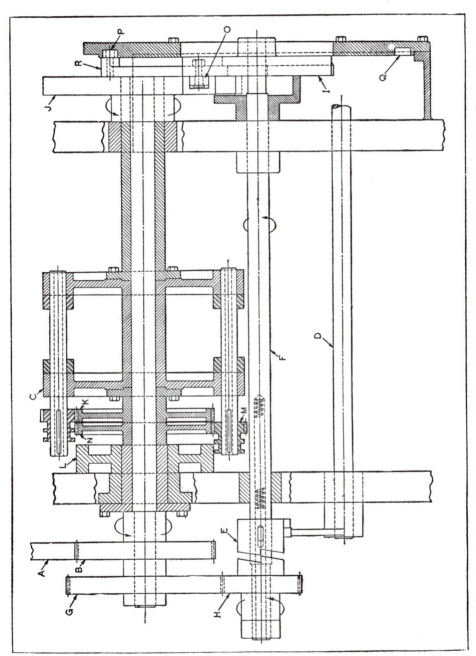

Fig. 15. Geneva Motion Mechanism which Indexes at Uniform Rotational Speed. Driving Member I Indexes Geneva Dial J One-sixth of a Revolution.

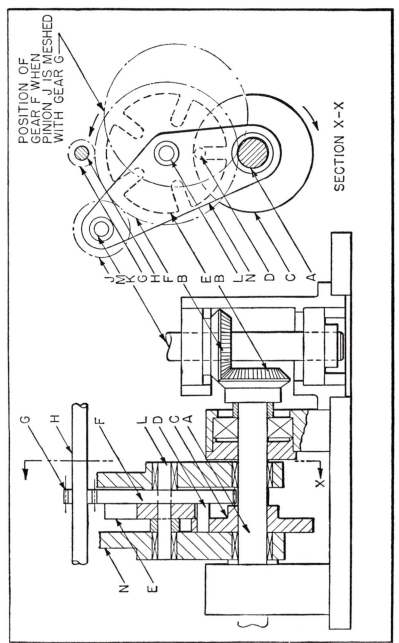

Fig. 17. Intermittent and Reversible Rotation of Shaft H is Obtained from Driving Shaft A by Means of Geneva Wheel E and Gearing.

lution of shaft *A*, wheel *E* is indexed through one-fifth of a revolution by pin *D*. Gear *F*, which is fixed to wheel *E* and rotates on the same shaft *L*, has sixty teeth of 24 pitch. Pinion *G*, having twelve teeth of 24 pitch, which meshes with gear *F*, is therefore driven intermittently through one revolution for each revolution of driving shaft *A*.

The direction of rotation of shaft *H*, on which pinion *G* is mounted, can be reversed by bringing gear *J* into mesh with pinion *G*. This angular movement disengages gear *F* from pinion *G*, as both shafts *L* and *M* are mounted in bracket *N*, which pivots about driving shaft *A*.

CHAPTER 4

Overload, Tripping and Stop Mechanisms

Mechanisms. which automatically operate to stop an operation when overload occurs, to trip and start a new sequence or operation when a certain position or part of a cycle is reached, or to bring an operation to a halt at the end of a given cycle or when a given amount of motion has occurred, are described in this chapter. Other mechanisms performing similar functions are described in Volumes I and II of "Ingenious Mechanisms."

Tripping Mechanism Operated by Revolving Shaft.— The mechanism shown in Fig. 1 is designed to operate the tripping lever A from the rotating shaft B. The object is to move lever A from the position shown by the full

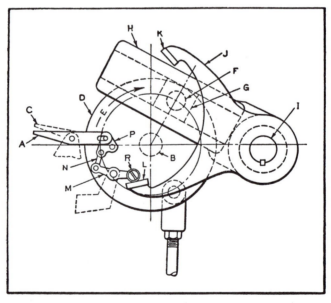

Fig. 1. Tripping Mechanism Operated by Rotating Shaft B which Shifts Lever from Position A to Position C and Back Again with Dwell Between Movements.

lines to that indicated by dotted lines at C, and back again, with an idle or rest period between movements. This is effected by means of disk D, which is keyed to shaft B, and linkage arrangement shown. Shaft B and disk D rotate in direction indicated by arrow E. A pin F in disk D fits a block G which is a sliding fit in a slot in lever H.

Lever H, being keyed to the shaft I, oscillates the latter shaft continuously as shaft B revolves. This oscillating movement is also transmitted to the fork-shaped member J, keyed or pinned to shaft I. Member J has two pads K and L which are alternately brought into contact with the roll R at one end of the lever M. The illustration shows the mechanism with lever H at approximately its highest point, and pad L in contact with roll R. With lever H in this position, the outer end of the trip-lever A is depressed to the position shown through the action of connecting link N and crank member P.

As shaft B continues to revolve, pad L moves away from roll R, allowing trip A to retain its position until pad K makes contact with the upper side of roll R, depressing it so that the outer end of trip A is raised to the position indicated by the dotted lines at C. Continued rotation of shaft B raises pad K from contact with roller R, so that the trip retains the position indicated by the dotted lines until pad L is again brought into contact with roll R.

Tripping Mechanism Operated by Revolving Shaft.— A modification of the design just described eliminates the large member J shown in Fig. 1.

As shown in Fig. 2, the lever H is provided with an integral extension piece C which acts upon roller R, causing it to transmit motion to the tripping lever A. Pad L is attached to lever H, as shown, and is located so as to come in contact with roller R when lever H is brought to its lowest position. The lever H has an oscillating movement obtained by means of shaft B, disk D, etc., as in the mechanism just described.

Two-Way Stop for Angular Movement of Shaft.—The mechanism shown in Fig. 3 comprises an arrangement for stopping the rotation of shaft A in two extreme angular positions by means of a single lever arm B attached to the shaft. The shaft C is connected to the lever arm by a pin, and is a slip fit through the trunnion D. The trunnion revolves around pin G and is held securely to the baseplate by the two bearing brackets E.

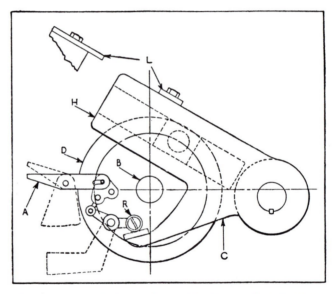

Fig. 2. Tripping Mechanism Operated by Rotating Shaft B which Shifts Lever A with Dwell between Movements.

The rubber shock absorber F on the end of shaft C stops against the trunnion block when shaft A is in either of the extreme positions X or Y. The position of the shock absorber is adjusted by means of the nut and lock-nut shown. By moving the trunnion block and its supporting brackets to either side of the center line, the extreme positions at which the angular movement of shaft A is stopped can be changed to suit a wide range of operating requirements.

Mechanism for Tripping a Rotating Lever.—The mech·· anism shown diagrammatically in Figs. 4 and 5 is designed to operate a trip-lever within a revolving housing at certain predetermined intervals, the exact time of each operation being controlled by a cam driven at the correct speed. The revolving housing A, Fig. 4, is mounted on sleeve B held on shaft C. The housing or head A, with its group of parts, revolves on sleeve B in the direction indicated by arrow D. The short lever E is free to pivot, and is pinned to shaft F, shown in cross-section. Lever E carries a roll G which comes in contact with a long radial dwell

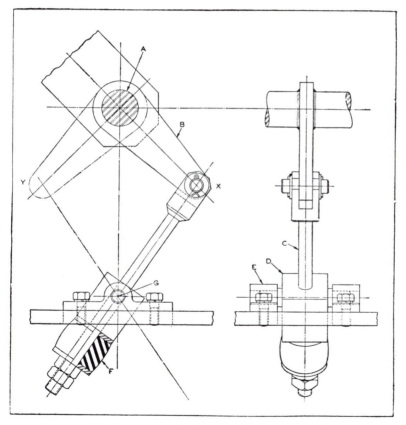

Fig. 3. Mechanism for Stopping Rotation of Shaft A in Two Extreme
Angular Positions.

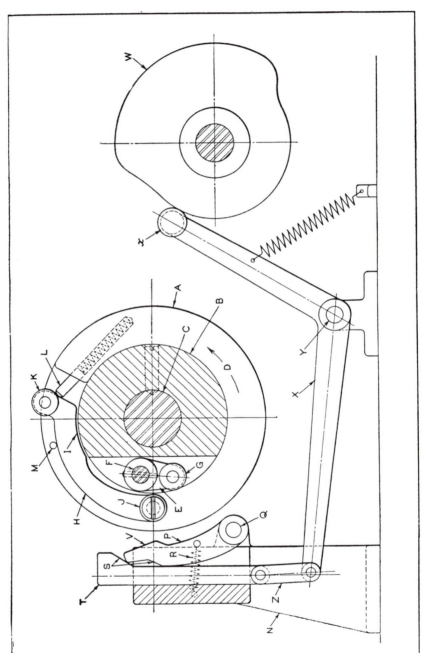

Fig. 4. Mechanism for Operating a Trip-lever on Shaft F within the Revolving Head A.

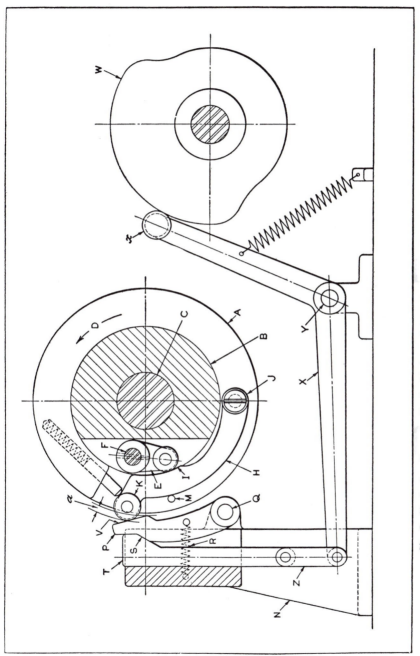

Fig. 5. Mechanism Shown in Fig 4 with Lever P Brought into the Operating Position by Cam W.

I on lever *H*, the latter lever being attached to the housing by fulcrum stud *J*.

The long lever *H*, with a roll *K* at the end, is held in the outward position, away from the center of the revolving housing by means of spring pin *L*, the outward movement being limited by a stop-pin at *M*. When roller *K* moves in and out a distance *a*, Fig. 5, pivoting about stud *J* as a center, the short lever *E* rocks shaft *F*. This rocking action, in turn, operates a trip mechanism (not shown) which is located within the machine itself. A spring within the machine serves to hold the short lever in its outer radial position.

The mechanism for operating the long lever *H* at predetermined intervals as it revolves past a non-revolving cam surface *V* on lever *P* is actuated by cam *W*. Referring to Figs. 4 and 5, bracket *N* is attached to the side of the machine and carries lever *P*, which is pivoted at *Q*. Spring *R* serves to hold lever *P* back against the beveled or cam surface *S* of a sliding member *T*. Member *T* is

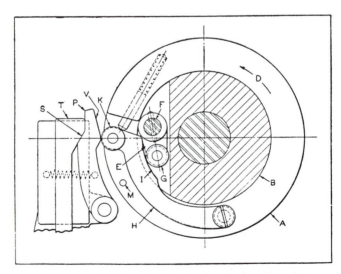

Fig. 6. Lever H is Forced Inward by Cam Surface V on Lever that Causes Cam I to Rotate Shaft F in Counter-clockwise Direction through Contact with Roller G.

free to travel up or down in a guide-way of the bracket. Cam surface S of the slide operates to push lever P to the right, so that roll K on the revolving lever H strikes the cam surface V at the correct time for operating the trip mechanism.

Actual timing of the trip is accomplished by cam W on a shaft which revolves in time or synchronism with the other operative units of the machine. Lever X, with a roller at x, pivots on a stud at Y. Lever X actuates slide T through connecting link Z; thus, when cam W and head A revolve in proper synchronism, the surface S on slide T is moved into engagement with lever P which, in turn, places surface V in position to push lever H over, as shown in Fig. 6, causing the short lever E to operate the tripping mechanism within the machine.

Overload Release Clutch Mechanism.—In the operation of an automatic machine that produces a formed wire product, it was impossible to prevent occasional jamming when changes in the wire size resulted in imperfect forming. To prevent serious damage to the machine, it was necessary for the operator to cut off the power immediately when the machine became jammed. As the operator was unable to maintain the close watch of the machine required to prevent damage, it was decided to attach an overload release clutch, as shown in Fig. 7.

The normal operating positions of the overload clutch members are shown in the left diagram of Fig. 7. Driveshaft A carries a collar B, which is keyed to it. Collar B is grooved on the periphery to receive a pad on the long arm of bellcrank lever D which swivels on stud E carried on gear C. Stud G on gear C carries lever F, the lower end of which is provided with a V-shaped cam surface. The lower side of this V-shaped cam is in contact with the rounded end of lever D. Spring H, attached to the short arm of lever D and the lower end of lever F, provides sufficient tension to hold the pad on lever D in

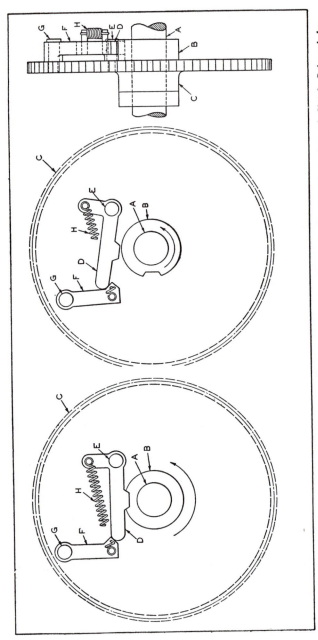

Fig. 7. (Left) Overload Release Clutch Mechanism with Driving Members Engaged; (Center) Mechanism of Clutch Released by Overload; (Right) End View of Clutch.

the groove of collar B during normal operation. The rotation of collar B is thus transmitted to gear C through lever D and stud E, the entire assembly rotating as a unit in the direction of the arrow, as shown in the left diagram of Fig. 7.

In the event that the clutch is overloaded, due to jamming of the work, the resistance to rotation of gear C overcomes the tension of spring H, causing the pad on lever D to be forced out of the groove in collar B, and thus disconnecting gear C from the source of power transmitted by shaft A. The movement of the end of lever D on the cam surface of lever F swings the latter member to the left; and as the rounded end of lever D passes over the high point of the cam surface on lever F, the upper edge of the cam surface on lever F acts on the lower edge of the end of lever D, lifting it completely out of contact with collar B, as shown in the center diagram of Fig. 7, and thus preventing lever D from returning to its normal driving position until it has been re-engaged by the operator.

An Overload Relief Device for Machine Protection.— An ingenious device designed to protect a driven machine mechanism from breakage in the event of accidental jamming of the work during normal operation of the machine is shown in Fig. 8. One feature of this device is that all forces are self-contained, and when it is tripped or released, no axial force is exerted on bearings or moving parts.

The mechanism consists primarily of a lever A which transmits a rocking movement to shaft B through suitable linkage. By itself, lever A is free to swivel about shaft B. It is held in one position on the shaft by a bearing at the right and arm C and collar F at the left. With plunger D seated in a conical socket in bushing E, however, motion is transmitted from lever A through arm C to shaft B, arm C being keyed to the shaft. Plunger D is held in the seated position by spring H.

The conical fit between plunger D and the socket of

bushing *E*, as well as the load provided by spring *H*, should be designed for transmitting only the desired amount of torque. Then, when this predetermined torque is exceeded, the tapered end of plunger *D* will ride up out of its socket and lever *A* will become disengaged.

To keep the parts disengaged is the function of plunger *G*. At the top point in the disengaging movement of plunger *D*, the deeper of two flat spots on the plunger comes opposite the end of plunger *G*. Plunger *G* then snaps to the right and holds plunger *D* in the disengaged position. This

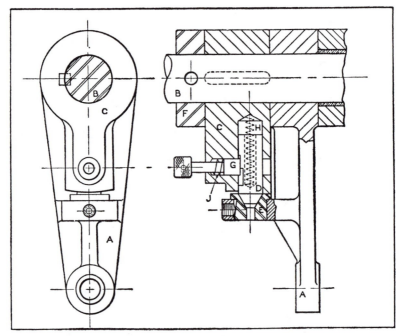

Fig. 8. Ingenious Device Designed for Protecting a Mechanism against Breakage if the Work Becomes Jammed.

permits the lever *A* to rock freely on shaft *B* until the device is manually reset, and power is, of course, no longer transmitted to shaft *B* through the mechanism.

To reset the device, the machine must first be cleared

of the obstruction, after which plunger G can be withdrawn to the left and plunger D reseated in the socket of bushing E. Plunger G is then free to return to the right, where it engages the shallower flat spot on plunger D. The purpose of this shallower flat spot is to prevent rotation of the plunger and to retain plunger D in position during disassembly of the mechanism.

This device is also applicable to rotary driving members, as well as to rocking members. In the case of rotating applications, the centrifugal force acting on plunger D would have to be considered as well as the load provided by spring H. The plungers D and G and bushing E should be hardened in order to insure satisfactory life.

Drive Unit with Overload Slip Mechanism.—The slip device shown in Fig. 9 was designed to prevent breakage of operative members, and is an important feature of the feeding mechanisms of certain machines. It is of the jump-pawl type, the pawl A being designed to operate with the radial plate Q. Pawl A is free to pivot upon stud B, although it is normally held in the position shown by the spring C.

The bellcrank lever D oscillates through the arc E, being actuated by the shaft F through pawl A, plate Q, and plate carrier R which is keyed to shaft F. Linkage levers G and H are moved up and down by means of connecting-rod J. Lever G is attached to a non-movable block at K, while lever H is attached at L to a sliding member M. The sliding member moves backward and forward in the direction indicated by the arrow N, once on the up stroke of the connecting-rod J and once on the down stroke, thus, in effect, producing a two-cycle movement of the member M for each reciprocation of rod J.

If slide M becomes jammed or is prevented from moving in a normal manner, pawl A will ride out of the notch in plate Q and travel over the radial surface of this member, thus disconnecting the drive.

new parts were made, the only work done consisting of a slight modification of the trigger elements of the drop-worm. Instead of being made practically square, to hold the drop-worm positively in mesh with its gear, the trigger faces A and B were made at an angle of 15 degrees, so that a downward thrust on B, resulting from an overload, would disengage the feed. Set-screw C compresses a spring, and can be adjusted to vary the pressure with which the trigger is held in the engaged position. The worm is lifted by knob D, and knob E operates the trigger to trip the feed. The feed-shaft runs in the direction indicated by arrow F.

Roll-Driving Mechanism that Stops Automatically in Case Material Breaks.

—The mechanism shown in Fig. 11 was designed for unwinding thin tissue from large rolls. The problem was to provide suitable means for feeding the delicate, thin material, which is easily stretched and torn by improper handling. The arrangement shown provides for stopping the feed if the material breaks, and will also slow up the speed of the roll if the feeding rate becomes greater than the rate at which the material is consumed. Thus, the roll of thin material is kept under control at all times, so that the tissue is subjected to a minimum amount of tension.

The two fork-shaped side frames A are bolted to a base-plate B. They support two feeding rolls C, mounted on shafts D, to which they are fastened by taper pins E. Rolls C are provided with molded rubber covers F to increase the driving friction. The roll of tissue G is kept in position by shaft H, which is fitted in bushings I. The bushings are arranged to slide freely in the vertical slots in the side frames A.

Mounted on each of the feeding roll shafts D is a spur gear J, fastened in place by a taper pin K. Bearing blocks L are so constructed that the feeding roll assembly can be removed by unscrewing bolts M. Driving gear N of the feeding control mechanism is fastened to stub shaft O by a taper pin. The driving gear meshes with the feeding roll gears J. Clutch member P is fastened to stub shaft O, in which there is a bore that supports the end of driving shaft Q. The driving shaft is thus supported between the stub shaft and a bearing in the side frame A. Clutch member R slides on shaft Q, guided by key S under the

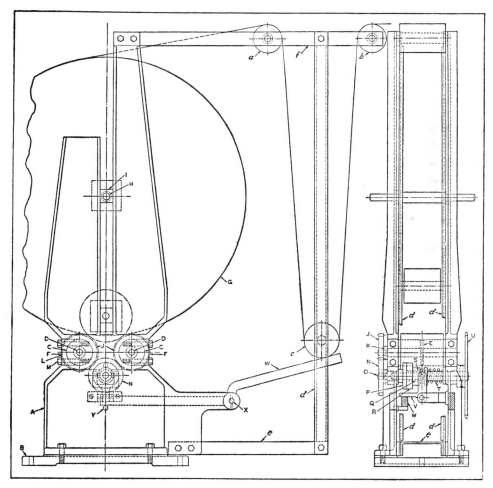

Fig. 11. Power-driven Mechanism for Unwinding Thin Tissue from Roll G which Stops if Material Breaks or if the Feeding Rate Exceeds the Rate at which the Material is Used.

action of spring T. Driving sprocket U is fastened to shaft Q.

Yoke V of the clutch-actuating mechanism is equipped with pins which fit into a groove in clutch member R. The yoke is supported by a bracket fastened to side frame A. Tripping bar W is pivoted on shaft X, which is supported in bearings provided in side frames A. The tripping bar is balanced, so that it rests on pin Y in the side frame.

The thin tissue material is guided over a series of three rolls, the positions of the two rolls a and b being fixed. The third roll c is free to move up and down in the side members or guides d. Guides d are fastened to horizontal members e and f which are, in turn, fastened to side frames A.

The operation of the unwinding mechanism is as follows: Sprocket wheel U actuates driving gear N through clutch members P and R, thus rotating feeding rolls C. The feeding rolls, in turn, rotate the roll of tissue, the entire weight of the material being supported by the feeding rolls. The tissue passes over guide roll a, under movable guide roll c, and up over the fixed guide roll b. In actual operation, movable guide roll c is raised from tripping bar W.

If the tissue is broken or ruptured, movable guide roll c comes to rest on tripping bar W, causing it to raise the opposite end from pin Y and make contact with yoke V. This action causes the yoke to draw clutch member R away from member P, thus stopping the feeding rolls. The same action takes place when the rate at which the material is being fed becomes greater than the rate at which the material is consumed. In such cases, movable guide roll c will fall on tripping bar W, causing the feeding rolls to stop and thus stop the rotation of the tissue roll G.

Clutch Equipment for Quick-Acting Brake.—The quick-acting brake in the double-ended jaw clutch shown in Fig. 12 was designed to permit the sudden stopping of a certain shaft E used for winding metal strip. The driving shaft A

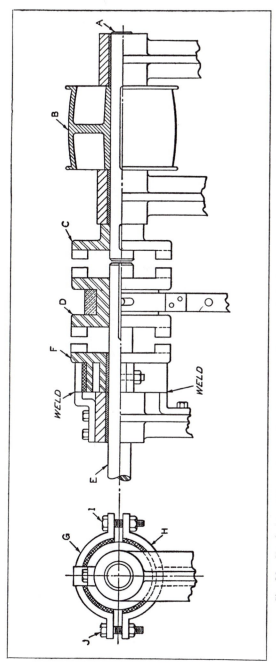

Fig. 12. Clutch Arrangement for Disengaging Shaft E from the Driving Shaft A and then Applying Brake to Shaft E.

is rotated continuously by a belt on the pulley B to which it is keyed. The double-ended clutch jaw D is free to move endwise along shaft E and slide along a key which compels it to revolve the shaft when engaged with the driving member C.

When jaw D is disengaged from member C, it engages the member F, which also acts as a brake-drum. Brake-shoes G and H, being tight at all times, will stop shaft E quickly upon the engagement of clutch D with member F. The brake-shoes are kept tight by means of bolts J and I. They are fastened to the bearing stand, which prevents them from turning. The brake linings are of standard width and can easily be replaced when worn out.

Safety Relief Mechanisms for Light Drives on Special Machines.—A relief mechanism of some type incorporated in the main drive provides adequate means of preventing breakdowns of many special light machines. It is sometimes difficult, however, to design or choose a simple effective device for this purpose that will entail a minimum of changes in existing mechanisms or machines. Figs. 13 to 18 show various pieces of mechanism that are capable of adaptation to almost any kind of special machine.

Practically all high-speed special-purpose machines have a number of connecting-rods, link arms, etc., one or two of which can be selected for modification. For example, the usual solid type connecting-rod can be replaced, as shown in Fig. 13, by one made of two parts A and B. Part B is tubular, and is threaded at one end with, say, twenty-six threads per inch to permit a fine adjustment of the operating arm by means of a knurled castellated nut G.

The other member A is a sliding fit in part B. A compression spring, located in part B, exerts pressure on member A. Two slots E are cut 180 degrees apart in B, and when this member is adjusted to the correct operating center distance by nut G, a pronged cotter-pin H is placed across one of the slots in the castellated head of nut G.

If the operating movement of the mechanism is obstructed, the arm simply collapses against the pressure of the spring until the mechanism has been freed. The amount by which the connecting-rod can be telescoped is limited by the length of slots E and the length of the prongs on cotter-pin H. The compression spring must, of course, exert sufficient pressure to operate the drive without yielding when the machine is working under normal conditions.

When the drive to a main camshaft or to a separate mechanism requires a safety relief, the one shown in Fig. 14 can often be used to advantage. On the main driving shaft C is a flanged sleeve D which is driven by key E, fastened to the main driving shaft. In the flanged sleeve there is an L-shaped slot F in which the projecting ends of the key are a sliding fit. A compression spring tends to force the bottom or end of the longitudinal leg of the L-slot into contact with the pin or key E. Thus the main driving gear or sprocket J is driven by means of pin E and its contact with slot F.

The flange also carries a clutch tooth K which is wedge-shaped, and at the instant the drive becomes excessive, the flange is forced against the spring and causes pin E to make contact with the other leg of slot F. The drive will then continue with the sleeve in an inoperative position until the obstruction is removed.

If a relief or safety device for a pusher member, slide, or folders is required, the simple attachment shown in Fig. 15 can be satisfactorily employed. The main cam arm should be replaced by an arm which has a slot Q at the operating end M. A small trigger-like arm N is pivoted about pin P. At the end of the trigger arm is a flared portion into which a recess is drilled to receive a compression spring. The compression spring tends to force the right-hand end of arm N slightly over the boss that carries the pusher, and under normal conditions has sufficient strength to transmit cam motion. If, however, the pusher strikes an obstruction,

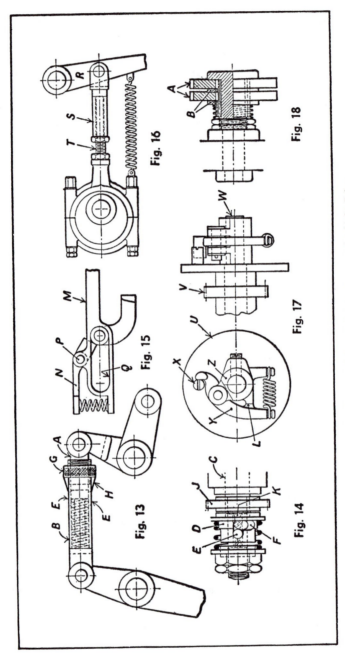

Figs. 13-18. Six Types of Safety Relief Mechanisms for Light Drives.

the pressure causes the right-hand end of arm N to move upward and allow the arm to ride in the slot until the obstruction has been removed.

In a positive cam-driven mechanism, there is always danger of bending an arm and ruining the mechanism. To avoid this, the drive can be changed or replaced by that shown in Fig. 16. In this case, the driven arm R may have a connecting-rod S, and the cam arm may have an adjustable rod T. To introduce a safety device, one of the ends of rod T is threaded and screwed into the cam arm while the other end is left plain and inserted in a long bore in piece S. The two parts S and T are made of steel and are hardened. The bore of part S should be 1/32 inch larger than the diameter of the plain portion of rod T.

The cam arm and the driven arm are pulled together by means of a tension spring having sufficient strength to transmit the necessary motion. If, however, the mechanism becomes jammed, the spring is stretched and the plain portion of rod T reciprocates in the bore of piece S until the obstruction is cleared and arm R is again pulled into its operative position.

The drive for a conveyor or stock-feeding mechanism of a box-making machine, for example, can be safeguarded against breakage by the addition of a slip drive like that shown in Fig. 17. A flange U is fastened to the main driving gear or sprocket V, the whole assembly being allowed to rotate freely upon the drive-shaft W.

Flange U carries one or more pins X. When the positive drive is in operation, one of the pins X engages the drive lever Y, which is pivoted on a pin carried by the driving boss Z, fastened to the main drive-shaft by a set-screw. The driving lever Y is forced into contact with pin X by a tension spring, and stops L are filed to permit proper functioning. If the drive is obstructed, the spring yields and the main drive-spindle is allowed to revolve freely without rotating the driving gear or sprocket.

If the drive is of a lighter nature, such as that for a sheet-feeding device or a dating mechanism, a safety arrangement like that shown in Fig. 18 can be employed. The driving gears A can be made a running fit on the drive-shaft or stud. Each of the side faces of the driving gears comes in contact with the friction disks B, which are preferably keyed to the shaft in such a manner as to prevent rotation and yet allow a sliding motion on the shaft. These gears are forced into contact with the side faces by means of a compression spring adjusted to exert the required amount of pressure for driving the mechanism. Lock-nuts are provided, as shown, for maintaining the required setting.

CHAPTER 5

Locking, Clamping and Locating Devices

Means of positively locking a mechanism, clamping a work-piece or part, and locating work in the proper position for some operation to be performed on it, or locating a carriage or table in the correct loading position, are described in this chapter. In some cases, the locking or clamping operation is performed automatically, while in others hand operation is required. Similar devices are described in Volumes I and II of "Ingenious Mechanisms."

Double Locking Lever Motion.—A lever-and-link motion used to operate a slide that must be locked against movement at both ends of its vertical travel is shown diagrammatically in Figs. 1 and 2. The linkage automatically pro-

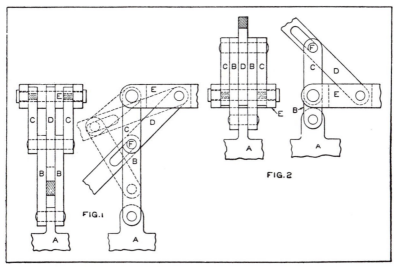

Figs. 1 and 2. Lever-and-link Motion which Locks Slide A at Each End of the Vertical Stroke.

vides the desired locking feature in both positions. The links *B* are connected to the slide *A* at their lower ends and are pivoted at their upper ends to the bracket *E*, attached to a stationary part of the machine. The hand-lever *D*, pivoted in the bracket *E*, passes between the links *B*, and is slotted to permit the pin *F* to pass through.

In Fig. 1, slide *A* is shown at its lower position. Links *B* and *C* have their centers in the same straight line, thus producing a dead center condition which locks slide *A* against movement in either direction. It will be noted that lever *D* is at an angle of approximately 45 degrees with links *B* and *C*. As the outer end of lever *D* is raised, the angularity of the slot in lever *D* causes pin *F* to be moved outward, so that links *B* and *C* are moved from their locked positions. Continued movement of lever *D* causes pin *F* to slide in its slot, and links *B* and *C* to swivel on their connecting pin, as shown by the dotted lines, Fig. 1, so that slide A is drawn upward.

Further movement of lever *D* causes links *C* to rotate 180 degrees on their upper pin, and links *B* to enter the spaces between links *C*. At the extreme upper position of slide *A*, as shown in Fig. 2, links *B* and *C*, again being in alignment, cause slide *A* to be locked against further movement. Thus the arrangement of the links is such that movement of slide *A*, when at its upper and lower positions, can be accomplished only through movement of lever *D*.

Lever Mechanism for Operating a Locking Pin and Clamping Bolts in One Movement.—The mechanism illustrated in Fig. 3 is employed on the swivel type fixture shown in Fig. 4 to withdraw a locking pin and release two clamping bolts in one movement of the hand-lever. The reverse operation of engaging the locking pin and tightening the clamping bolts is similarly accomplished by a simple return movement of the lever.

Referring to Figs. 3 and 4, it will be noted that hand-lever A is located near one corner of the fixture in a convenient operating position and at some distance from locking pin K and the two clamping bolts B and C which it actuates. The clamping bolts act on both ends of the

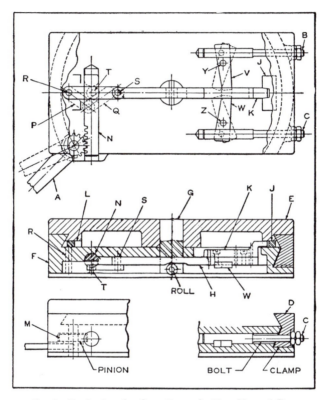

Fig. 3. Mechanism for Operating a Locking Pin and Two Clamping Bolts in One Movement of a Hand Lever.

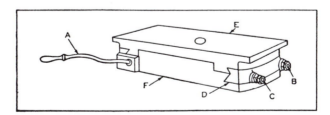

Fig. 4. Swivel-type Fixture Equipped with Locking Mechanism Shown in Fig. 3

dovetail segment D, as shown in detail at the lower right, lever A being the clamping medium.

The purpose of the entire unit is to bind upper portion E of the swivel base to the lower portion F. As shown in Fig. 3, stud G, which passes through the entire unit, is slotted and has a roll at its lower end. Bar H passes through the slot in stud G and rests on the roll. Stud G serves as a pivot on which section E can be swiveled. Swivel section E is located on the base by locking pin K which fits the slot in plate J attached to member E, while another plate at L serves to locate the swivel section when it is reversed or revolved through an angle of 180 degrees.

Pinion M, attached to operating lever A, meshes with a rack bar N. There are two toggle levers P and Q, lever P being pivoted at one end on pin R, fixed in base F. The other end of this lever is connected to one end of lever Q, the outer end of which is connected to bar H by pin S. Pin T acts as a fulcrum for the toggle joint.

At this fulcrum point, the ends of the levers are yoked into a cross-slot in the rack bar, so that, as the rack is moved into the position shown in Fig. 3, the toggles force bar H into the clamping position. When lever A is in the position shown by the dotted lines, the toggles are likewise in the positions shown by dotted lines, which results in withdrawing the clamping bar from plate J and unlocking dovetail clamp D, at which time two levers (shown at V and W) assume the positions shown by the dotted lines.

Levers V and W engage a slot in the lower side of bar H and pivot on pins Y and Z. These levers also have yoke connections with the two clamping bolts B and C, so that a solid metal-to-metal contact is provided. In effect, this forms a quick-operating, toggle-action mechanism by means of which the simple action of moving lever A is all that is necessary to unlock and unclamp upper member E.

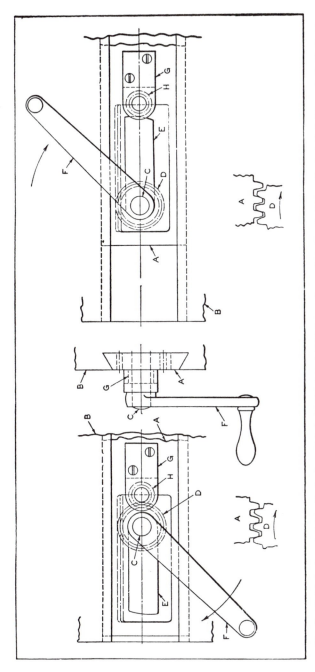

Fig. 5. (Left) Rack-and-pinion Mechanism with Rack Member A Withdrawn from Clamping Position; (Center) End View of Mechanism; (Right) Member A Locked in Clamping Position by Bar E.

in block *E*, and is keyed to shaft *F*, as is handle *G*. Block
E is free to slide in a recess in the base; this recess and
the contacting edge of the block are machined at an

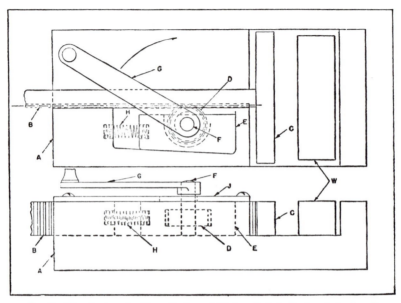

**Fig. 6.   Vise Equipped with a Rack and Pinion that is Designed to Permit
Rapid Clamping and Positive Locking in Any Position.**

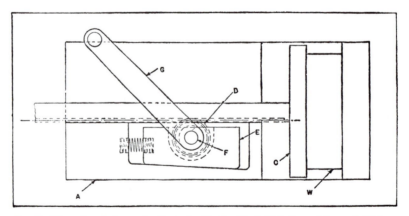

**Fig. 7.   When Jaw C Comes in Contact with the Work, Rotation of Lever G Wedges
Block E Between the Rack and the Wall of the Recess, Thus Locking the Work in Place.**

angle to the pitch line of the rack. Spring *H*, which is in compression, exerts pressure against the block and holds it against the right-hand wall of the recess with a force sufficient to overcome the frictional resistance of the rack to movement. Plate *J*, shown in the front elevation view, serves as a retainer for the assembly.

In operation, as handle *G* is turned in the direction of the arrow, the rotation of the pinion causes the rack to move to the right. This movement continues until jaw *C* comes in contact with the work-piece *W*, when further movement is prevented. Continued turning of the handle causes block *E* to be moved to the left, as shown in Fig. 7, by the force exerted by the pinion. Owing to the wedging action of the block, the rack is firmly locked against any reverse movement. However, when handle *G* is turned in a counter-clockwise direction, the block returns to its original position and the rack is freed for releasing the work.

A Non-Reversible Rack and Pinion Motion.

A rack and pinion motion is ordinarily reversible, so that rotation of the pinion will cause straight-line movement of the rack, and power applied to the rack will produce rotation of the pinion. In the assembly illustrated, movement of the rack can be obtained only by rotation of the pinion, and the mechanism is locked to prevent movement of the rack against the pinion.

Referring to Fig. 8, the bearing *A* supports shaft *B*, which carries the hand-crank *C*, pinned to the shaft, and pinion *D*, which is free to rotate. Collar *E* is also pinned to the shaft and serves to retain pinion *D*. Bearing *A* is counterbored to receive the closely wound spring *F*. The two ends of the spring are turned in, as shown in Fig. 9, which illustrates the counterbored end of bearing *A*. The clearances are exaggerated in the illustration for the sake of clarity.

Spring *F* is made to a diameter which requires that it

be pressed into the counterbore of bearing *A*, so that a certain amount of frictional resistance to movement will be applied. Part *G* is a disk, keyed to shaft *B*, from which a segment-shaped section has been removed. The

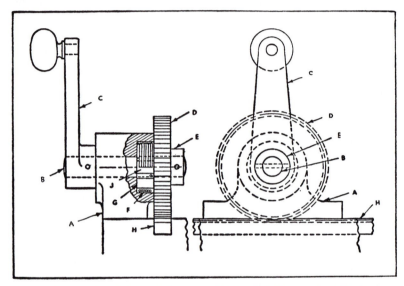

Fig. 8. A Rack-and-pinion Motion Assembly in which Movement of the Rack can be Obtained Only by Rotating the Pinion.

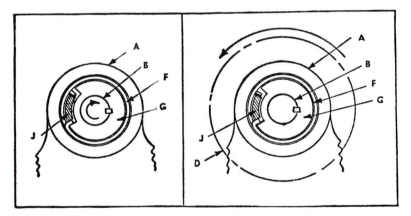

Fig. 9. (Left) View of Counterbored End of Bearing A Indicating Rotation of the Pinion in the Direction Shown by Arrow, which Transmits Motion to the Rack; (Right) View of Counterbored End of Bearing A Illustrating Operation of Locking Action which Prevents Movement of the Rack.

shape of disk G is shown clearly in Fig. 9. Pinion D has on its inner side a segment-shaped projection J, which enters the counterbore between the ends of spring F and the segment-shaped section which has been removed from disk G.

The two diagrams in Fig. 9 illustrate the operating principle. In the left-hand diagram, shaft B is being rotated by the hand-crank in the direction indicated by the arrow. As disk G is keyed to shaft B, it rotates with the shaft, one side of the cut-away section contacting one end of spring F. Since spring F is a tight fit in the counterbore of bearing A, the end of the spring at first offers resistance to further turning of disk G, but continued turning effort applied to the crank C and transmitted through disk G to the end of the spring, causes the latter to be wound tighter, thus slightly reducing the diameter of the spring. When this occurs, spring F is free to turn with disk G.

As projection J on pinion D is in contact with the end of spring F, and as pinion D is free to rotate on shaft B, pinion D is likewise carried around with disk G, the motion thereby being transmitted to rack H. When crank C is turned in the opposite direction, a similar effect is produced. Motion may, therefore, be transmitted to rack H by turning crank C in either direction.

The locking action operates as follows: If force is applied to rack H to rotate pinion D in the direction indicated by the arrow in the right-hand diagram of Fig. 9, the turning effort of the pinion will be applied to the end of spring F through projection J in the reverse direction to that shown in the left-hand diagram of Fig. 9. This causes spring F to unwind, thus increasing the diameter of the spring, so that the pressure against the walls of the counterbore in bearing A is increased. This pressure becomes greater as the force applied to the rack is increased, preventing movement of H in either direction.

In both diagrams of Fig. 9, the space between the end of spring F and the cut-away portion of disk G is greatly exaggerated. It is only necessary to provide sufficient clearance to permit a slight movement of the ends of the spring. The required clearance can be determined by trial.

Rack-and-Pinion Mechanism with Self-Locking Non-Reversing Feature.—In the operation of a wire-forming machine, it is frequently necessary to change the adjustment of a guide manually. The adjustment must be made quickly and must be maintained without any additional manual locking operation. The rack-and-pinion mechanism shown in Fig. 10 was designed to provide the required

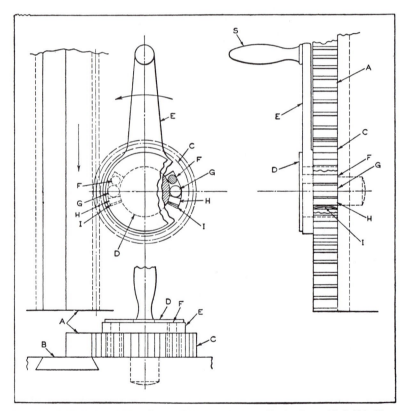

Fig. 10. (Left) End and Plan Views of Rack-and-pinion Mechanism with Self-locking Feature; (Right) Side View of Mechanism.

speed and ease of adjustment with the necessary locking feature. With this arrangement, the wire guide can be adjusted in either direction by turning handle *S* in a clockwise or a counter-clockwise direction. The mechanism is automatically locked in position.

Referring to the plan and end views of the arrangement shown at the left of Fig. 10, rack *A* is attached to slide *B*, which carries the wire guide that is to be adjusted. Stud *D*, locked on a stationary part of the machine, carries pinion *C* which meshes with rack *A*. Pinion *C* has grooves in the bore at two points, in which rollers *G*, blocks *H*, and springs *I* are assembled. The grooves in pinion *C* are cut deeper toward one end, so that rollers *G* will be free at the deeper end but will wedge tightly toward the center at the shallow end, forming the conventional type of free-wheeling clutch which permits free movement in one direction and locks or transmits motion in the other.

Blocks *H* are backed up by springs *I*, which tend to force rollers *G* toward the shallow end of the grooves, thus insuring a positive wedging action. It will be noted that the wedging action in the two grooves takes place when the driving member is rotated in opposite directions, so that gear *C* is normally locked against rotation in either direction.

Lever *E*, which is free to swing on stud *D* above gear *C*, carries two pins *F* which project into the grooves in gear *C* behind rollers *G*. A slight clearance is provided between pins *F* and rollers *G*, so that lever *E* will have a small amount of free movement. When lever *E* is moved in the direction indicated by the arrow, pin *F* on the left comes in contact with roller *G*, moving it toward the deeper section of the groove in gear *C* and thus eliminating the wedging action on that side. As the wedging action on the opposite side takes place in the opposite direction only, gear *C* is now free to rotate with lever *E*, causing rack *A* and slide *B* to be moved in the direction shown by the arrow. When lever E is rotated in the opposite direction to that indicated in Fig. 10, the pin F on the right-hand side comes in contact with roller G, eliminating wedging action on that side and freeing gear C to rotate with lever E.

When lever E is released, the wedging action of both rollers G again becomes effective in locking pinion C against rotation in either direction. As very little movement of rollers G is required for effective locking, rack A is locked against movement in either direction with almost imperceptible backlash, except when lever E is moved.

Locking and Releasing Mechanism for Traveling Carriage.

—The stitching mechanism of a machine used for stitching a wire through a fabric must be firmly locked in position during the stitching operation, but must be immediately freed for resetting when the operation is completed. These requirements were effectively met in a simple manner by the mechanism shown in the accompanying illustrations. In Fig. 11 are rudimentary views of the machine with the stitching mechanism omitted to show the driving mechanism, which automatically locks and releases the stitching mechanism in its various positions.

The stitching mechanism is supported on carriage A, which rolls freely on two rails B, and is operated by gear H, which is also mounted on carriage A. Gear H is rotated by square shaft G, supported by bearings C. The motive power is furnished by belt F, operating on pulley D, which is keyed to shaft G. Loose pulley E is provided for disconnecting the power. Plate J, attached to bearing C at the right-hand end of the machine, is used for operating the disconnecting mechanism, to be described later.

In Fig. 11, belt F is shown on the loose pulley E. With the belt in this position, no power is applied to shaft G, and carriage A is free to be moved longitudinally to any position in its range of travel. Square shaft G passes through a square hole in gear H, to which it transmits

motion. The hole in gear H is somewhat larger than shaft
G. During idle period of the stitching head, shaft G is
located symmetrically within the square hole in gear H, as
shown at upper left, Fig. 12, clearance space permitting free
longitudinal movement of carriage A, Fig. 11.

When belt F is shifted to the tight pulley D, causing
shaft G to rotate, the latter carries gear H with it, the

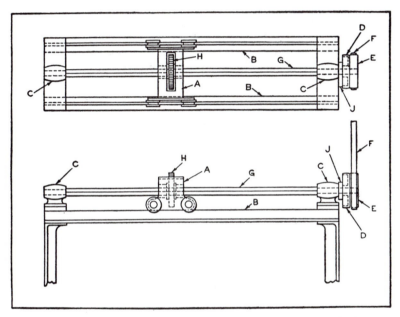

Fig. 11. (Top) Plan View of Wire-stitching Machine Frame and Carriage;
(Bottom) Front View of the Machine.

corners of shaft G acting as driving members in the manner shown at the upper right of Fig. 12. The frictional
resistance to rotation of the stitching head produces a
sufficiently powerful wedging action of shaft G in the
square hole in gear H to lock carriage A firmly against
longitudinal movement. This condition exists as long as
shaft G continues to rotate.

The mechanism that disengages carriage A from shaft
G to permit longitudinal movement is shown in Fig. 12.

This illustration shows driving pulley D as viewed from the side next to the machine. The internal hub of pulley D carries lever I, which is clamped to it with sufficient tension to support its own weight, so that the lever and the pulley tend to rotate as a unit. However, the tension is light enough to be easily overcome by the driving power exerted by belt F on pulley D.

The split bore in lever I is lined with leather to prevent scoring the hub of pulley D. Plate J, the position of which is indicated by dotted lines in this view, is fastened rigidly to bearing C, shown in the lower or front view in Fig. 11. The outer end of lever I carries pin K which passes through a slot in plate J.

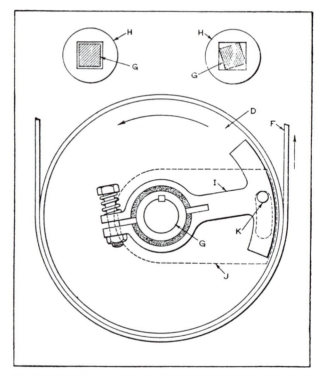

Fig. 12. (Top) Diagrams Showing Shaft G in Unlocked and Locked Positions in Gear H; (Bottom) Mechanism for Unlocking Shaft G.

In operation, belt F causes pulley D to rotate in the direction indicated by the arrow. The friction of lever I on the hub of pulley D tends to rotate the lever; but as pin K comes in contact with the upper end of the slot in plate J, further rotation of lever I is prevented, as shown in Fig. 12, this position of lever I being maintained throughout the stitching operation, or as long as pulley D rotates.

When belt F is shifted to loose pulley E, Fig. 11, the unbalanced weight of lever I causes pulley D to reverse its direction of rotation until pin K comes in contact with the lower end of the slot in plate J, at which time shaft G has again returned to its unlocked position in gear H, as shown at the upper left of Fig. 12. In this manner, carriage A is automatically freed for repositioning at any point of its cycle.

Mechanism for Operating a Floating Jaw Vise.—Fig. 13 shows a vise used on a routing machine for holding a wood block A while a cavity is being routed out, as indicated by the dot-and-dash lines at B. The jaws C and D of this vise are operated by a mechanism that permits them to float to suit the work, locking firmly when the work is finally gripped. The work A is located on two pins E which enter two previously drilled holes. The jaws C and D prevent the wood from bulging under the pressure of the cut where it closely approaches the sides of the piece.

As the routing must be accurately positioned relative to the holes and as the width of the work varies considerably, it is necessary that the jaws C and D have a floating action. That is, the jaws must grip the work in the position in which it is located by the pins E before they are clamped to the fixture bed F. The jaws C and D are slidably mounted on base F to provide the required floating action. The base F is slotted to receive keys machined on the under side of the jaws. Retaining plates G and H are screwed to the under side of the keys to hold the slidable jaws in place. The blocks I and J, located in slots in the jaws, swivel on pins at their upper end, and are held at the back of the slots by means of springs K. Blocks I and J carry at their lower ends the swivel-pins L and M, which are threaded to fit the right- and left-hand threads on the rod P. Rod P is equipped with a handwheel W which is turned clockwise for clamping the work. The blocks I and J are T-shaped, being wider at the bottom, as shown in the end view to the right. The lower ends of blocks I and J are therefore wider than the slot in base F.

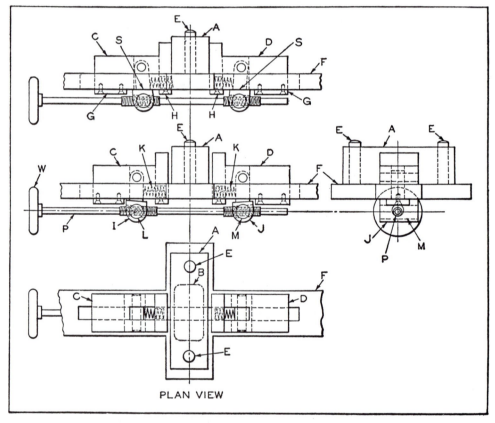

PLAN VIEW

Fig 13. Mechanism that Causes Floating Jaws C and D to Grip Work A and then Grip Base F through the Clamping Action of Blocks I and J.

The two lower views in Fig. 13 show the vise in the open position, with work A located on pins E. In this position, the jaws and their blocks I and J are bound together by rod P and slide together as one unit on base F. When handwheel W is turned clockwise, the action of the right- and left-hand threads on rod P will throw swivel-blocks I and J toward each other. As pins L and M are prevented from swiveling by the resistance of springs K, jaws C and D are also thrown together.

When either jaw comes in contact with the work, its movement ceases, and the action of the screw is transmitted to the other jaw until they are both in contact with the work. Further turning of the handwheel causes blocks I and J to swivel on their upper pins, compressing the springs, so that at this point, work A is held by spring pressure. Continued turning of handwheel W causes blocks I and J to swivel still further until the enlarged or T-sections at their lower ends come in contact with the under side of base F, as indicated at S in the upper view, thus locking jaws C and D firmly in position.

Mechanism for Clamping and Releasing a Spring-Actuated Tailstock Center.—A mechanism designed for releasing a spring-actuated center and for clamping the center in the work-holding position is shown in Fig. 14. The fixture on which this mechanism is used is designed for holding the thin wooden part G, on which a routing operation is performed. The pointed ends of the work are supported on centers, the rear center being located in the block H. Although there is slight variation in the length of the wooden pieces, they must be held firmly in position, but without sufficient pressure to distort them.

Referring to the view to the left in Fig. 14, the rack A meshes with the gear B which turns freely on the flanged bushing D, mounted on stud C. The bushing D is slidably keyed in the bearing member J which supports the assembly. Stud C is threaded into the collar F which is fastened to bearing J. Handle E is pinned to stud C and serves to advance it into the threaded collar F. The work G is supported by the block H at the end of rack A. The opposite end of the work is similarly supported by a stationary block. A spring L furnishes the required pressure to support the work.

To place a piece of work between centers, the handle E is turned to the left into contact with the pin K in gear B, so that the gear turns with the handle. Then movement of gear B moves rack A to the right. Next, the

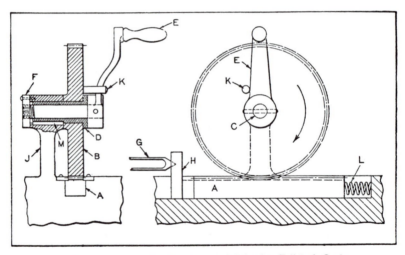

Fig. 14. Mechanism for Clamping and Releasing Tailstock Center.

work is located between the centers and handle E turned to the right, as indicated by the arrow. This permits the pressure of spring L, acting on rack A, to support the work with the maximum permissible pressure. Although this pressure will support the work, it is not sufficient to resist the pressure of the cut and some method of positively locking rack A is necessary.

Continued movement of handle E in the direction indicated by the arrow causes stud C to be screwed deeper into collar F, so that the hub of handle E, acting on the flange of bushing *D*, locks gear *B* against the end of bearing *J*. As gear *B* now is locked, rack *A* is prevented from moving. Thus the support *H* holds the work *G* in position under the pressure exerted by the cutting tool. As bushing *D* is restrained from rotating by the key *M*, the rotary movement of handle *E* is not transmitted to bushing *D*, and the endwise pressure applied to the work when it is placed between centers consists only of that exerted on rack *A* by spring *L*.

Mechanism for Adjusting Arc-Shaped Levers around Rotating Cylinders.

—The mechanism shown in Fig. 15 is designed to enable the arc-shaped members *B* of levers *A* to be closed around the rotating work *X* in such a manner as to smooth out a covering material placed on the cylindrical surface.

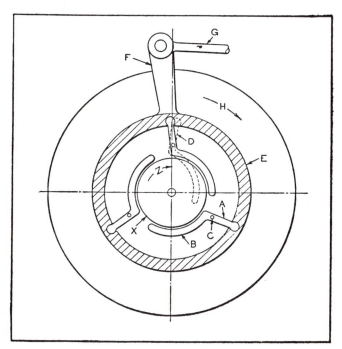

Fig. 15. Mechanism for Closing Arc-shaped Members B around Rotating Cylinder X.

The three smoothing levers A have their extensions B pivoted about pins C as centers, and thereby operate on the work to smooth out the covering material. The levers are shown at their outer positions, the dotted lines at D indicating the contracted position which would be assumed if work X were not in position. The levers are brought together under spring tension in the following manner: Each of the three levers A has its outer end located in a socket in a revolvable flange E having an extension arm F. Through the medium of the connecting-rod G, the flange is rotated in the direction indicated by arrow H to close the jaws. At the pulling end of the connecting-rod there is a tension spring, not shown. Thus when the smoothing fingers come in contact with the work, which revolves in the direction indicated by arrow Z, pressure is applied under spring tension to the levers B.

Milling Machine Spindle Brake and Circuit-Breaker Mechanism.

—The purpose of the mechanism shown in Fig. 16 is to provide a means for locking the spindle of a milling machine positively when changing the milling cutter, in order to insure safer operation. When the machine is running, the milling spindle A is rotated by means of disk B. Disk B is provided with six holes C for the locking pin D which serves to prevent rotation of the spindle when it is raised to engage one of the holes in the disk. Pin D is actuated by means of lever E and the operating lever F which has a handle G.

A spring H presses pin D against disk B, so that it automatically engages one of the indexing holes, as shown in the middle view of Fig. 16. The braking lever I has the same center as lever F, but is independently fixed on a hollow shaft which terminates in a bevel gear K, shown in the lower view of Fig. 16. This gear meshes with another bevel gear Z on shaft M. The latter gear is connected with a crank disk N. Between the rotating pin U and another fixed link is attached a brake-band which

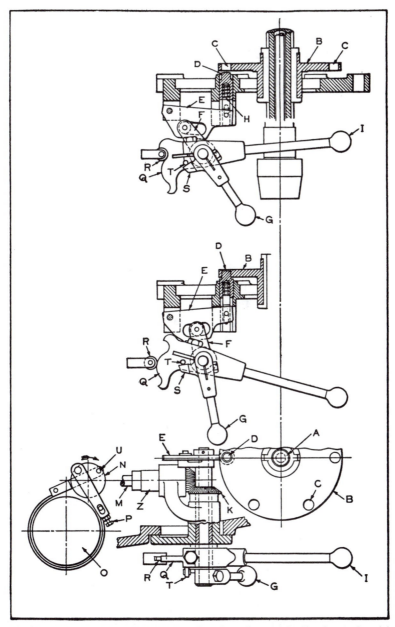

Fig. 16. Milling Machine Spindle Brake. (Top) Operating and Braking Levers G and I Disengaged to Allow Spindle to Rotate; (Center) Levers Positioned to Lock Spindle against Rotation; (Bottom) Top View of Mechanism and Side View of Brake.

acts upon the brake-disk O. The brake-band can be adjusted by means of a screw P. On the free end of the braking lever is a formed cam Q which actuates the limit switch of the electric motor by means of a roller R. A pin T on lever I and a projection on lever F provide the necessary interconnection.

When the machine is running, both levers are disengaged, as illustrated in upper view of Fig. 16. When brake-lever I is moved downward, roller R is pushed backward by the action of cam Q so that the current supply is interrupted. When lever I is moved downward further, the brake-band is applied, the index lever F remaining in its first position. However, with lever I in this position, it is possible to actuate lever F. Lever F can now be brought into such a position that the nose S comes into contact with pin T. The various members of the mechanism now occupy the relative positions shown in the middle view of Fig. 16, the spindle A being locked to prevent rotation by pin D entering hole C of the disk B.

If the brake-lever band is released by operating lever I, lever F is also returned to its first position by means of pin T. A further lifting movement brings lever I into an almost horizontal position, switching on the current. A false or unintentional movement of the levers is made impossible by having the length of lever F so short that it is first necessary to move lever I.

Automatic Work-Locating Mechanism for Milling Machine.—A large number of pieces similar to the one shown at the left of Fig. 17 are end-milled as indicated at M after the groove N is milled. The operation is performed on a vertical hand miller by setting the cutter to the proper depth and sliding the work into a clamping fixture, the central groove permitting the end-mill to enter, as indicated at the right of Fig. 17. The work-table is then moved to the right and left for cutting, and returned to the central position for removing the work.

As the central groove is but slightly wider than the diameter of the end-mill, it is necessary to stop the table in the position for loading and unloading with a fair degree of accuracy. Fig. 18 shows the construction of an attachment that locks the table positively in the loading position and permits the required travel for cutting.

Referring to Fig. 18, top view, the bar C is fastened to the machine table B, and carries the sliding block D, which is drawn to the left by the spring S. Bar C is notched in the center, as shown more clearly in the middle view of Fig. 18. Block G is fastened to the knee of the

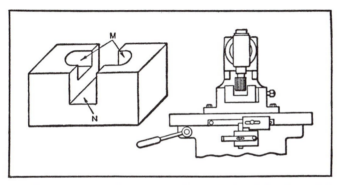

Fig. 17. Automatic Work-locating Mechanism. (Left) Work is End-milled as Indicated at M; (Right) Milling Machine with Fixture and Locating Attachment.

machine and carries the sliding key E, which is actuated by the lever F. The block D is reduced in width at the left-hand end. The travel of the table B is limited by the usual stops, not shown.

Referring to Fig. 18, middle view, which is the starting point of the cycle, the table B is locked by the key E being located in the notch in bar C. After the work fixture has been loaded, the lever F is depressed, withdrawing key E from the notch in bar C and permitting block D to be drawn to the left by the spring. When the lever F is released, key E comes to rest on the step on the end of block

D, as shown in Fig. 18, top view. The table *B* is then moved to the left, the block *D* being restrained from movement by key *E.*

When the travel to the left has been completed, and the table *B* is moved to the right, block *D* is permitted

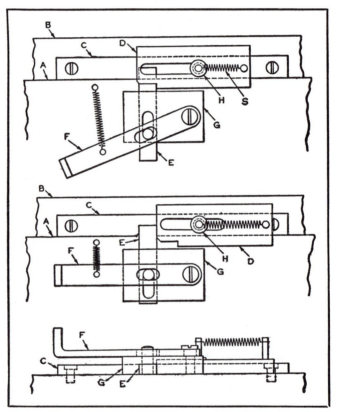

Fig. 18. (Top) Work Locating Attachment with Block D Drawn to Left by Spring S; (Center) Locating Attachment at Starting Point of Cycle; (Bottom) Bottom View of Attachment.

to slide toward its original position until the end of the slot comes in contact with the stud *H.* Continued movement of table *B* causes block *D* to be withdrawn from the end of key *E,* which then slides onto the solid portion of the bar *C,* the key *E* having been carried over and to the right of the notch by the block *D*. When table *B* reaches the end of its right-hand travel, it is returned to its central position, the key *E* resting against the end of block *D* and sliding on bar *C* until it reaches the notch in bar *C*. Key *E* then enters the notch in bar *C*, again locking the table in the loading position.

Work-Locating Mechanism for Milling Machine.—The mechanism shown in Fig. 19 is designed to accomplish the same purpose as that just described. It is very accurate, can be operated rapidly, and has a wider range of adjustment. Machine tables *B* that do not have bolt slots in the front face can be fitted with a slotted plate *C* to which stops *D* are clamped by bolts *F*. Plate *C* is held in place on the table by two machine screws. It will be noted that a shallow double-angle or beveled slot *J* is machined in the center of plate *C*. This slot is engaged by the point of latch *G* through pressure exerted by spring *I*. This arrangement stops or locates the table in the correct position for loading and unloading the work without inter-

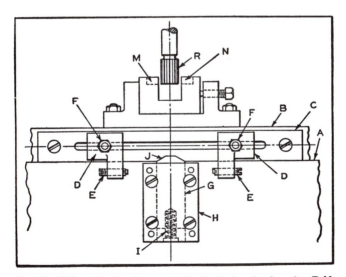

Fig. 19. Milling Machine Equipped with Mechanism for Locating Table in Correct Position for Loading.

ference of the cutter R with the sides of the slot in which the cavities are to be milled.

After the work is clamped in the fixture, cavities M and N are machined by the end milling cutter R. The pressure exerted by spring I is light enough to enable the machine or hand feed to disengage latch J. The machine feed can be used to within 1/32 inch of the final stops, after which the table is fed by hand until the points of set-screws E come in contact with latch casing H on the base A.

CHAPTER 6

Reversing Mechanisms of Special Design

Described in this chapter are various arrangements for obtaining reversal of motion. These include mechanisms for reversing a chain driven table; for providing a dwell at each reversal; for reversing a carriage traversing screw; for automatic stroke reversing in a variable speed traversing motion mechanism; for reversing a cable winding machine; for reversing a cross-head feed screw at any predetermined point; for varying the point of reversal; and a positive type of clutch that releases readily under heavy load. Other reversing mechanisms are described in Chapter 6 of Volume I and Chapter 7 of Volume II of "Ingenious Mechanisms."

Reversing Mechanism for Chain-Driven Table.—The accompanying diagrams illustrate the operation of a table reversing mechanism designed for use on a wire fabricating machine in which the table is driven by a roller chain. As shown in Figs. 1 and 3, the table is reversed or reciprocated by alternately disengaging and engaging toothed dogs M and N with the upper and lower portions of the driving chain at opposite sides of the sprockets. The mechanism, while not applicable to very heavy drives, possesses the advantages of simplicity and the ability to function when long table traversing movements are necessary. A front elevation of the machine and an end view on section XX are shown diagrammatically in Fig. 1. The table A is reciprocated by the chain B, its length of travel being controlled by the positions of the adjustable dogs C and D.

Referring to Figs. 2, 3, and 4, the vertical slide E is carried between two blocks F and G, which are attached to a supporting member H, Fig. 1. Blocks F and G are grooved to carry the horizontal slide built up of the plates I and J, which are spaced to accommodate springs U and V, as well as the levers Q and R. Plates I and J are held together by rivets. The plates K and L retain the horizontal slide in the blocks F and G.

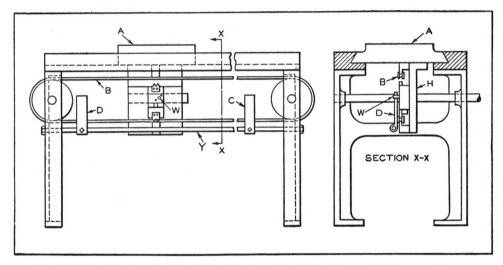

Fig. 1. Table is Reciprocated by Alternately Engaging Toothed Dogs with Upper and Lower Portions of Driving Chain.

Vertical slide E is grooved to permit the horizontal slide to pass freely through it, and carries at its upper and lower ends the toothed dogs M and N, which are shaped like the teeth of a standard roller-chain sprocket. These toothed sections are alternately inserted between the rollers of the chain, and serve to transmit the movement of the chain to the table A. The chain B is supported by the guide blocks O and P, which are attached to the supporting member H, Fig. 1, and therefore travel with the table A. The levers Q and R are supported freely on pins between plates I and J of the horizontal slide, and are normally held in contact with the pins S and T by the springs U and V. The pin W, carried by the horizontal slide, serves to operate the mechanism when it comes in contact with the adjustable dogs C and D, Fig. 1. It will be noted, by referring to Fig. 3, that lever Q projects out of the horizontal slide into the widened groove in the vertical slide E, while lever R projects into the recess formed in block G.

Fig. 3 shows the mechanism in the normal driving position, the teeth of dog N being engaged between the rollers of chain B, which is supported at that point by the guide P. Dog N cannot be disengaged from chain B, as the

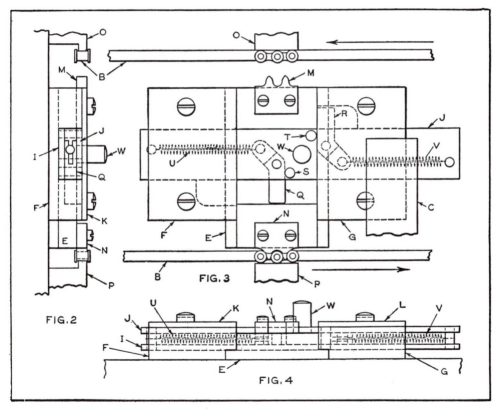

Figs. 2, 3, and 4. Three Views of Table Mechanism in Normal Driving Position with Teeth of Dog N Engaging Lower Portion of Roller Chain.

position of lever Q prevents slide E from moving in its groove. As that side of chain B into which the teeth of the dog N are engaged is moving in the direction indicated by the arrow, table A is caused to travel with it to the right until pin W makes contact with the fixed dog C. This restrains the horizontal slide from further movement with the assembly.

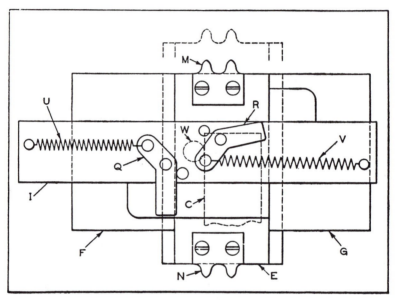

Fig. 5. Table Reversing Mechanism that Serves to Raise Slides E to Position Shown by Dotted Lines.

Continued movement of table A causes lever R to be turned on its pivot pin and enter the groove in slide E, as shown in Fig. 5. In this view, plates K, L, and J have been removed and the positions of pin W and dog C are shown by dotted lines. Lever R is shown held against the upper edge of the groove in slide E by the increased tension of spring V produced by the swinging movement of lever R. However, no movement of slide E takes place, as it is still locked in place by lever Q.

Further movement of the table causes lever *Q* to pass beyond the lower edge of the groove in slide *E*, at which time the tension of spring *V*, acting against the upper edge of the groove in slide *E* through the lever *R*, causes slide *E* to be raised quickly to the position shown by the dotted lines. The teeth of dog *M* entering between the rollers of the upper length of chain *B* carry the table in the opposite direction, or to the left.

Table *A* continues its movement to the left until pin *W* comes in contact with dog *D*, locked in position on the bar *Y*, Fig. 1. At this time, lever *Q* acts on slide *E*, again bringing about the engagement of dog *N* with the lower part of chain *B*. This causes table *A* to move to the right again.

Reversing Mechanism that Provides for Dwell at Each Reversal.—The mechanism shown in Fig. 6 is used to transmit motion from shaft *A* to shaft *I* in such a manner that when the direction of motion is reversed, there will be a definite, specified lag or dwell in the transmission of the motion to shaft *I*, regardless of the point in the cycle at which the reversal takes place. This mechanism is used on a machine for fabricating a wire product.

Referring to the illustration, shaft *A*, which carries pinion *B* keyed to it, rotates in the direction indicated by the arrow, transmitting motion in the reverse direction to gear *C*, which rotates freely on the hub of the machine member. Gear *C* carries the pin *D*, which comes in contact with the extended arm of the sector *E*, the latter member likewise rotating freely on the hub of the machine member. Lever *G*, which also rotates freely on the same hub, carries the pin *F*, one end of which enters the slot in sector *E*. The opposite end of pin *F* is extended so as to make contact with the lever *H*, which is keyed to the driven shaft *I*.

In operation, gear *C*, rotating in the direction indicated by the arrow, transmits motion to sector *E* through pin

D. Sector *E,* in turn, transmits motion to levers *G* and *H* through pin *F.* Lever *H* transmits motion to shaft *I.* On the reversal of the motion, and in practically one complete revolution of gear *C,* the pin *D* will come in contact with the opposite side of the extended arm of the sector *E,* causing the latter to rotate with gear *C* in a counter-clockwise direction.

When the end of the slot in sector *E* comes in contact with pin *F,* the lever *G* is likewise caused to rotate with gear *C.* On the completion of the second revolution of gear *C,* pin *F* is in contact with the opposite side of lever *H,* the position of the parts at this point being as shown by dotted lines in the illustration. At this point, the shaft *I* begins to rotate in unison with gear *C.*

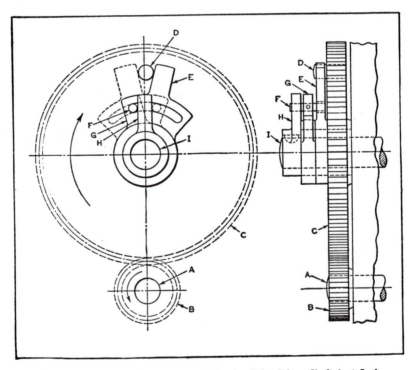

Fig. 6. Gear Drive Mechanism that Provides Dwell for Driven Shaft I at Each Reversal of Driving Shaft A.

It will be noted that in order to change the position of pin F from one side of lever H to the other, two complete revolutions of gear C are required, shaft I remaining stationary during the cycle. As the ends of the slot in sector E control the exact time at which shaft I begins to rotate, slight changes in timing can be accomplished to suit different applications by varying the length of this slot.

Reversing Rotating Motion Mechanism for Carriage Traversing Screw.—A simple reversing-motion mechanism constructed to operate a carriage traversing screw on a wire fabricating machine is shown diagrammatically in Figs. 7 and 8. This mechanism is so designed that there is no lag at the end of the traverse motion. A single-thread traversing screw is used, and there are no springs, sliding keys, or gears. Only moderate speeds are possible, however, due to the momentum developed at high speeds.

On the drive-shaft A, Fig. 7, are mounted the mutilated gears B and C, rotating in the direction indicated by the arrow. The double pillow block F carries the driven shaft H and the idler shaft G. Shaft H supports the gear E, which meshes with the gear B over one-half of its face and over the other half of its face with gear D. Gear D, on shaft G, has one-half of its face in mesh with gear E and the other half of its face in mesh with gear C.

Gears B and C each have a few more than half their teeth removed, and are keyed to shaft A with their toothed portions in diametrically opposite positions. The length of the gap between the two portions of gears B and C is governed by the center-to-center distance between shafts G and H, and must be such that the leading tooth of one gear comes in contact with its mating gear just as the trailing tooth of the other gear terminates its contact.

Referring to Fig. 7, gear B, rotating in the direction indicated by the arrow, meshes with gear E, causing it and the driven shaft H to rotate in the opposite direction,

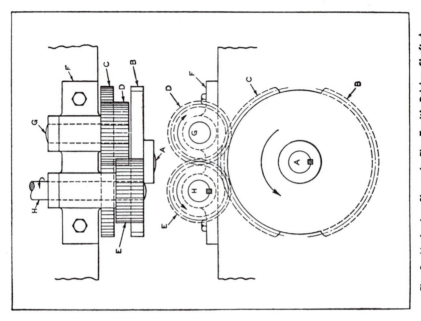

Fig. 8. Mechanism Shown in Fig. 7 with Driving Shaft A Rotating Driven Shaft H in a Counter-clockwise Direction.

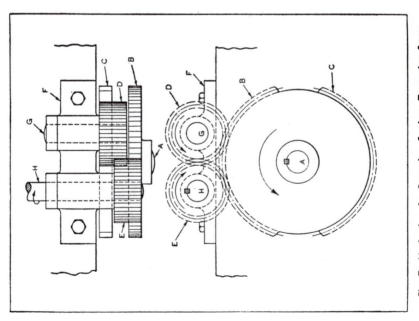

Fig. 7. Mechanism for Reversing Carriage Traversing Screw Shown with Driving Shaft A Rotating Driven Shaft H in Clockwise Direction.

as indicated by the arrow. At this point, the toothless portion of gear *C* is passing under gear *D*, and there is no connection between the two, but as gear *D* meshes with gear *E*, which, at this point, is being driven by gear *B*, gear *D* is caused to rotate in the direction indicated by the arrow. In rotating in this direction, however, no useful work is performed, the gear simply acting as an idler. Thus, the motion of shaft *A* is transmitted in the opposite direction to shaft *H* through gear *E*, while gear *D* and shaft *G* idle, or rotate without transmitting any motion.

Referring to Fig. 8, shaft *A* is shown in the position it occupies after one-half revolution, which brings the teeth of gear *C* into mesh with gear *D*. Gear *E*, meshing with gear *D*, is rotated in the direction shown by the arrow and, being keyed to shaft *H*, serves to drive the latter member. At this point, the toothless portion of gear *B* is passing under gear *E*, there being no connection between these two members. Thus the rotation of shaft *A* is transmitted in the same direction to shaft *H* through gears *D* and *E*. The end teeth of gears *B* and *C* are modified in shape as required to provide the clearance necessary for practically instantaneous reversal of the direction of rotation.

Automatic Stroke-Reversing and Variable-Speed Traversing Motion Mechanism.—The mechanism shown in Fig. 9 was designed in connection with the development of a machine for grading food products. It transmits a continuous reciprocating motion to the rod *A* at the rate of thirty strokes a minute, although provision is made for obtaining any desired traversing speed from one to sixty strokes a minute. The maximum length of the reciprocating stroke is 5 inches. The mechanism runs smoothly and is practically noiseless in operation.

The drum *B* is driven at a speed of 300 revolutions per minute by a motor which transmits power to the mech-

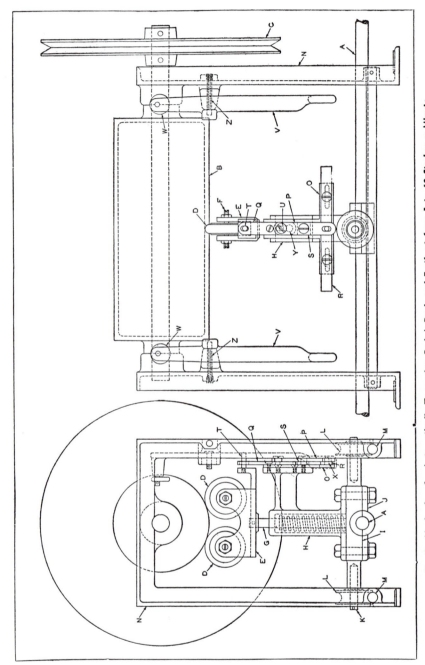

Fig. 9. Mechanism for Automatically Traversing Rod A Back and Forth at from 1 to 60 Strokes a Minute.

anism by means of a belt through the grooved pulley *C*. The drum *B* transmits motion to the rod *A* through contact with the friction-driven wheels *D* mounted in bracket *E*. The wheels *D* are equipped with hardened steel centers and revolve freely on conical screw points *F* which are adjusted and locked in the bracket *E*. The wheels *D* are equipped with semi-hard rubber or rawhide rings for friction driving, which operate smoothly.

Spindle *G*, fitted in bracket *E*, turns freely in the guide casting *H*. The assembly, consisting of the guide casting *H* and castings *I* and *J*, is supported by the grooved wheels *L*, mounted on pins *K* in the casting *I*. The guide wheels roll on the rods *M*, secured to the under frames *N*. The reciprocating rod *A* is clamped between the two castings *I* and *J*.

Bearing plate *O* is fastened to member *H* with countersunk machine screws. Assembled on plate *O* are levers *P* and *Q*. These levers have a compound action, and are actuated by the push-lever *R* which is held in place by screw pins acting in slots in plate *O*. The spacer *S* provides working clearance for levers *R*, *P*, and *Q*. Lever *Q* is pivoted on bearing plate *O*, and at its upper end has an elongated hole which engages the pin *T* in bracket *E*. The lower end of lever *Q* is slotted, as shown at *U*, to engage a pin on lever *P*. The lower end of lever *P* also has an elongated hole which engages a pin *X* on lever *R*.

When drum *B* is rotated, the traction wheels *D*, with their horizontal center lines parallel with each other and with the horizontal center line of drum *B*, revolve at speeds governed by the size and speed of the drum. In the position shown in the illustration, no traversing movement by the traction mechanism takes place, but any turning movement of bracket *E*, with spindle *G* as its axis, immediately causes the bracket *E* and its assembled members to travel to the right or left, as the case may be, and at a speed governed by the angle at which the wheels D are set in relation to the horizontal center line of drum B.

The travel of the traverse assembly, either to the right or left, brings lever R in contact with one of the levers V. Continued movement of lever R causes the eccentric rollers W to move in toward the revolving drum. Contact between the drum and rollers W causes the latter to revolve so that the high side of the rollers pushes back the reverse lever V with a rapid movement. Lever V thus moves lever R in the opposite direction. This action transmits, through levers P and Q and pin T, a turning movement to member E, which changes the angle at which wheels D make contact with drum B, thus reversing the direction of travel of rod A. This series of reverse movements is repeated when lever R makes contact with the lever V at the opposite end of its travel.

The enlarged end Y of the slot U locks the traction wheels D when the pin on lever P moves into this part of the slot. Pressure applied at any point above point Y when the pin on lever P is in the enlarged end Y of the slot will not unlock levers P and Q. The slightest pressure below point Y, however, particularly when applied to lever R, immediately unlocks the mechanism and places the wheels D in the reverse position. The springs Z serve to eliminate vibration of levers V and to keep them in their proper operating positions.

Reversing Mechanism for Cable-Winding Machine.—
An automatic reversing mechanism for a cable-winding machine which is designed to allow reversal of the winding guide at any desired point from zero to maximum, so that reels of various widths can be wound on the same machine, is shown in Fig. 10. The winding of various pitches is accomplished by using pick-off gears. An interesting feature of this mechanism is the use of two free-wheeling flywheels to store up energy for completing the

automatic clutch-engaging movements at each reversal of the lead-screw P which drives the winding guide.

The mechanism is driven by the gears A and B, which rotate in opposite directions. These gears mate with gears C and D, which are pinned to the shafts E and F, formed with clutch slots G. For simplicity, only one slot is shown, but the clutches are of the usual multi-tooth type.

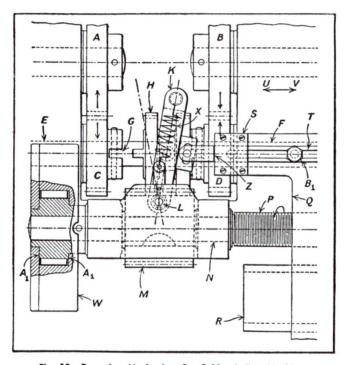

Fig. 10. Reversing Mechanism for Cable-winding Machine.

A combination gear and spool H runs freely on shafts E and F. The spool is shifted from side to side by the pins in the fork K, which is pivoted at L. It should be noted that the spool H remains constantly in mesh with the gear M, which is keyed to the shaft N. This shaft is formed integral with the lead-screw P which actuates the winding-guide casting Q. The casting Q is guided by the way R. The guide head carries the member S in which slides the slotted rod T. This rod is connected to the fork K as indicated. On the end of shaft N a free-wheeling flywheel W is pinned, which locks when the lead-screw rotates in the direction indicated, but unlocks in the opposite direction and runs free. On the other end of shaft N a second free-wheeling flywheel (not shown) is pinned, which operates in the opposite way to W. Spring X keeps the clutch in mesh with gear C or D until acted on by fork K.

In operation, spool H, being in mesh with gear D, causes lead-screw P to operate in the direction shown. This causes guide Q to move in the direction U until it. comes in contact with the shoulder Z on rod T. The guide then pushes fork K, causing the spool to move out of the clutch slot until it is entirely disengaged. At this point lead-screw P would normally stop revolving. However, the energy stored in the free-wheeling flywheel W continues to turn the lead-screw until fork K passes through and slightly beyond the dead center position, whereupon spring X causes the spool to snap over against the face of gear C and into the slot. At this instant, the lead-screw reverses and flywheel W commences to free-wheel, while the opposite flywheel is driven to store up energy for the next reversal. It was found necessary after a trial run to make the surfaces A_1 in the free-wheeling flywheel a friction fit, so that after a short interval the energy would be dissipated, and it would then tend to drop down to a speed below that of the shaft. Thus, when again actuated, it could immediately pick up in the proper direction.

Upon the reversal of the lead-screw, guide Q travels in the direction V until member S strikes the stop B_1, when reversal again occurs.

Mechanism for Reversing Cross-Head Feed-Screw at Any Predetermined Point.—The purpose of the mechanism shown in Fig. 11 is to reverse the direction of rota-

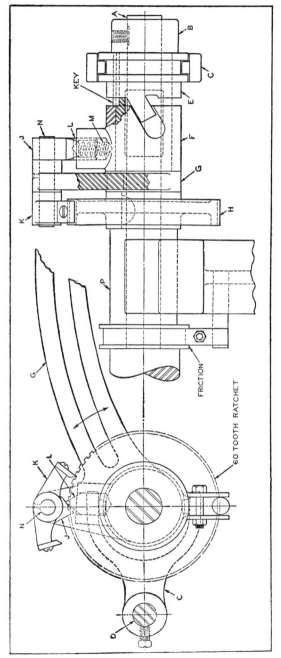

Fig. 11. Feed-screw Reversing Mechanism Using Double-pointed Pawl which can be Positioned to Revolve Ratchet Wheel Forward or Backward.

tion of a cross-head feed-screw when the cross-head has reached any predetermined point in its traverse. Adjustment of the cross-head speed is accomplished by a slotted pawl-arm. Reversal of the feed is obtained by a double-pointed pawl which can be positioned to revolve the ratchet wheel forward or backward. The mechanism has been applied to a machine for evenly distributing wire or cord on a reel or spool 50 inches in diameter, and also to a machine for a 30-inch spool.

The feeding and reversing mechanism consists of a single-threaded screw A, Fig. 11, supported in frame bearings P; pawl-arm G, which is free to rotate on screw A between ratchet H and collar B; the pawl-reversing cam F, placed over the hub of pawl-arm G; grooved sleeve E having a finger that engages a slot in the pawl-reversing cam F; fork C, secured to guide rod D for sliding sleeve E to the right or left when a cross-head (not shown) moves guide rod D through contact with a collar (not shown).

The collars are set to reverse the direction of rotation in accordance with the length of traversing movement desired. The sliding movement of sleeve E, Fig. 11, tips or rotates the reversing cam F which, in turn, depresses plunger L until it has passed the point of lug J. Spring M then forces plunger L outward against the side of lug J secured to shaft N. Shaft N, mounted in the short arm of pawl-arm G, is thus rotated, causing the pawl K to engage the ratchet on the other side of the arm, so that the lead-screw is revolved in the opposite direction. Pawl-arm G is oscillated by an eccentric on the drive-shaft. Should lead-screw A tend to back up, a simple friction brake can be applied.

Reciprocating Devices that Vary Their Points of Reversal.—Gradual variations of one or both reversal points of a reciprocating slide or shaft can be obtained automatically by using the star-wheel mechanism shown in Fig. 12 in combination with a proper intermediate drive, four types of which are illustrated in Figs. 13 and 14. Eight different movements can be obtained with the various combinations; and any combination can be instantly converted to obtain a constant point of reversal by simply swinging the pawl F to its upper position so that it does not engage the star-wheel.

These movements are particularly adapted for mixing, valve-grinding, and textile machinery, where the stroke

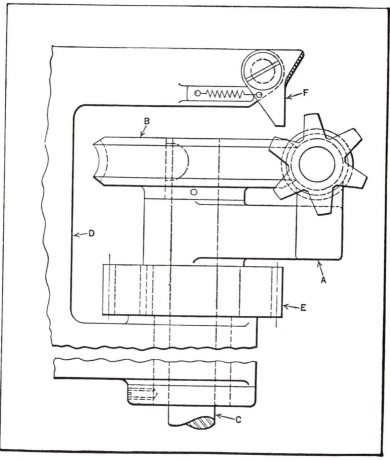

Fig. 12. Device Used with Any One of the Drives in Fig. 13 or Fig. 14 for Gradually Varying the Reversal Points of a Shaft or Slide.

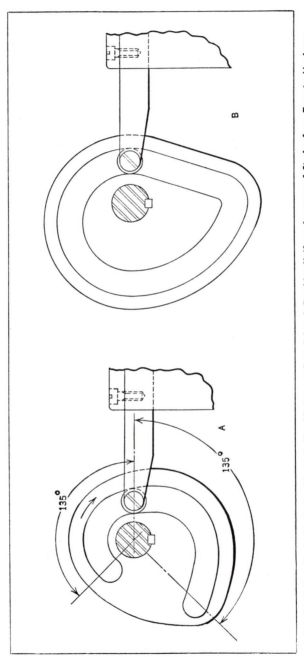

Fig. 13. When Used with Device Shown in Fig. 12, Intermediate Drive A Provides Uniform Increase of Stroke from Zero to Maximum. Intermediate Drive B Varies Time at which Reversal of Slide Occurs.

of a slide or a reciprocating shaft movement must be gradually increased or diminished, or where the positions of both points of reversal of the slide or the shaft must change gradually without changing the distance traveled between these two points.

The star-wheel mechanism consists principally of the arm A, Fig. 12, on which is mounted a shaft carrying the star-wheel, together with a worm meshing with the worm-wheel B. The worm-wheel is keyed to the shaft C, which has for its bearing a long sleeve, integral with the hub of arm A. This sleeve is free to turn in a stationary bearing on the bracket D, and has a spur gear E keyed to it. The driving member for this mechanism may be either a reciprocating or a continuously rotating gear, depending upon the movement to be transmitted through the intermediate drive; this gear (not shown) meshes with gear E.

Suppose that it is required to transmit movement to the slide at A in Fig. 13, so that the length of the stroke increases uniformly from zero to the maximum travel, the position of the point of its reversal at the left remaining constant. The cam shown here would be mounted on shaft C (Fig. 12), and the driving member for the star-wheel mechanism would be a reciprocating gear or rack. Now, as shaft C is locked to the arm by the worm and worm-gear, an oscillating movement is imparted to it by the driving member. However, during each cycle, the oscillating shaft will be advanced or retarded (according to whether the worm has a right- or a left-hand thread) by the action of pawl F on the star-wheel.

Assume now that the angular movement of the shaft C is 135 degrees, and that the cam is about to rotate in the direction of the arrow (Fig. 13). The cam-roll will follow the concentric portion of the cam groove and the slide will have no movement. During the next oscillating cycle of the cam, the position of both reversing points with respect to the cam will be slightly advanced in a

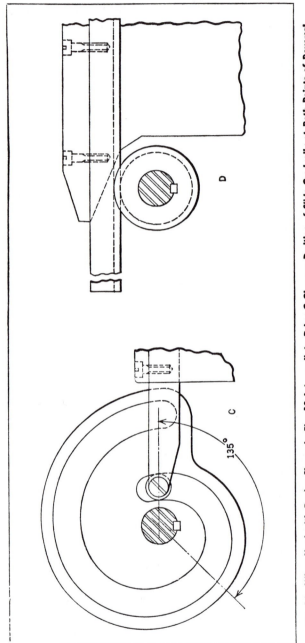

Fig. 14. When Used with Device Shown in Fig. 12 Intermediate Drive C Changes Position of Slide Gradually at Both Points of Reversal. Intermediate Slide D Produces Practically the Same Movement.

clockwise direction, due to the action of the star-wheel mechanism. This action will cause the roll to pass a short distance into the irregular portion of the cam groove, thus moving the slide a very short distance.

After each oscillation of the cam, the roll enters the irregular groove farther, uniformly increasing the stroke of the slide until it has reached its maximum travel. To prevent the roll from coming into contact with the end of the cam groove and damaging the mechanism, automatic dogs are usually provided on the machine for discontinuing the operation of the star-wheel mechanism; or in some cases, provision is made for automatically returning the cam to its starting position.

It should be noted, however, that owing to the fact that the roll passes over part of the concentric portion of the groove before the slide has reached its maximum travel, the slide will have a dwell at the beginning of each stroke. This dwell will decrease as the stroke increases, until finally, when the slide has reached its maximum travel, the dwell will equal zero.

To gradually vary the time at which the reversal of a slide occurs relative to the movements of the rest of the machine, the position of the reversal points remaining constant, the intermediate drive shown at B is used. In this case, the gear in the star-wheel mechanism is rotated continuously by another gear. Thus the angular movement of shaft C (Fig. 12) will be uniformly advanced or retarded (depending on whether the worm thread is right- or left-hand) relative to the movements of the rest of the machine members. Consequently, the time at which each reversal of the slide occurs will also be advanced or retarded relative to the movements of the other machine members.

When it is required to change the position of a slide gradually at both points of reversal, the drive shown at C, Fig. 14, may be employed. The cam, as before, is keyed to the shaft of the star-wheel mechanism, which, in this case, is driven by a reciprocating gear or rack meshing with gear E, Fig. 12. Assume, as in case A, that the angular movement of the shaft is 135 degrees. The rise of the cam groove is uniform. Hence, no matter what part of the groove the roll follows, the stroke of the slide will remain constant.

However, owing to the action of the star-wheel mechanism, the cam lags or advances after each cycle (depending upon whether the worm thread is left- or right-hand). Therefore, although the stroke of the slide remains constant, the positions of both reversal points of the slide change uniformly until the roll approaches the end of the cam groove. To prevent the roll from striking the end of the groove and damaging the mechanism, the same provision is usually made as mentioned for case A.

Another type of drive for producing practically the same movement as that just described is shown at D, Fig. 14. In this case, a pinion and sliding rack are used instead of the cam and roll. This is perhaps a simpler and more economical design, and it has the advantage of varying the position of the reversal points of the slide over a longer distance than would be practical with the cam type of drive.

To vary both points of reversal of a reciprocating shaft uniformly, no intermediate drive is required, because the shaft on the star-wheel mechanism—when driven by a reciprocating gear or rack—has this movement. All other reciprocating shaft movements corresponding to those of the slides shown in Figs. 13 and 14 can be obtained by combining the star-wheel mechanism with the proper intermediate drive and mounting a rack on the slide. The rack, meshing with a gear on the part to be reciprocated, will transmit the required movement.

Positive Type Reversing Clutch.—To design a positive type clutch that will release instantly under a heavy load without causing too much strain on the clutch fork or operating levers, usually necessitates considerable experimenting in order to determine the correct angle for the driving side of the teeth. When the proper angle is employed, the teeth will retain a sufficient grip to perform their work without releasing under the pressure of the load, yet the operator will be able to release the clutch with little effort.

The working conditions change, of course, with the method of operation, especially when the clutch is employed to control the movements of a slide weighing 2000 pounds. For example, the operating conditions are somewhat different when the slide is advanced by a series of short quick movements from when the slide is advanced smoothly and continuously at a uniform rate. Whether the movement is controlled by hand or by power feed, and whether or not oil or grease is employed to lubricate the ways on which the slide travels, are also factors to be considered. If the slide is fitted with a gib which must be kept fairly tight to prevent side play, as in the case of a wheel-slide on a surface grinder, this requirement must also be considered. Under these operating conditions, it is necessary that all the teeth make good contact so that the load will be distributed evenly, and thus permit the clutch to release instantly when reversing the direction of the slide movement either automatically or by hand.

The reversing clutch shown in Fig. 15 was first made with teeth cut straight without any releasing angle. With this type of teeth, it was found impossible to reverse the clutch when the slide was in motion and carrying a full load. It was, therefore, necessary to grind the cut teeth to an angle on the driving side. This was done by setting up an indexing head to hold the clutch on the table of a surface grinder and traversing the table back and forth by hand.

After several trials, an angle of 14 degrees, as indicated by view J, was decided upon as the most satisfactory for

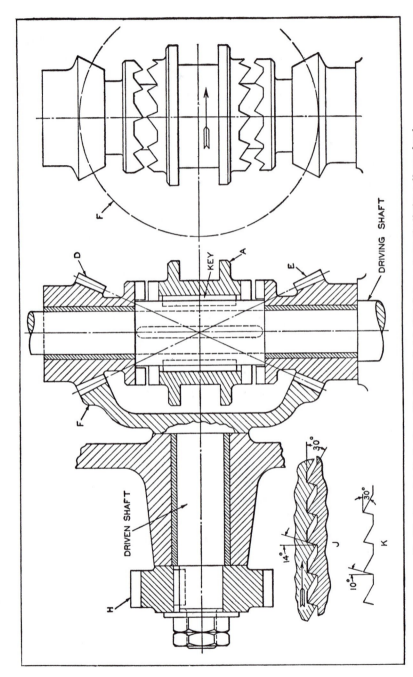

Fig. 15. Reversing Mechanism with Positive Clutch that Releases Instantly Under a Heavy Load.

the contact faces of the clutch teeth for moving a wheel-slide weighing 2000 pounds. For a slide weighing 1000 pounds with exactly the same operating conditions, an angle of 10 degrees, as at K, was found to be most satisfactory.

Various oils and greases were tried as lubricants for the wheel-slide, with the result that grease under pressure was found to give the slide the best working conditions. Oil, even in the heavy grades, would squeeze out on the sides of the ways due to the weight of the slide, which would then stick or freeze so badly that it was almost impossible to start the slide after it had remained idle for a short time. The grease, however, was found to stick to the ways, giving them a film that made it possible to move the slide easily.

It was found advantageous to have an odd number of teeth in the clutch face to facilitate milling the teeth. A clutch with an odd number of teeth will permit an over-running milling cutter to enter a tooth space on the opposite side of the clutch, instead of cutting into a tooth.

The clutch member A, which does the driving, has right-hand teeth cut in it at one end and left-hand teeth at the other end. The pinion bevel gears E and D have clutch teeth to match those cut on the clutch member A. The large bevel gear F meshes with the teeth in pinions E and D. A spur gear H meshes with a rack that moves the wheel-slide in or out.

CHAPTER 7

Reciprocating Motions Derived from Cams, Gears and Levers

Among the reciprocating mechanisms described in this chapter are: linkage arrangements for imparting reciprocating motion in a straight line and for operating a slide and plunger simultaneously; a chain and sprocket mechanism for providing a long stroke at uniform speed with a rest period at the end of each stroke; an ingenious arrangement for providing a reduction of speed at one end of a reciprocating movement without using a cam or changing the speed of the driving member; a simple means of providing reciprocating motion suitable for instrument use; an arrangement which causes a slide to be reciprocated from three different positions; a rack and intermittent gear mechanism which provides a uniform speed on both forward and reverse strokes; a mechanism designed to give various ratios of strokes to revolutions when converting rotary into reciprocating motion and for varying the phase of reciprocation continuously with respect to the rotary motion; a double-contact cam which provides reciprocating motion with positively locked rest periods; three types of sliding block mechanisms for converting rotary into reciprocating motion; a combination driving crank and Geneva dial mechanism for obtaining a positive, accurately timed reciprocating motion with a dwell at the end of each stroke and a means for converting reciprocating motion in one plane to similar motion in a plane at right angles to it.

Other reciprocating mechanisms based upon the action of cams, gears and levers are described in Chapter 9 of Volume I and Chapter 9 of Volume II of "Ingenious Mechanisms."

Straight-Line Reciprocating Mechanism for Disk Saw.—
A mechanism consisting primarily of a link suspension
frame for the rotating spindle of a saw for wood and
similar materials is shown in Fig. 1. The interesting
feature of this arrangement, which restricts the hand feeding movement to a horizontal motion along the straight
line X—X, is the large amount of space it allows between
the saw S and the supporting links A. There are two supporting links A and two links B. These four links are

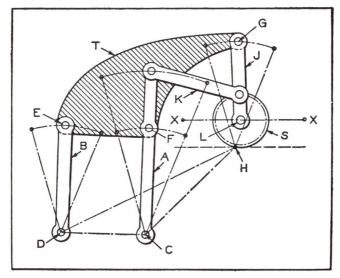

Fig. 1. Mechanism for Guiding Saw S Along Straight Line X—X.

mounted on bearing studs on the base at C and D. This
arrangement permits links A and B to swing through the
arcs indicated by the dot-and-dash lines.

The connecting bar T is triangular in shape. Triangles
E, F, G, and D, C, H, are equal and are located in corresponding parallel positions. Link J is connected to T at
G and to a rod K which is attached to the extension of
link A. Thus the end L of lever J which supports the
rotating saw S has a nearly straight-line motion. When

member J is moved by hand, the saw S is given a straight-line motion, as required for cutting material. A mechanism similar to the one described is used on electric saws.

Straight-Line Reciprocating Motion.—The device shown diagrammatically in Fig. 2 was developed to obtain a movement of the point B in a straight line from D to E when point A of the fork H is moved from F to G. This, however, is only one of several possible motions that can be obtained by applying the same general principles to the construction of a motion-transmitting device of similar design.

Referring to the construction of the device, fork H is free to oscillate about pivot J on slide K. This slide is free to move along the fixed rod L. Rods M and N, fixed in fork H at angle x, are free to slide through members

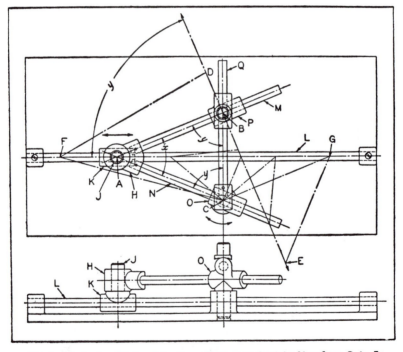

Fig. 2. Mechanism Designed to Move Point B in Straight Line from D to E when A is Moved from F to G.

P and *O*, respectively. Member *O* is pivoted at a fixed point in relation to rod *L*.

Rod *Q* is fixed in member *O* at the angle *y* of the isosceles triangle *ABC*. The rod *Q* is free to slide through member *P* at the fixed angle *y* in relation to rod *M*. As *A* is moved along rod *L* triangle *ABC* changes its altitude, but always remains isosceles. The triangle oscillates about *A* in relation to slide *K*. As it oscillates about *A* the point *B* travels in a straight line *DE* at the angle *y* in relation to rod *L*.

The movement of *A* is uniform with that of *B* in the proportion given in the equation:

$$\frac{\text{Movement of } A}{\text{Movement of } B} = \frac{\text{Leg } AB}{\text{Base } BC}$$

When point *A* of slide *K* has reached point *G* on rod *L*, the point *B* will be at point *E*. Various intermediate positions of rods *M* and *N* are indicated by the light dot-dash lines.

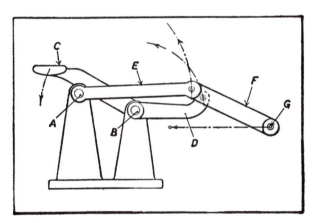

Fig. 3. Simple Foot-operated Mechanism for Obtaining Straight-line Motion.

Simple Straight-Line Reciprocating Motion Mechanism.—

Fig. 3 shows a mechanism designed to withdraw a slide horizontally in a straight line to permit a part to be dropped through the opening into a chute of a conveying system

For the long stroke required, this device is quite efficient and can be adapted to other mechanisms. It is foot-operated and can be constructed at small cost. The paths through which the various members travel when the pedal is depressed are indicated in the illustration.

Pedal C is simply a continuation of lever D, which is pivoted so that it can be rotated freely about the center of stud B. Lever E pivots about the center of stud A. The other ends of arms D and E are connected by pivot-pins to the operating lever F. It will be noted that the inner of these two connecting points does not lie on the same center line as the end connections of lever F. At G, on the lower end of arm F, is a stud for connecting it to the slide arrangement. In use, the pedal is depressed, causing G to move in a relatively straight line toward the left, the motion being controlled by the two links D and E, which move in the paths indicated by the dot-and-dash lines. Two of these mechanisms are employed, one being mounted at each end of the slide.

Linkage Mechanisms for Operating Slide and Plunger Simultaneously.—The linkage arrangements shown in Figs. 4 and 5 are designed for advancing a slide and a plunger simultaneously during the first part of their movements, and then stopping the slide while the plunger continues to advance. The object of the devices illustrated is to advance the lower slide A for a distance B and then continue to advance plunger F. Slide A carries a receptacle C containing a sheet of paper D from a cut-off position to an operating position within the machine.

Resting on top of the slide is a cylindrical-shaped piece E which is carried along simultaneously by the plunger F, attached to another slide unit G. Plunger F also pushes piece E through the paper-holding receptacle C, and in doing so, causes the paper to be wrapped about the piece. The chute H, Fig. 4, leads from a hopper which furnishes a continuous supply of the parts to be wrapped.

A portion of the machine frame is shown at *J*. A bracket *K*, mounted on the frame, carries the dual slide mechanism. A lever *L* is employed to operate the slides. This lever is actuated by a cam *M* and a co-acting lever *N*, mounted on the rocker shaft *P*. Up to this point the mechanisms illustrated in Figs. 4 and 5 operate in a similar manner.

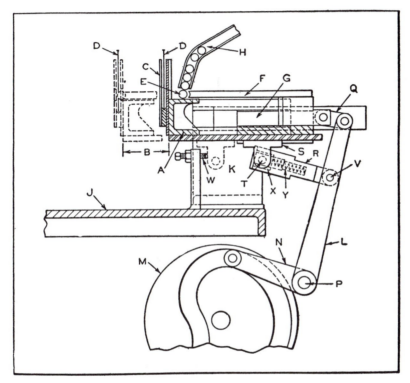

Fig. 4. Link Mechanism Designed to Operate Slide A and Plunger F Simultaneously.

Referring to Fig. 4, the upper end of lever *L* is attached to a link *Q* which moves the slide or plunger *F*. Attached to the lower sliding unit of which *A* forms a part, is a spring cushion link *R*. This link is connected to a block *S*, the pivot points being at *T* and *V*. When slide *A* is advanced a distance B, block S comes in contact with the stop-screw W, which prevents the lower slide from advancing beyond this position. At this point the spring cushioning mechanism comes into effect and permits the lever L to continue advancing. The continued movement of the upper slide and plunger F carries the cylindrical piece into the machine.

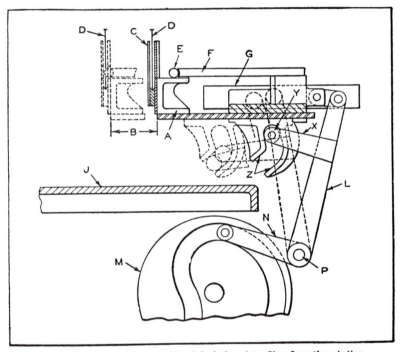

Fig. 5. Mechanism Shown in Fig. 4 Redesigned to Give Smoother Action.

The spring cushion consists of a movable block X in which the stud T pivots. This block slides in the connecting link R, while the spring at Y holds the sliding block in the position shown. As lever L advances, block X slides along the connecting link R, thereby compressing spring Y. Interrupted advance linkages requiring the compressing of a spring of sufficient stiffness to advance a slide mechanism introduce into the machine what may be termed a harsh movement and one that may pound and wear out the mechanism quickly, with an attendant loss of power. To eliminate this action, the lower slide connection was redesigned as shown in Fig. 5.

The lever L, Fig. 5, is fitted with an extension X which supports a roll Y. Fastened to the under side of the slide bracket is an open cam Z. The contour of this cam is so designed that the roll will follow the angular portion of the cam indicated by the arrows at Z without advancing the lower slide after it has been advanced the required amount, although the upper slide and work pusher or plunger F continue to advance. It is obvious that the smoother action of the open cam Z will permit the moving parts to travel at a faster rate than is the case with the construction shown in Fig. 4.

Long-Stroke Reciprocating Motion for Guiding Wire on Reel.

—Fig. 6 shows diagrammatically a reciprocating motion mechanism used for guiding wire on a broad reel. The mechanism is designed to give a long stroke, with a uniform rate of traverse speed and a period of rest at each end of the stroke.

Rod E, which is supported in bearing F, carries two pins P at its outer end for guiding the wire. Sprocket A drives sprocket B through the block chain C. The special link D, inserted in the chain in place of one of the standard links, carries the rod E on the pin G. The length of link D is such that pin G travels along the center line between the two sprockets. Link D, in traveling from sprocket to sprocket, transmits a uniform reciprocating motion to rod E.

At each end of the stroke, pin G remains stationary while the link D rotates half way around the sprocket, providing a period of rest which is determined by the diameter of the sprocket. On extremely long strokes, it is necessary to support the chain to avoid excessive sag.

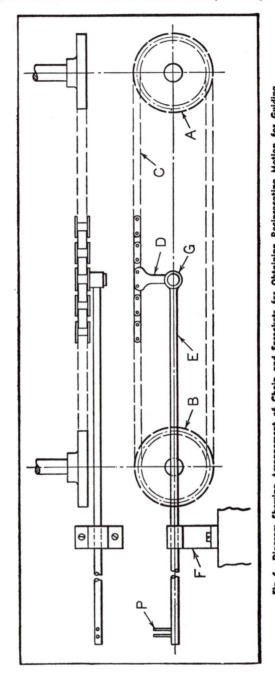

Fig 6. Diagram Showing Arrangement of Chain and Sprockets for Obtaining Reciprocating Motion for Guiding Wire on a Long Reel.

Reciprocating Motion Used on a Wire-Forming Machine.—Fig. 7 shows the design of a reciprocating motion used on a wire-forming machine, which employs a unique method of providing for a reduction of speed at one end of the movement. This is accomplished without the use of a cam and without changing the speed of the driving member. The lower view shows a plan of the arrangement, and the two upper views show side elevations. The part A is a sliding fit in a dovetail slot in the stationary part B, and carries the lever C, which is connected at its upper end to the rod F, from which it receives its motion. The lever D is keyed to the same shaft as lever C, and carries at its lower end the roller E, which rides on the plate G.

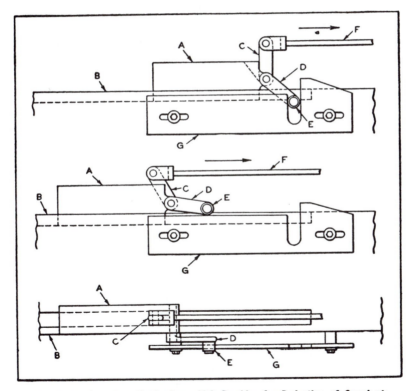

Fig. 7. Reciprocating Mechanism which Provides for Reduction of Speed at One End of Movement.

The part A is recessed, as indicated, to receive lever C, the recess being of a depth that will permit the roller E to ride on the upper edge of the plate G without any lost motion.

In the central view, rod F is shown as having completed its movement to the left and as moving to the right. As levers C and D are keyed to the same shaft and lever D is prevented from rotating downward by the roller E resting on plate G, there is no movement of either C or D relative to part A until roller E reaches the slot in plate G. The motion of A up to this point is directly transmitted to it by rod F. As roller E reaches the slot in plate G, it rides downward, as shown in the upper view. At this point, levers C and D, being fastened to the same shaft, operate as one lever, with roller E acting as the fulcrum. As levers C and D are of the same length, the speed of movement of part A from this point is reduced to about 50 per cent that of rod F. The length of travel at this point is likewise reduced. Plate G is slotted to provide adjustment.

Simple Reciprocating Mechanism for Chart Recording Pen.—A very simple means for producing reciprocating motion was discovered while developing a special instrument. Aside from the simplicity of the device, which is shown in Fig. 8, there is the desirable feature of lightness, with a minimum amount of power loss due to friction.

The mechanism consists of a cam A, fabricated from spring wire. The two legs B of the spring cam cross and pass through a hole in bar C, while screw D serves to hold the spring cam in place in the instrument case.

Cams E, acting on spring cam A, cause the position of the point where legs B cross to change as cam A is depressed and released. In changing the position of the crossing point, bar C is made to move back and forth. The upper view of the illustration shows the device at the starting position, with spring cam A completely relaxed, while the lower view shows the relative position of the members of the mechanism when the maximum depression of spring cam A has been accomplished to displace bar C a distance equal to F.

Bar C is intended to support a pen for marking a continuous chart, and it is for this reason that lightness and easy operation are desired. The two cams E are not of

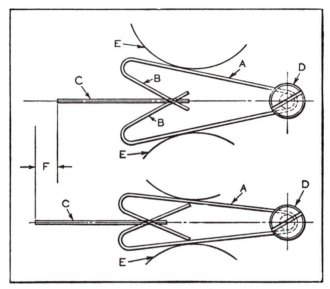

Fig. 8. Simple Method of Producing a Reciprocating Motion for a Chart Recording Pen.

the full rotating type. The motion of the cams about their pivots or centers is dependent upon the action of a bellows that is subject to pressure pulsations.

Pivoted Holder for Cam Follower Rolls Provides Unusual Reciprocating Motion.—The design of an interesting mechanism used on a wire-forming machine is shown in Fig. 9. This mechanism is used to operate a slide G which carries a forming tool, the slide being caused to reciprocate from three different positions. The lower cam F imparts the reciprocating movements to the slide, while the upper cam E intermittently locates the slide G in three different positions.

Referring to the illustration, the shafts A and B carry the meshing gears C and D which rotate in the directions indicated by the arrows. Shaft A is the driving member,

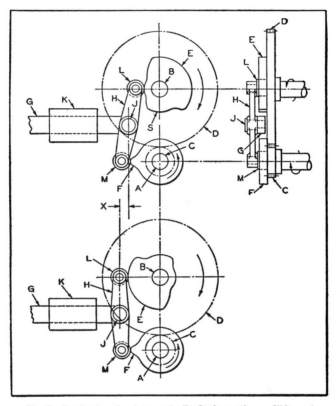

Fig. 9. Mechanism for Automatically Reciprocating a Slide and Locking it in Three Different Positions.

shaft B running idle in its bearings. Cams E and F are attached to gears D and C and operate with them. The speed ratio between gears C and D is 4 to 1. The slide G is supported in bearing K. Slide G carries the stud J, upon which lever H fulcrums at its center. Lever H carries the rollers L and M, which engage the cams E and F, respectively, being held in contact by a spring (not shown) which draws slide G to the right.

Cam E is designed with three points of rest, one of which is for a period of 180 degrees, while the other two are for periods of 90 degrees each. The single-lobed cam F is designed to give a quick rise and fall. In the upper view to the left, the roller L is shown on the low point of cam E, while the roller M is positioned on the high point of cam F. As the lobe of cam F passes under the roller M, the lever H fulcrums on the center of the roller L, transmitting to the slide G a movement equal to one-half the rise of cam F.

As gear D rotates in the direction indicated by the arrow, the intermittent step S of cam E is brought under roller L, causing lever H to fulcrum on the center of roller M, thus moving the slide G to the left an amount equal to one-half the rise of step S on cam E, slide G now being in its second or intermediate position. The lobe of cam F again actuates the lower end of lever H, giving the slide G its second reciprocating movement from its new position.

Continued rotation of gears C and D brings the high point of cam E under roller L, as shown in the lower view. As compared with the upper view, the position of the stud J has moved to the left a distance equal to X. As the high point of cam E includes 180 degrees, the lobe of cam F acts twice on the roller M during this period. Thus the slide G is given one oscillating motion at each of two positions and two motions at the third position. The roller L then returns to the low point on cam E, completing the cycle. The slide bearing K, it will be noted, is not shown in the side view of the assembly.

Rack and Intermittent Gear Mechanism for Obtaining Uniform Reciprocating Motion.—The reciprocating motion produced by the experimental mechanism shown in Fig. 10 differs from that of the crank-driven movement so commonly used, in that member C moves at a uniform speed on both the forward and reverse strokes. No sliding gears or clutches, such as are usually employed in gear-driven reciprocating motions, are used in this mechanism, the motion being obtained by two modified gears and a rack.

Referring to the illustration, the intermittent driving gear A rotates constantly in the direction indicated by the arrow. The teeth of gear A mesh alternately with the teeth of rack C and gear B. In the position shown in the illustration, the leading tooth on gear A is about to make contact with the first tooth of gear B, causing the latter to rotate in the direction indicated. The motion of gear B is transmitted to the rack C, causing it to move to the right.

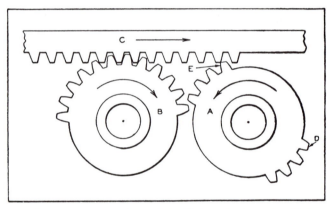

Fig. 10. Mechanism for Transmitting Uniform Reciprocating Motion to Rack C from Rotating Intermittent Gear A.

As the last tooth E in gear A completes its contact with the mating tooth in gear B, the tooth D in gear A is ready to engage the mating tooth in rack C. When this occurs, the direction of movement of rack C is reversed. At this point, the teeth on gear B are in the clearance space of gear A. Gear B is then driven by rack C until tooth D of gear A meshes with gear B. The alternate driving of rack C by gear A and gear B produces a uniform reciprocating motion of the rack.

It will be noted that the number of teeth in each tooth section of gear A is two less than one-quarter of the number of teeth that would be carried by a full gear of the same size and pitch. Although this is necessary to prevent clashing, the effect is the same as that produced by a quarter turn of a gear having the full number of teeth. It will also be noted that the two end teeth in each toothed section of gear A are modified to provide sufficient clearance to permit the reversal of rack C. On large gears, it will be necessary to modify the succeeding tooth as well. Also, it may be necessary to remove a small amount of material from the leading face of the leading tooth in each group of teeth in gear A to permit a slight lag in the reversal of rack C. The degree of modification of the teeth necessary for the proper functioning of the mechanism will vary with the size of the gears and their pitch.

The length of travel of rack C is determined by the diameters of gears A and B. The number of teeth in these gears must be divisible by 4. Although this mechanism can be used only where moderate speeds are employed, due to the necessity for suddenly arresting the momentum of rack C and gear B, its simple construction recommends its application wherever the speeds are within the permissible range.

Mechanism for Converting Rotary into Reciprocating Motion and a Single-Chain Differential Drive.—The mechanism shown in the upper diagram of Fig. 11 is used for converting rotary into reciprocating motion. It can be designed to give various ratios of strokes to revolutions and to cause the phase of the reciprocating motion to vary continuously with respect to the rotary motion. Such variation is desirable for lapping and polishing. For example, it can be built to have sleeve T make 0.565 reciprocation per revolution of sprocket C. The reciprocating motion is continuous and positive and takes place without shock.

The stationary housing K supports a bushing D, held in place by a screw E. The shaft A rotates in bushing D and is restrained from endwise motion by the hub of sprocket C and the base of the truncated cylinder F. At L is another truncated cylinder similar to the one at F.

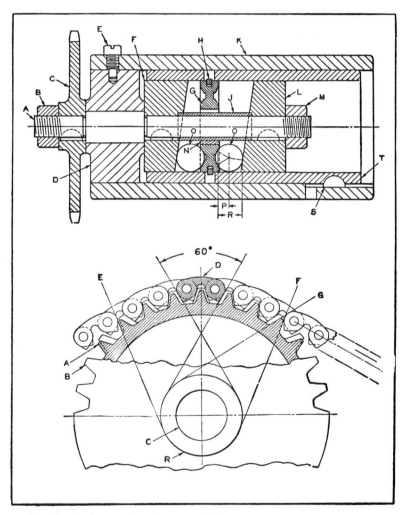

Fig. 11. (Top) Mechanism for Converting Rotary into Reciprocating Motion. (Bottom) Single-chain Differential Drive with Standard Sprocket A and Special Sprocket B.

These cylinders are keyed to shaft A with their truncated faces parallel and spaced by the sleeve J. The nut M serves to clamp the members F, J, and L together.

Sleeve T is a sliding fit in housing K, and is restrained from rotating by the key S. The internal collar or race G is held in sleeve T by the snap-ring H. In dismantling, the snap-ring may be reached through the holes in sleeve T. The two balls O are positioned by the ball cage N which straddles the internal collar G, so that the line joining the ball centers is parallel to the axis of shaft A. The cage N bears on the faces of the collar or race G and thus holds the balls in line.

Rotation of driving sprocket C rotates the truncated cylinders or face-cams F and L, thus causing the balls O to roll on the cam surfaces and on the internal collar G which they cause to reciprocate or shift from side to side. The reciprocating motion is transmitted from collar G to sleeve T through ring H.

If the faces of collar G were not grooved but constructed with two plain surfaces, the balls and their cage would make one revolution about the axis of shaft A to every two revolutions of shaft A. This gives two revolutions of A for each complete reciprocation of sleeve T. With the faces of the collar G grooved as illustrated, however, the ratio is somewhat different, as shown by the equation:

$$\frac{\text{Number of reciprocations of } T}{\text{Number of revolutions of } A} = 1 - \frac{P}{R}$$

where P and R are the horizontal distances shown in the upper diagram of Fig. 11.

It is not necessary to groove the faces of collar G in order to get a varying-phase relation between A and T. Any creep of the balls or the collar itself, however little, will cause a continuous though slight change in the phase relations. The ratio of revolutions of the driving sprocket C to the number of strokes of sleeve T can be further modified by mounting housing K so that it can be rotated about the axis of shaft A.

The chain mechanism shown in the lower diagram of Fig. 11 provides a convenient means for driving shaft A and housing K at slightly different speeds. Ordinarily, two concentric sprockets would be driven by two separate chains, each with its own take-up mechanism. The mechanism shown, however, drives the two concentric sprockets at slightly different speeds by means of a single chain which is wide enough to engage both sprockets. Referring to the lower diagram of Fig. 11, B is a standard sprocket, the teeth of which are shown in unshaded outline, and D a silent chain. The special sprocket A, the teeth of which are shown in shaded outline, has one tooth less than sprocket B, but has substantially the same outside diameter and root diameter as sprocket B. Its teeth are cut with the same cutter as is used for sprocket B, but the sprocket is indexed and recut to make the tooth thickness less than normal, so that the teeth A do not interfere with chain D and lift it off sprocket B.

The approximate tooth outline can be determined as follows: Draw the circle R tangent to the 60-degree faces of the standard gear tooth, and also draw a standard chain in contact with sprocket B, making the angle of wrap as small as possible without sacrificing proper functioning of the driving chain. Next, draw the line E tangent to the circle R and to the chain tooth which is about to engage the standard sprocket B. This establishes the driven face of the sprocket tooth on sprocket A. Next draw line F tangent to circle R and to the chain tooth that has started to leave the standard sprocket B to establish the opposite face of a tooth on sprocket A. Draw the line G tangent to the circle R and a whole number of tooth spaces from the line E, as measured on sprocket A. The space included between F and G will be the approximate tooth on sprocket A. It may be easier to place a chain on the sprockets and cut the teeth of the second sprocket back by trial cuts until there is no interference than it is to lay out the tooth as described.

The operation of the mechanism is as follows: The chain, coming in wedge-like contact with a tooth of sprocket A along the line E, moves sprocket A a little faster than sprocket B. The narrow teeth on sprocket A permit it to move faster than sprocket B without interfering with the chain. One tooth at a time on sprocket A takes all the drive. If necessary, these teeth can be made shorter than standard teeth, as shown in the illustration.

The sprocket A shown in the lower diagram of Fig. 11 has 27 teeth, while the sprocket B has 28 teeth, so that the relative speeds of the two sprockets are inversely proportional to their number of teeth.

In using the single-chain differential drive on the mechanism shown in the upper diagram of Fig. 11, the standard tooth sprocket B was used to drive the housing K, which should be mounted so that it can rotate but is restrained from endwise movement, while the special 27-tooth sprocket A took the place of sprocket C. Thus when housing K and collar G have made 27 revolutions, the cams F and L will have made 28 revolutions. The cams, therefore, gain one revolution on collar G for each 27 revolutions of the housing. Since it requires approximately two revolutions of the cams with respect to collar G to produce one reciprocation, it will require 2 × 27 or 54 revolutions of housing K to produce one reciprocation when the collar G has plain faces. The sleeve J may be omitted and the nut M pulled up until the balls just grip collar G. The nut can then be locked with a cotter-pin or check-nut. A light spring could be used under the nut M, but this was not found necessary for the particular mechanism shown.

Care should be taken not to make the angle of the truncated face of the cylinders F and L too large. The limiting angle may be found by experimenting with a ball placed between two lubricated surfaces. The angle used for the mechanism shown was 8 1/2 degrees. This angle was found satisfactory for balls 3/8 inch in diameter.

Mechanism for Converting Rotary into Reciprocating Motion.—In Fig. 12 is shown a mechanism for converting the rotary motion of shaft A into a reciprocating motion of the sleeve which is composed of the two members B and C. This mechanism is similar to the one just described, the principal difference being in the construction of the reciprocating sleeve.

In the mechanism shown in the upper diagram of Fig. 11, the snap-ring H for transmitting the drive from collar G to the sleeve member was assembled by springing it into the groove in member G. Member G was then pushed into the sleeve until the ring reached the groove on the inside of the sleeve. The ring then expanded into the groove in the sleeve. In order to permit ring H to be compressed so that collar G could be removed, it was necessary to drill two holes through the reciprocating sleeve, as shown in the cross-section view in the upper diagram of Fig. 11. The diameter of these holes was somewhat larger than the width of the snap-ring and the holes were drilled diametrically opposite each other.

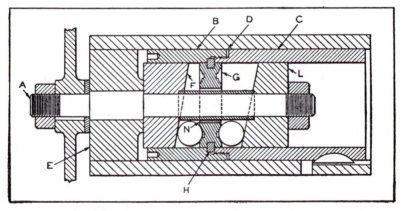

Fig. 12. Mechanism for Imparting Reciprocating Motion to Sleeve Members B and C.

With the design shown in the accompanying illustration, a heavier ring H is used and the necessity for drilling holes through the sleeve is eliminated. The ring H is fitted into the groove in collar G, which is then inserted in member B of the reciprocating sleeve so that ring H fits snugly into the counterbore. Threaded sleeve C is then screwed into sleeve B, a clearance of about 0.025 inch being left at D so that ring H can be clamped securely in place.

The operation of the mechanism is the same as the one just described and shown in Fig. 11. Shaft A, with the two truncated cylinders F and L keyed to it, revolves, causing the balls in contact with the truncated faces and the race of collar G to transmit a reciprocating movement to the sleeve members B and C. The balls are kept in alignment by a cage N which straddles the collar G.

Ball thrust bearings placed on both sides of the stationary bearing E might be a worth-while improvement.

Mechanism for Converting Rotary to Constant-Velocity Reciprocating Motion.—The mechanism shown diagrammatically in Fig. 13 was devised to impart a reciprocating motion to the sliding member A from the rotating shaft B. This reciprocating motion was required to operate the bellows of an artificial respiration machine in such a manner as to simulate the inspiration and expiration of natural breathing.

In natural breathing, the expanding movement of the ribs, accompanied by a downward pull of the diaphragm, produces a suction power for inhaling air which operates immediately at its full velocity and continues to do so to the end of the inhalation. It is apparent, therefore, that the accelerating and decelerating motion produced by a simple crank mechanism such as is usually employed to convert rotary to reciprocating motion cannot be used to drive the reciprocating bellows-operating member A.

The special mechanism devised for this purpose, like the natural mechanism of the chest, immediately starts the flow of air into the lungs at full velocity and continues the flow at a uniform rate to the end of the air injecting motion, after which it causes the air to be withdrawn from the lungs in a similar manner.

To duplicate the rate of natural breathing, the driving motor C was equipped with reduction gears to give shaft B a uniform speed of 16 R.P.M. The reciprocating motion is imparted to member A in the direction indicated by arrow D from the rotating shaft B through arm E, flat-faced wheel F, and a flexible steel band G in the manner to be described.

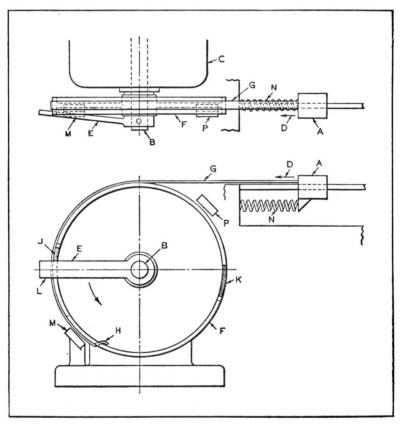

Fig. 13. Constant-velocity Reciprocating Motion Mechanism Designed to Operate the Bellows of an Artificial Respiration Machine.

Shaft B, with the arm E secured to it by a pin, rotates in a counter-clockwise direction, causing arm E to ride on the rim of wheel F, which is a free running fit on shaft B. One end of the flexible steel band G is fastened to wheel F at H, and the other end is secured to the bellows-actuating member A.

The arm E is made of spring material, and is so designed that it will automatically drop into slots or notches J and K in the rim of wheel F as it is rotated in a counter-clockwise direction. Fig. 13 shows arm E in slot J, in which position it acts as a driver for wheel F, causing the wheel to rotate counter-clockwise.

This rotation of wheel F serves to wind steel band G over the face of the wheel and thus pull the bellows operating member A to the left, in the direction indicated by arrow D. The movement of member A continues at a uniform speed until the outer projecting end L of arm E comes in contact with the cam-faced dog M, which lifts the arm out of the driving notch J in wheel F.

As soon as arm E is lifted from the driving slot, the spring N, which has been compressed by the movement of A to the left, instantly expands and forces member A to the right, thus completing the return stroke. This return stroke of member A causes wheel F to revolve clockwise back to its original starting position, where it is stopped by a device not shown.

The arm E then continues to ride or slide on the rim of the wheel until it drops into the slot K. The movement of member A to the left is then repeated in the manner described until another cam-faced dog P disengages arm E and allows the spring N to return the member A again to the starting point. Member A travels at a constant velocity during its stroke to the left because the rotating speed of wheel F is constant and therefore provides a constant speed for winding the flexible band G around the wheel face.

If a power return motion is required for member A in place of the spring-actuated return motion, gears can be employed to operate the mechanism in the reverse direction. Since the bellows are only a secondary part of the mechanism, they are not shown.

The notches J and K in wheel F are deep enough to seat arm E. One end of each notch is beveled or sloped to allow arm E to drop into the slot until it strikes the driving end of the notch which is at right angles to the rim. The dogs M and P are fixed, and are so positioned that they lift arm E out of the notches in wheel F at the proper time as it traverses around with the wheel rim, and then allow it to drop back into contact with the rim. It then slides along the edge and in contact with the rim in a counter-clockwise direction while the wheel is being reversed and being moved in a clockwise direction by spring N, until it drops into the next notch in the rim edge, which again immediately reverses the motion of wheel F.

Reciprocating Motion with Positively Locked Rest Periods.—A double-contact cam, such as is shown in Fig. 14, can be used to secure a reciprocating motion with a rest period for the sliding member. This mechanism is positive, self-locking, and has no backlash. The cam of this mechanism is keyed to the driving shaft at A. The contour of the cam is composed entirely of arcs of circles, and is constructed as described in the following: Let e equal the angular magnitude of the two rest periods of the cycle. Then 180 degrees minus angle e equals the angle of the two action periods of the operating cycle.

Next, triangle ABC is constructed with the angle at A equal to angle e as shown in Fig. 14. The lengths b and c are then selected for two sides of the triangle, so that $b + c - a$ equals the length of the reciprocating stroke S of the yoke. Then, with the vertex of the smallest angle of the triangle as a center, say at B, and with an arbitrary radius r, construct the arc DE; next, with C as a center, construct the arc *EF*, and with *A* as a center construct the arc *FG*. With *B* as a center, construct arc *GH* and with *C* as a center, construct the arc *HJ;* finally, with *A* as a center, construct the closing arc *JD*.

The distance between any two parallel tangents to the cam is equal to the constant distance $2r + a + b - c$. Thus, the yoke follower maintains two contact points with the cam surface throughout the cycle, and operates without any backlash, except the small amount necessary for working clearances.

It will be noted that angles *B* and *C* are unequal, a fact that results in a different rate of acceleration or timing of motion on the forward and return strokes. Motion distribution or timing can be varied still further by changing radius *r*. Identical forward and return motions can be secured, if desired, by making angle *B* equal to angle *C*.

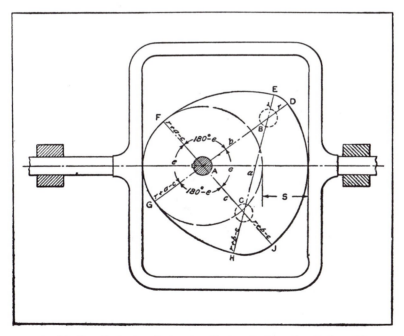

Fig. 14. Cam Mechanism Designed to Produce Reciprocating Motion with Positively Locked Rest Periods.

The shaft could have been keyed to the cam at points *B* or *C* as well as at *A*. In that case, however, a different rest period and a different travel of the follower would have been obtained with the same cam and follower, the rest periods corresponding in magnitude to the three angles *A, B,* and *C* of the triangle. The cam will operate with a rocker-arm follower as well as with a sliding type follower.

Sliding-Block Mechanisms for Converting Rotary into Reciprocating Motion.—Rotary motions are frequently converted into reciprocating movements by means of sliding-block mechanisms. Such mechanisms generally include a

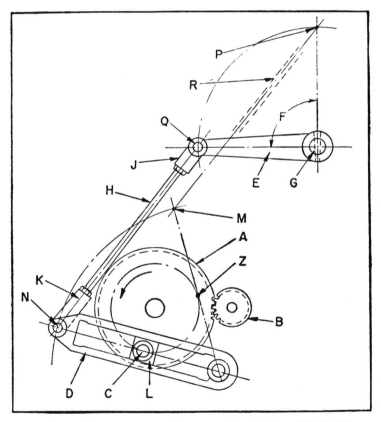

Fig. 15. A Long Eccentric Block Movement in which Lever E is Quickly Raised to a Vertical Position, then Slowly Lowered to Position Shown.

box-shaped cast-iron lever arm in which a bronze block slides. Three different applications of this type of mechanism are shown in Figs. 15, 16 and 17.

A long eccentric block movement is shown in Fig. 15. In this construction, the large gear *A* is driven by pinion *B*. Sliding block *L* is free to pivot about stud *C*, which projects from the face of the gear. When the gear is rotated in the direction indicated by the arrow, the sliding block will move lever *D* from the lower position shown to that indicated by the center line connecting points *M*

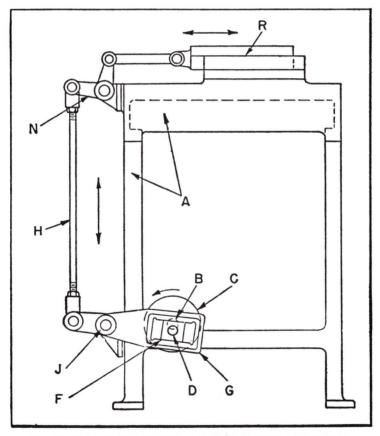

Fig. 16. Sliding Block F Mounted on Eccentric Stud D is Employed with Connecting Linkage to Reciprocate Member R on Table of Machine.

and Z. M represents the upper position of connecting pin N, and Z indicates the common center of the stud and block when the lever is in its uppermost position.

As block L slides back and forth in member D, the lever E and shaft G to which it is pinned are reciprocated through arc F to operate a mechanism not shown. Connecting-rod H and adjustable links J and K join levers E and D. Point P represents the upper position of pin Q, while the center line joining points P and M, and broken lines R, indicate the uppermost position of the connecting-rod.

One feature of this design is that only a little more than one-quarter of a revolution of gear A is required to lift lever E to its vertical position. To return the lever to the position shown requires almost three-quarters of a revolution. In other words, the slow motion of lever E and shaft G is obtained while work is being performed on the machine, and the quick motion returns them to their starting positions.

A contrasting application of the same mechanism is the so-called "short eccentric block movement," Fig. 16. In this case, sliding block F is free to pivot about an eccentric stud D, which projects from flange C on driven shaft B. The shaft is mounted between frames A of the machine. Block F slides within the slotted box-like lever G, pivoting the lever about shaft J. Connecting-rod H joins one end of lever G with bellcrank lever N, which, in turn, reciprocates member R in ways provided on the table of the machine.

An interesting variation of the sliding-block movement is the cam-operated mechanism shown in Fig. 17. In this case, the lever M containing the sliding block P is on the driven rather than the driving end of the mechanism. Roll B, which rotates about a stud extending from lever C, fits in the grooved face of cam A. Lever C is hinged on pin D, so that it oscillates through arc H as the cam is rotated.

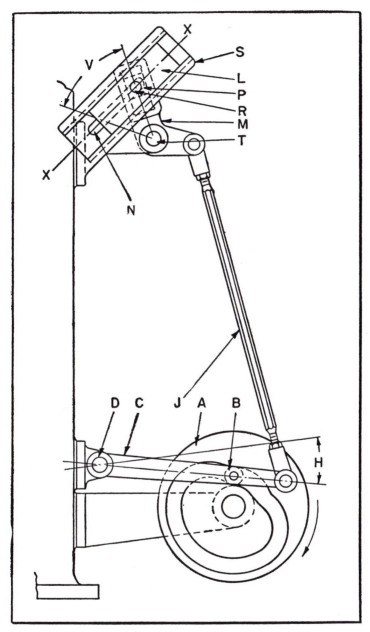

Fig. 17. Cam-operated Sliding Block Mechanism which is Employed to Reciprocate Slide L in Ways Provided on Bracket S.

This motion is transmitted to the bellcrank lever M by connecting-rod J. The bellcrank lever is free to pivot about pin T, and its longer arm is slotted to fit sliding block P. Stud R, which passes through the sliding block and is secured to slide L, is confined to movement along center line XX by an elongated slot N in bracket S. As the slotted arm of the bellcrank lever is pivoted through arc V, slide L is reciprocated along center line XX in ways on bracket S.

In making such sliding-block mechanisms, it is sometimes advantageous to have the slot in which the block travels open at one end. This type of forked construction facilitates assembly, and is generally satisfactory when the travel of the block is small. In cases of longer travel or high-speed operation, it is usually necessary to increase the strength of the forked member or resort to a box type lever.

Positive Reciprocating Mechanism.—Machines of certain types sometimes require feeding slides that are given a positive, accurately timed, reciprocating motion with a dwell at each end of the stroke. In the case of the mechanism shown in Fig. 18, the feeding slide S is required to be positively locked in the position indicated by the full lines while driving shaft T, rotating continuously in one direction, completes a certain portion of a revolution, following which the feeding slide S is moved to the position at the right indicated by dotted lines. The slide then remains locked in this position for a certain portion of a revolution of shaft T, after which it is returned to the position shown by the full lines. This cycle is repeated continuously as long as the driving shaft T continues to rotate.

While it would be possible to obtain a reciprocating and dwell motion of this kind by the use of simple cams, the usual cam and follower arrangement would not insure the necessary positive locking action, accurate positioning during the dwell period, and freedom from lost motion or backlash provided by the mechanism illustrated. As shown in

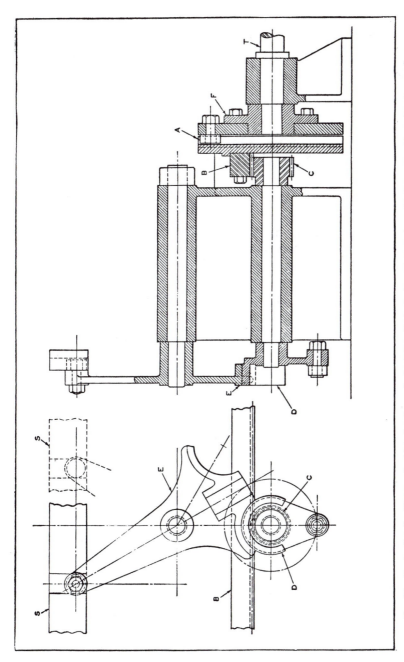

Fig. 18. Mechanism by Means of which Continuously Rotating Shaft T Causes Slide S to Reciprocate with a Dwell Period at Each End of Stroke.

the view to the right, the mechanism consists of two units—the driving crank unit shown at the right-hand end of the assembly or section view, and the Geneva dial mechanism indicated at the left-hand end.

Members D and E are laid out as a six-station Geneva motion, which allows D to revolve through an angle of 240 degrees while E remains stationary in either one of the two positions. This provides for the rotation of D through an angle of 120 degrees while transferring E from one dwell position to the other.

Reversal of member D after completing one revolution, as required to give slide S the reciprocating movements and dwells described, is accomplished by a crank with a roller A which revolves continuously with shaft T. Roller A engages a vertical slot in a member attached to rack B, as indicated in Fig. 19. Rack B, in turn, meshes with a pinion C on the shaft on which member D is mounted.

Thus continuous rotation of shaft T causes rack B to reciprocate, driving D first in one direction and then in the other. Rack B rotates driver D one full turn during

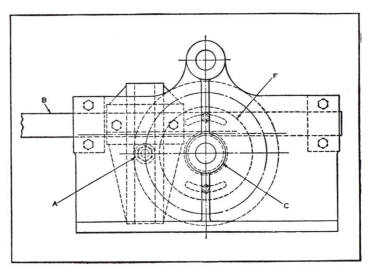

Fig. 19. End View of Mechanism Illustrated in Fig. 18.

one-half of a revolution of shaft T, and then reverses and drives it one full revolution in the opposite direction. Flange F, to which the disk carrying the driving roller A is attached, has elongated slots for the fastening studs in the flange to permit adjusting the angular position of roller A.

Swivel Joint Mechanism for Changing Direction of Movement.—Although the mechanism shown in Fig. 20, designed for changing the direction of movement transmitted by a lever, is not suitable for use where considerable rigidity is required, it nevertheless has much to commend it where the force applied is comparatively small.

The illustration shows one corner of the machine table on which the mechanism is used. The rod A has a reciprocating movement, rocking lever B about shaft C as a fulcrum. Shaft C is supported by two brackets M, the one at the front end of the shaft not being shown. This motion is transferred by the mechanism to a horizontal reciprocating slide D. There is a swivel link unit at E on the end of lever B. Lever H, being an integral part of the hub of lever B in the form of a separate arm, is also rocked back and forth by rod A.

Swivel joint F, of which S is a connecting-rod having a second swivel joint at G working in conjunction with the swivel link K, transmits motion in a horizontal plane to a rocker lever L. The horizontal lever L pivots on stud N, transmitting motion to link P, which, in turn, transmits the required reciprocating movement to slide D. Slide D is held on the base of the machine by means of guide plates Q. The opposite end of slide D serves to actuate transfer plates in the proper sequence of operations performed by the machine.

The principle involved in the swivel connections at E, F, and G is the same in all instances. The link E carries a pivot pin R, held in place by means of a washer and cotter-pin. This permits an up and down rocking movement of rod *A*. It also permits a rocking action of the joint forward or back, as well as sidewise, in the horizontal plane. The joint *G* at the end of the connecting link *S* allows a

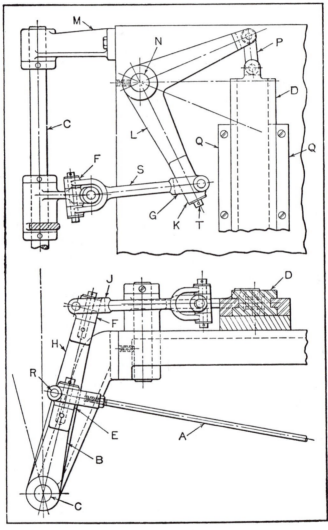

Fig. 20. Swivel Joint Mechanism by Means of which Reciprocating Rod A Operates Slide D.

swiveling motion in the horizontal plane, while pin T permits a swiveling motion in the vertical plane.

Therefore, regardless of the inclination of the connecting links or the variations in height due to the radial action of the several rocking levers, movement is transmitted from the rod A to the slide D. While the mechanism appears somewhat complex because of the many angles involved, it nevertheless often provides the simplest arrangement for obtaining the desired results.

CHAPTER 8

Crank Actuated Reciprocating Mechanisms

Among the special features of the reciprocating mechanisms described in this chapter are: operation of two slides alternately from one shaft; providing straight line reciprocating motion without support by ways or slides; synchronizing of horizontal and vertical motion; operation of two reciprocating slides from a single crankpin; provision of crank motion with dwell or rest period and obtaining reciprocating motions in two different positions in each cycle.

Other crank actuated reciprocating mechanisms are included in Chapter 9 of Volume I and are covered in Chapter 8 of Volume II of "Ingenious Mechanisms."

Mechanism for Operating Two Slides Alternately from One Shaft.—Figs. 1, 2 and 3 show the construction of a mechanism used on a packaging machine. Two slides B and S are operated alternately from the rotating shaft D, one of the requirements being that one slide start its movement when the other stops. Figs. 1 and 3 show the end and plan views.

The stationary part A of this mechanism is dovetailed to hold the two slides B and S on opposite sides, as shown in Fig. 1. Bearing C supports shaft D, which rotates in the direction indicated by the arrow. Lever E is free to turn on shaft D, and carries at its upper end the gear K which runs free on its stud. Gear K receives its motion from gear L which is keyed to shaft D. Connecting-rod F is carried on the stud on the upper end of lever E and the stud on slide B. Connecting-rod J is carried on a stud on gear K and a stud on slide S. Slides B and S are slotted, pin G passing through part A and acting as a stop for

both of these slides. Spring *H* serves to draw lever *E* to
the right.

Taking Fig. 3 as a starting point in the cycle, slides **B**
and *S* are held against pin *G* by spring *H* acting on lever
E. The rotation of gear *L* in the direction indicated by
the arrow causes gear *K* to rotate in the opposite direction; this, in turn, causes rod *J* to draw slide *S* to the left,
slide *B* remaining stationary until slide *S* is again returned
to its resting point against pin *G*. In Fig. 3, the dotted

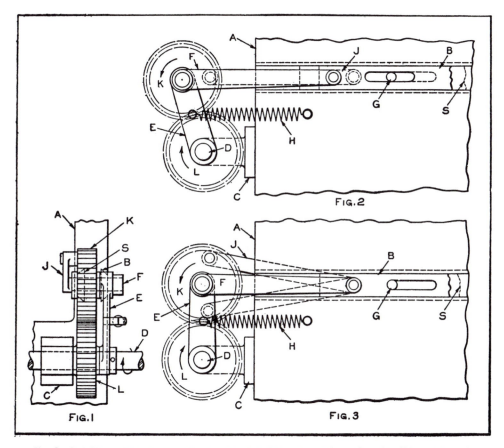

Fig. 1. End View of Mechanism, Showing Dovetail Slides B and S on Opposite Sides of Part A.
Fig. 2. Plan View, with Slide B in Central Position and Slide S in Right-hand Position. Fig. 3. Slides
B and S in Extreme Right-hand Positions.

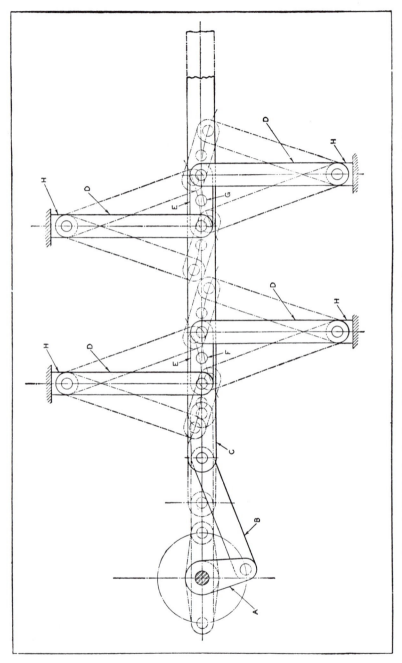

Fig. 4. Mechanism for Imparting Straight-line Motion to Ram C without Support by Ways or Slides.

outline of lever J indicates its position after slide S has completed its movement. As slide S is restrained from further movement to the right, continued rotation of gear K brings rod J to the center position, as in Fig. 2, thus causing lever E to swing to the left against the resistance of spring H. The movement of lever E draws slide B to the left, returning it to its resting point against pin G as rod J passes the center position, thus completing the cycle.

Straight-Line Reciprocating Motion for Link-Supported Ram.—A horizontal ram located in an atmosphere laden with abrasive material was required to have a straight-line reciprocating motion without being supported by ways or slides. The linkage devised to meet these requirements is shown in Fig. 4. Crank A and connecting-rod B impart the required reciprocating movement to ram C. In order to confine the motion to a straight path, four links D and two links E are utilized. These links are connected to ram C at the pivoting points F and G, the whole linkage mechanism being swiveled at the four stationary bearings H. All bearings are sealed to protect them from the abrasive material.

Links D and E are of such proportions that their centers F and G move in a straight line; and since the ram is connected at these points, it also moves in a straight path. Links E are approximately 38 per cent as long as links D, and the stroke of the ram does not exceed 65 per cent of the length of links D. This limitation on the length of the stroke with respect to the length of the links is necessary because points F and G only move in a straight line within a certain distance, beyond which they begin to move in a curved path.

Gear and Link Mechanism for Synchronized Horizontal and Vertical Motion.—The mechanism shown in Fig. 5 was designed to move work horizontally to a position where a vertical lift is applied. Referring to the illustration, the

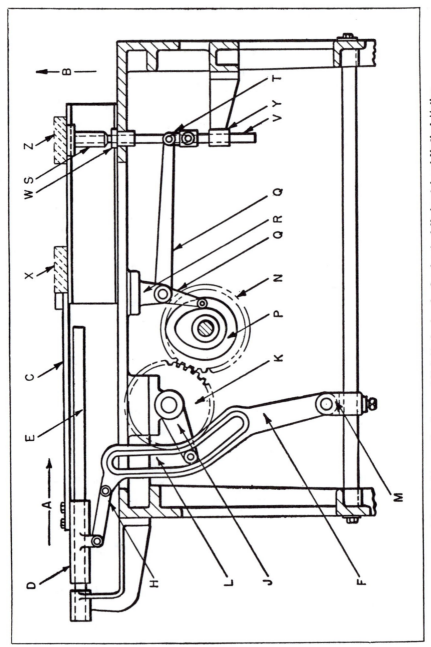

Fig. 5. Schematic Diagram for Gear and Link Mechanism for Synchronized Horizontal and Vertical Motion.

work X is to be moved in the direction indicated by the arrow A. Upon reaching position Z, the work is raised, as shown by the arrow B.

The horizontal pushing plunger C is attached to a slide D, which moves along two rods E when pulled by the lever F through the medium of the connecting link H. A crank J, rotating with gear K, operates in a slot L in the lever F, thus moving the lever so as to actuate the plunger slide to the right and left. Lever F is secured by means of a pivot to a stationary sleeve M, which is attached to a tie-rod in the base of the machine.

Gear K is driven by gear N in which there is a cam groove P. The purpose of the cam groove is to operate the bellcrank lever Q, which is attached by a pivot to the fixed bracket R. The movement of the bellcrank lever causes the lifting plate S to be carried up or down on the rod V, which is connected to the bellcrank lever by means of the linkage shown at T. Bearings at W and Y guide the lifting rod.

Timing of the horizontal and vertical movements relative to each other is accomplished by proportioning the slot L in lever F so that the forward movement of the work in the direction of arrow A is accelerated and then decelerated gradually as the crank J travels in the slot. A quick return stroke is provided by the angular section of slot L. The cam groove P in gear N is so positioned that lifting of the work at Z takes place during the return movement of lever F and plunger C.

Mechanism with Two Reciprocating Slides Driven from One Crankpin.—On a wire forming machine, a reciprocating motion was imparted to a sliding member by a crank disk to which it was connected by a pitman rod. Owing to a change in the design of the product, it became necessary to add another sliding member, which would carry the original sliding member. The added sliding member had to be given a reciprocating motion relative to the original

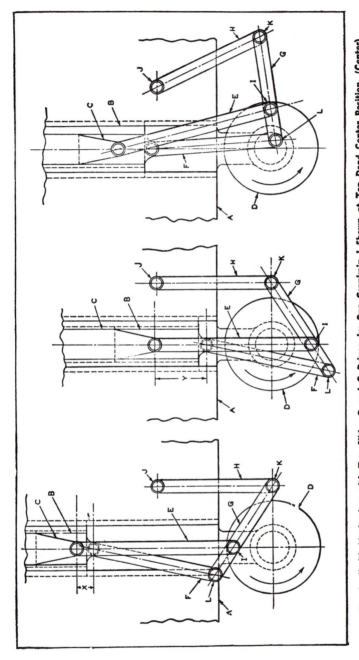

Fig. 6. (Left) Mechanism with Two Slides B and C Driven by One Crankpin I Shown at Top Dead Center Position. (Center) Mechanism with Crankpin I Shown at Bottom Dead Center Position. (Right) Mechanism Shown with Slides in Position Occupied After Crankpin I has made Three-fourths Revolution from Position Shown in Left View.

sliding member, both motions being transmitted from the same crankpin. This was accomplished in a comparatively simple manner, as shown in the accompanying illustrations.

Referring to the left-hand diagram in Fig. 6, the disk *D*, rotating in the direction indicated by the arrow, imparts a reciprocating motion to the slide *C* through the crankpin *I* and the pitman rod *E*. This constituted the original movement before any changes were made. The slide *B*, which was added, has a dovetailed base which is a sliding fit in a dovetail groove in member *A* of the machine, and carries the part *C* which is similarly mounted in slide *B*.

The lever *G*, pivoted on crankpin *I*, is connected at one end to the link *H* by the stud *K*, and at the other end to the pitman *F* by the stud *L*. Pitman *F* is connected to slide *B*, and pitman *E* to slide *C*, each transmitting the motion received from lever *G*, which is actuated by crankpin *I*. Link *H* is pivoted on stud *J*, which is stationary on member *A*.

In the left-hand diagram of Fig. 6, crankpin *I* is shown at the top dead center position. At this point of the cycle, slides *B* and *C* occupy the relative positions indicated by the dimension *X*, measured through the centers of the studs. In the center diagram of Fig. 6, crankpin *I* is shown at the bottom dead center. As lever *G* is operating in the third order, the greatest movement takes place at the work end, or at stud *L*. By virtue of this action, slide *B* is given a greater downward movement than slide *C*. The relative positions of the slides *B* and *C* are shown by the dimension *Y*, which is greater than dimension *X*, Fig. 6, left-hand diagram, indicating the difference in the movements of the two slides *B* and *C*.

In the right-hand diagram of Fig. 6, crankpin *I* is shown in the position it occupies after completing three-quarters of the cycle. This illustrates the action of link *H*, the purpose of which is to provide a floating fulcrum for lever

G without resorting to a slotted member or other form of track for stud K.

Crank Motion with Rest Period.—A crank motion with a dwell or rest period, designed for use on a wire forming machine, is shown in Fig. 7. This mechanism imparts a reciprocating movement to a part in the usual accelerating and decelerating cycle manner, except that a rest period is provided at one end of the stroke.

Referring to the upper left-hand view, the disk A, rotating in the direction indicated by the arrow, carries the bar B in a slot in which the bar slides freely. Bar B is retained by a cover plate (not shown), and is held in its extreme outer position by the spring F, carried on a rod that passes through the block G. The nut on the end of this rod restricts the movement of bar B, and is used for

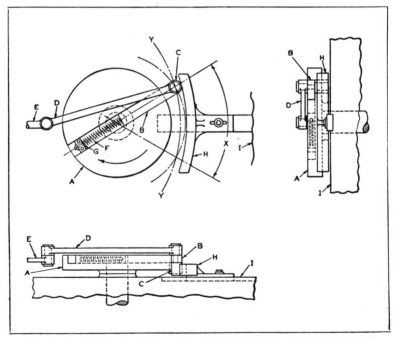

Fig. 7. Crank Drive for Reciprocating Slide that Provides Dwell at One End of Stroke.

making slight adjustments. Bar B is connected at its outer end with the rod D, which, in turn, is connected to the rod E to which the reciprocating motion is to be imparted by rotation of disk A.

The roller C, lower view of Fig. 7, carried on the under side of bar B, makes contact with the guide H during a portion of the cycle. Guide H is adjustably mounted on the stationary part I of the machine. The contact surface of guide H is formed to an arc of a circle having a radius equal to the center-to-center distance of the studs on rod D plus half the roller diameter.

Again referring to the upper left-hand view of Fig. 7, the rotation of disk A normally carries the center of roller C in the circular path indicated by the dot-and-dash circle concentric with disk A. However, as roller C contacts guide H, it cannot continue its normal path, but follows the path of the face of guide H. As this surface is formed in the arc of a circle that has its center on the center of the stud connecting rods D and E, rod D at this point rotates on this center, and there is no movement of rod E until roller C again leaves guide H. This causes a shortening of the stroke equal to the distance shown between the two arcs, and produces a rest period for rod E equal to the angle X.

Positive Crank Motion that Causes Slide to Dwell at Both Points of Reversal.—The use of a cam of ungainly size was avoided in the design of a wire-forming machine by employing a rather ingenious crank motion that produces a dwell at each end of its stroke. The mechanism employed is shown in Fig. 8. It imparts a relatively long stroke (6 inches) to a slide, and produces a dwell at each end of the stroke. The movement is positive and smooth. The compact nature and simplicity of the design may suggest its application to other types of machines requiring a similar movement.

The crank indicated at A is mounted on the drive shaft B, the latter being supported in the stationary bearing C.

A roll D engages the two grooves E and F in the driven slide G, which is mounted on the machine frame H. Positive action of the roll in passing from one groove to the other is assured by the switching arm J. This arm is secured to stud K, which is a free turning fit in the slide. It will be noted from Fig. 8 that the face of the roll is

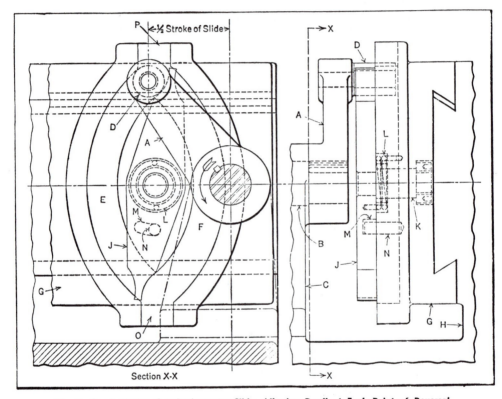

Fig. 8. Crank Motion that Reciprocates Slide, Allowing Dwell at Each Point of Reversal.

long enough to engage both the groove and the arm. Normally, the arm is held in the position shown by the torsion spring L, which forces the right-hand end of the elongated slot M in the arm against the pin N in the slide.

As the drive shaft rotates in the direction indicated by the arrow, the roll is guided by the arm into groove E.

Since the radius of this groove coincides with the radius of the path of roll D as crank A is rotated by the drive shaft, no motion of the slide will result and, consequently, the slide will have the required dwell at this end of the stroke. When the roll reaches the lower end of this groove, however, it pushes the arm J around in a counter-clockwise direction until the left-hand end of slot M engages pin N. In this way, the lower end of the arm forms a part of a continuous groove leading from groove E into the vertical groove O; and as the crank continues to rotate, the slide is carried toward the right to the end of its stroke. At the middle of this stroke, the roll is at its lowest position, at which time the lower end of the arm clears the roll. This allows the spring L to force the arm back to the normal position shown, so that a continuous groove leading from groove O to groove F results.

When the slide has reached the end of its stroke toward the right, the roll enters the concentric groove F, allowing the slide to dwell until the roll reaches the top of this groove. At this point, the roll engages the arm, forcing its upper end toward the left, so that it serves as a guide for transferring the roll to the vertical groove P. Continued rotation of the crank causes the slide to move toward the left to the position shown, the switching action of the arm being identical to that which took place at the bottom of the slide when the roll entered and left groove O.

It is possible that this mechanism would operate without the switching arm, but the action would not be certain. That is, the roll would be likely to jam at either point of reversal should the slide be accidentally displaced. Moreover, the impact of the roll, at each point of reversal against one corner of each of the grooves O and P would certainly damage the face of the roll or the groove corner. In addition to these disadvantages, a jerky action would be obtained at the points of reversal, which is absent when the switching arm is used.

Double Reciprocating Mechanism with Displaced Operating Positions.—The mechanism shown in Fig. 9 was developed to produce two distinct reciprocating movements of the feeding rod S. The limits of each of the reciprocations are identical, the difference being in the relative positions of the rod when the reciprocations take place. The action is best described by referring to the diagrams

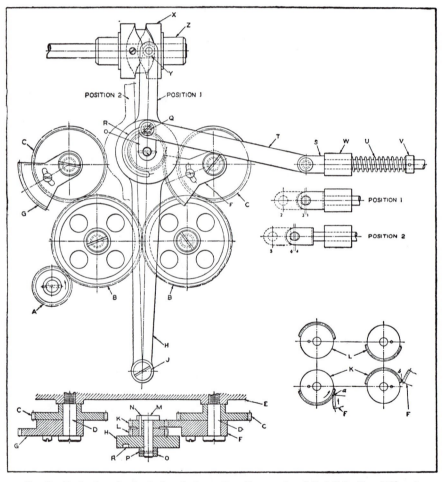

Fig. 9. Mechanism for Producing Reciprocating Movements of Rod S in Two Different Positions. At Lower Right are Shown Form and Positions of Gears K and L at Two Moments in Position 1.

which show the movements in the two reciprocating positions. The diagram below rod S designated Position 1 represents one reciprocating movement starting at a point marked 1, extending to 2, and finally returning to 3. When the mechanism is displaced, so that the upper end of rocker H occupies the position indicated by dotted lines in Position 2, the reciprocation starts at 4, extends to 5, and comes to rest at 6. This sequence of movements may be likened to the movements of a crank whose entire structure can be displaced or moved from Position 1 to Position 2, thereby giving the same reciprocating movement at two distinct points or positions separated by a definite distance. Provision is also made for time adjustment between the reciprocations.

The mechanism is actuated by gear A, which drives idlers B, the latter, in turn, driving the gears C on the studs D, in the main plate E of the machine. Gears C are provided with gear segments F and G. A slot in the gear segments permits adjusting their positions on gears C to produce a definite relation between the positions of the two segments in their continuous rotation with the gears.

The rocker arm H is pivoted on the stud J and is free to oscillate between gears C. Two mutilated gears K and L are fastened to the shaft M by pin N. In order to retain shaft M in place on the rocker arm H, the link or crank-arm O is fastened to the shaft M by pin P. Crank O is provided with the guide pin Q which fits into the concentric slot R in the rocker-arm head. Crank O is connected to the reciprocating rod S through the connecting-rod T.

The spring U furnishes the necessary tension between a collar V, fastened to rod S, and the bearing block W, which is a part of the main plate E. The cam X actuates rocker arm H between the gears C. A roller Y at the end of the rocker arm fits in the slot of the cam and produces a simple oscillating motion of the rocker arm. The bearing blocks Z support cam X.

The functioning of the mechanism is controlled by cam X. The cam is so designed that one revolution is necessary for a complete cycle of the reciprocating member S. One cycle consists of starting in Position 1 and returning finally to the same position. The cam can be given any irregular shape or form necessary to give any time cycle required of the reciprocating mechanism.

The mutilated gears K and L are alternately brought into contact with the gear segments F and G, which are displaced in such a manner as to make contact with their respective mutilated gears. Fig. 9 shows the form and relative positions of the mutilated gears at two moments in Position 1. First, the gear segment F is in contact with gear K at a ready to rotate the unit. The position of gear L should be noted at this time. The next view shows the gear segment F after it has completed its driving movement of gear K and is passing away from it at b. The changed position of gear L at this point should be noted. Gear L is now in such a position that when the rocker arm H is moved to Position 2, gear segment G can be engaged with it and made to produce a reverse rotation of gear L, returning link O to the upper part of the rocker-arm head and thus completing the reciprocation for Position 2.

It will be noted that points 1 and 3, as well as points 4 and 6, shown in the two position diagrams do not coincide. This is due to the fact that the slot R, subtending an angle of 180 degrees, combines with the angularity of rocker H to produce a slight difference in positions. However, this difference is necessary in the process to which the mechanism is applied. It is interesting to note that by slightly extending the slot at each end, the difference can be eliminated, and the points 1 and 3, as well as 4 and 6, can be made to coincide.

An interesting feature of the mechanism is that it provides a wide range for the timing of the interval between reciprocations. The timing is controlled by cam X, which is so interconnected with gears C that their positions are in a definite relation to the position of the cam X at all times. With the cam rotating intermittently, so that there is a sufficient pause in each position of rocker arm H to give the segment gear time to complete its function, there will be a definite time interval between reciprocations, dependent upon the relative positions of the two gear segments on their respective gears C. The interval can be varied by alternating the gear segments and bringing them closer together or spacing them farther apart.

CHAPTER 9

Variable Stroke Reciprocating Mechanisms

Means of adjusting the length, speed or timing of the reciprocating stroke are provided in mechanisms described in this chapter. Part of the stroke may be accelerated or retarded or the reciprocating motion may have a constantly varying length of stroke. Variation in the length of successive strokes may be obtained in one of the mechanisms described, while another provides a means of adjusting the length of stroke during operation. A mechanism for varying the stroke of a toggle-lever press is described. Another interesting arrangement is a double-lever mechanism for obtaining a variably controlled range of action from a single cam.

Adjustable Oscillating Mechanism for Reciprocating Movement.—The length of stroke of a reciprocating member on a wire-forming machine required occasional adjustment to suit changes in the product. The reciprocating movement on this machine was obtained by the cam-operated oscillating lever shown in Fig. 1 at B, rod J being used to impart the required movement to the reciprocating member. The cam D rotates in the direction indicated by the arrow, the roller C on lever B being in contact with the cam surface.

Lever B fulcrums on stud G, which is located on the upper end of the short lever A, the latter, in turn, fulcruming on stud H. The lower end of lever B has an extension which is acted upon by the spring F. Normally, this spring serves to keep levers A and B in alignment. The stop-screw E in a lug on a stationary part of the machine limits the movement of lever A.

214

In the view to the right, the roller C is shown in contact with the middle portion of the lobe on cam D, the rise from the lower lobe of the cam being accomplished while levers A and B operate on stud H as a unit. As the upper lobe of cam D reaches roller C, continued movement causes screw E to make contact with lever A, preventing further movement of the latter lever, so that lever B then fulcrums on stud G, as indicated in the view to the left. The change in the ratio of leverage increases the movement of rod J,

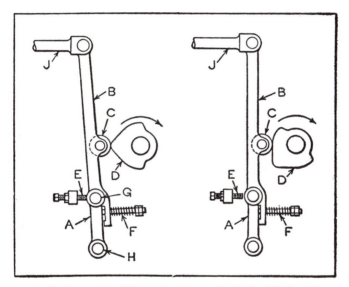

Fig. 1. Cam-operated Oscillating Lever with Stroke Adjustment, Used to Obtain Reciprocating Motion.

as controlled by the setting of stop-screw E. Roller C is held in contact with cam D by a spring which acts against the working end of rod J.

Mechanism for Varying Speed of Slide.—On a certain type of forming machine it was necessary to modify the action of the reciprocating slide carrying the metal-tape forming tools, so that a part of the forward and return stroke of each cycle of slide movement could be increased

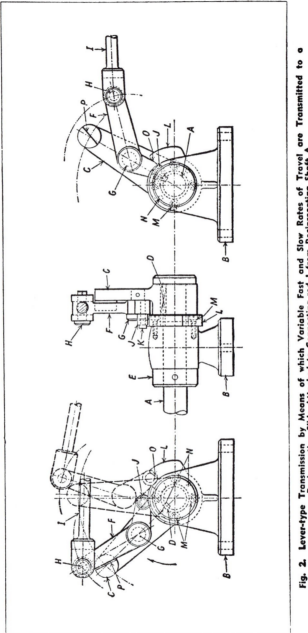

Fig. 2. Lever-type Transmission by Means of which Variable Fast and Slow Rates of Travel are Transmitted to a Reciprocating Slide Attached to Connecting-rod I from Reciprocating Shaft A.

considerably in speed. In addition, the starting and finishing points of such accelerated speeds were to be variable within specified limits, and a simple, inexpensive means of making such changes in speed was needed.

Fig. 2 shows the simple, smooth-running mechanism designed to effect the changes specified. The left-hand view shows the relative positions of the levers, etc., at the beginning of the cycle of operations, while the right-hand view indicates their positions at the end of the forward stroke. In the middle is an end elevation view.

The drive to the slide was transmitted from an oscillating drive-shaft A having a regular unvarying angular velocity and a fixed amplitude, while the total length of travel of the driven slide had to be kept constant. Shaft A is made a free running fit within the bore of the pedestal bearing bracket B. Secured by key D to this shaft is the lever C, which is prevented from endwise movement beyond due working clearances by the head of the shaft at one side of the lever and the retaining collar E.

Attached to the side of the lever C is the bellcrank lever F, which swings easily on the fixed-head stud G that is secured to lever C. The upper end of lever F is pivoted on stud H, which passes through the slotted knuckle end of the connecting-rod I. The latter member is connected at its other end (not illustrated) with the end of the driven reciprocating slide.

To the lower and shorter arm of lever F is attached a hardened and ground steel roller J, fulcrumed to revolve easily on the stud K, the head of which serves to retain the roller in correct endwise location.

Fitted on a hub machined at the right-hand side of pedestal B and concentric with the bore in which shaft A oscillates is the disk-shaped cam-plate L, secured by the screws M and two dowel-pins N to withstand radial thrust. The cam is of the slotted type, the slot O being wide enough to permit roller J to pass readily. The slot is situated in a radial position, and is deep enough to insure free working of the roller as bellcrank lever F swings on its fulcrum G.

It will be seen that the rim of the cam-plate is built up in diameter for a short distance at the right-hand side of the slot O. The roller J is normally held in contact with the rim portion of the cam-plate at the left of the slot, being maintained in this position by the stepped portion P of the boss at the end of lever C. The straight side of this step is so arranged as to bear against the left-hand side of the bellcrank lever F. In this position, sufficient clearance must be left between the step and the side of the lever to permit the roller to rotate freely.

In operation, commencing the cycle of movements from the position shown in the left-hand view, shaft A is oscillated toward the right, carrying with it the lever C, and this, in turn, carries along at the same speed the bellcrank lever F and the connecting-rod attached to the slide. Throughout this stage the roller J runs along the circular rim of the cam-plate L in advance of the slot O. Thus the driven slide is actuated at the same slow speed as the driving shaft.

This slow speed will continue until the roller reaches the slot O in the stationary cam-plate. As soon as this point is reached, the roller passes down the slot, and continued movement of lever C causes the bellcrank lever F to be swung round on its fulcrum stud G, thereby imparting to the connecting-rod a movement at an increased speed—greater than that possessed by the rod during the first stage of its movements. The point of the start of this rapid rate of speed is shown in heavy broken lines superimposed upon the left-hand view. When the roller is in radial line with the slot O, the built-up side of the slot insures the correct entrance of the roller, which might not occur if the length of the slot sides were identical.

The position of the levers at the termination of the clockwise stroke of the shaft is shown in the right-hand view,

from which it will be noted that the roller has passed well down the slot *O*, and the bellcrank lever *F* is accordingly swung around its fulcrum stud. Thus the first half of the complete cycle of movements of the driving shaft *A* imparts a slow speed to the reciprocating slide, followed by a much quicker rate of travel throughout the later portion of its stroke.

When shaft *A* starts the second half of its cycle by moving backward toward the left, the driven slide will start off at a fast rate, followed by a slower speed as the roller *J* leaves the slot. Thus each cycle of motions affords to the slide two slow and two fast movements.

By providing a simple means of adjusting the position of the cam-plate radially on its hub bearing, thereby altering the position of slot *O* relative to the vertical center of the arrangement, the point of the beginning of rapid travel can be varied within appreciable limits. This provision can be secured by passing the retaining screws *M* through elongated slots in the bracket *B*, thus enabling the cam-plate to be shifted. This will entail the omission of the dowel-pins *N*, which can be used when the cam-plate does not have to be moved.

This mechanism gives extremely smooth running and shockless transmission. It is simple, both in construction and design, inexpensive, and has simple means for changing the timing and duration of different speed rates for the driven slide.

Variable Reciprocating Motion.—A reciprocating mechanism employed on a wire fabricating machine to convert a uniform motion into a variable motion is shown diagrammatically in Figs. 3 and 4. The reciprocating motion was required to meet conditions of operation resulting from a change in the design of a certain part of the machine. Originally, the reciprocating slide, traversing a stationary part of the machine, was given a uniform motion, coming to rest at a predetermined point.

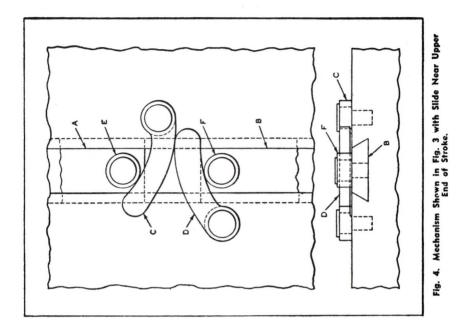

Fig. 4. Mechanism Shown in Fig. 3 with Slide Near Upper End of Stroke.

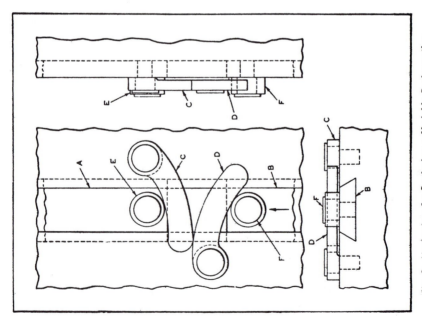

Fig. 3. Mechanism for Producing a Variable Reciprocating Motion.

Owing to the change in the part referred to, which reduced the clearance space available, it became necessary to retard the movement of the slide during a portion of its travel without changing the relative time at which the slide reached its ultimate position. This was accomplished by dividing the slide into two parts A and B, and introducing the levers C and D.

Referring to Fig. 3, slides A and B are dovetailed into a stationary part of the machine. Slide B, which is given a uniform reciprocating motion, carries the roller F which contacts with lever D, mounted on a fulcrum stud at the left. Slide A carries the roller E, which contacts with lever C mounted on the fulcrum stud to the right. Levers C and D are in contact at one point of their convex edges.

As slide B moves in the direction indicated by the arrow, its motion is transmitted through roller F to lever D, which is caused to pivot on its fulcrum stud. The motion of lever D is transmitted to slide A through lever C and roller E. It will be noted that D acts as a second-class lever on the end of lever C, which, in turn, acts as a second-class lever on roller E. Since it is a characteristic of the second-class lever to transmit the motion of the power member to the working point at a reduced speed and length of travel, it is obvious that slide A, at this point, is moving more slowly than slide B, thus providing the required clearance at the far end of the slide A.

As continued movement of slide B causes levers C and D to change their relative positions, the lengths of the lever arms are constantly changing, so that slide A is given an accelerated motion which, however, is less than the movement of slide B. As levers C and D change their relative positions, they eventually arrive in positions in which the contact points between rollers E and F and the levers lie on the same straight line. At this point, slide A is momentarily traveling at the same speed as slide B, since the motion is transmitted by direct contact, there being no lever action at this point. Further movement of slide B results in a continued acceleration of the speed of slide A, which is then moving faster than slide B. This is due to the fact, as shown in Fig. 4, that C and D are acting as third-class levers.

It will be noted that there is no change in the relative positions of slides A and B at the two extreme positions, indicating that the length of travel of slide A is equal to that of slide B. The ultimate effect of this arrangement is that of a continuous slide, with the added advantage of a retarded movement at the point where clearance space is required. Slides A and B are returned to their original positions by a spring (not shown).

Variable-Stroke Mechanism for Graduating Feed-Screw Dials.—The variable-stroke mechanism shown in Fig. 5 moves the head T up and down in synchronism with the indexing movement of the drum D, which stops the downward movement of the head as required for the production of the three different lengths of graduation lines on the work or dial A. The dials thus graduated are used on the cross-feed screws of turret lathes.

Referring to the illustration, the cutter-head T is actuated by the lever G and the connecting link J, which enters the sleeve H, in which it is securely held against the spring L by means of the spring K. The roll P at the lower end of sleeve H contacts with cam M, mounted on the main drive-shaft O. With this arrangement, the cutter-head T makes a full stroke at each revolution of the drive-shaft O and cam M.

The lengths of the graduation lines on dial A are controlled by the hardened steel circular drum or anvil D which comes in contact with the projection F on the sliding head T and thus interrupts the downward movement. While the cutter C is clear of the work, the anvil D is indexed by means of cam S on the main drive-shaft. Ten strokes of the cutter-head are required to complete the lines on one main division. Ten teeth, such as shown at R, are provided for indexing

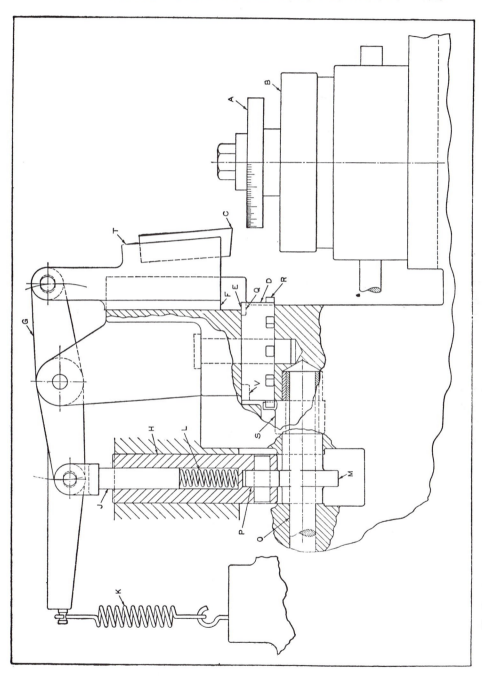

Fig. 5. Dial Graduating Mechanism which Automatically Varies Length of Stroke to Suit Different Lengths of Graduation Lines.

the anvil one complete revolution. Four of the strokes are interrupted at the short lines by the top face E of the anvil.

At the fifth indexed position, a slot Q in the face of the anvil receives the projection F, thus lengthening the stroke for the intermediate graduation line. A slot V, similar to the one at Q but diametrically opposite, allows the projection F to descend far enough to cut the longest graduation line.

The compression of spring L compensates for the various lengths of line. This spring also forces the cutter-head downward as sleeve H is lifted by cam M. Spring K serves to return the cutter-head to its uppermost position after the completion of the downward stroke. The faceplate B is indexed by a separate mechanism (not shown), which is properly synchronized with the variable-stroke mechanism.

Irregular Reciprocating Motion.—On a machine for manufacturing a wire product, it is necessary to guide the wire over a given path at an irregular rate of speed which decelerates toward the end of its travel. Although the required motion could readily be secured by the use of a cam, it was desired to obtain it from a crank used to operate another unit of the machine. The four views E, F, G, and H in Fig. 6 show how this was accomplished.

The crank A rotates in the direction indicated by the arrow, a reciprocating motion being transmitted to the slide-block C through the connecting-rod B. Gear teeth, cut in the large end of rod B, mesh with similar teeth on lever D, which is carried on block C. The wire being guided passes through a hole in the upper end of lever D.

In operation, the mechanism starts with lever D in the position indicated in view E. The angular movement of connecting-rod B causes lever D, by means of the gear teeth, to oscillate on its stud, so that the movement of the upper end of lever D is added to the movement of the block C as produced by the rotation of crank A, thus causing lever D to assume the position shown in view F. Continued rotation of crank A carries rod B to the horizontal position, causing lever D to return again to a vertical position, as indicated in view G, the movement of the upper end of lever D being deducted from the movement of block C.

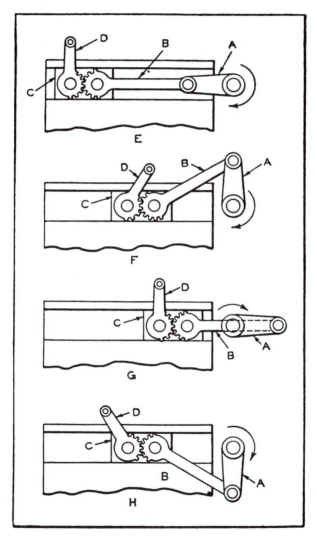

Fig. 6. Crank-operated Slide with Segment Gear Lever which is Given Irregu'ar Reciprocating Motion Through Segment Gear on Connecting-rod.

A comparison of views E, F, and G indicates that during the first quarter revolution of crank A, the movement of the upper end of lever D is considerably greater than in the second quarter. In view H, which shows the mechanism at the end of the third quarter revolution, the condition is the same as in view F, but in the reverse direction.

Cam Drive with Variable-Stroke Mechanism.—In order

to obtain a variable-lift motion from a simple cam mechanism, the length of the follower lever D was extended and equipped with an eccentric stop G, as shown in Fig. 7. With this arrangement, the stroke of lever D is lengthened or shortened as required by adjusting stop G. Baseplate H carries rotating cam B, which is in contact with follower roller C, mounted in the locking lever D. Lever D rocks or swings about pivot E and extends beyond follower roller C, terminating in a spherical end F, preferably made of fiber or laminated plastic material. End F is located opposite an adjustable eccentric G.

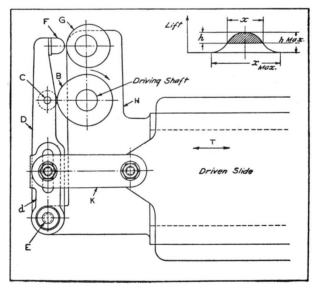

Fig. 7. Mechanism for Obtaining a Variable-lift Stroke by Adjusting Eccentric Stop G.

The extension d connected to lever D is provided with a long slot along which connecting-rod K can be adjusted and clamped. Rod K connects lever D with slide T, which represents the driven member. The position of slide T during the time in which the reciprocating motion is imparted to it by rod K can be changed by adjusting the driving pin in the slot in extension d. The length of the stroke can be changed by adjusting eccentric G.

This effect is shown diagrammatically in the upper right-hand corner of the illustration. The maximum and minimum lifts of the cam are represented by dimensions h, and the maximum and minimum angular movements during which the lift movement occurs are represented by dimensions x. By adjusting the stop, the values of these dimensions can be changed from maximum to minimum and, in extreme cases, they can be reduced to zero.

Variable-Stroke Mechanism for Toggle Lever Press.—
Toggle lever presses and eccentric presses of the usual type have no provision for adjusting the stroke, and only by adjusting the length of the connecting-rod can the position of the ram at the end of the stroke be changed. With the new mechanism shown in Fig. 8, however, it is possible to make any desired adjustment of the stroke length while the press is running.

The ram I is actuated in the usual manner by a connecting-rod H attached to lever G. The length of lever H can be varied by means of a screw, so that the ram can be located at any desired position at the end of the stroke. Between the free end of lever G and the swinging lever D which has a fixed bearing in the press body, there is a short member F and a longer member C. To obtain a positive movement, point J on member C is compelled by member E to follow a circular path.

A rotating crank A drives member C by means of a connecting-rod B. The stroke adjustment is accomplished by means of the lever E. Lever E has the form of a quarter sector provided with holes that correspond with other holes in the press body. By changing the bolt K on which lever E swings to one of the other holes in the press body, the length L of the ram stroke will be changed. The position of the ram at the end of the stroke, however, remains the same.

The members D, C, and F being located in nearly the same vertical plane, as shown in the illustration, then exert the maximum pressure on the ram. In the actual design, sector E is replaced by a worm-wheel which permits regulating the position of bolt K on which lever E is pivoted. With this arrangement, the bearing or bolt K can be moved on a circle around the center of the worm-wheel which coincides with point J.

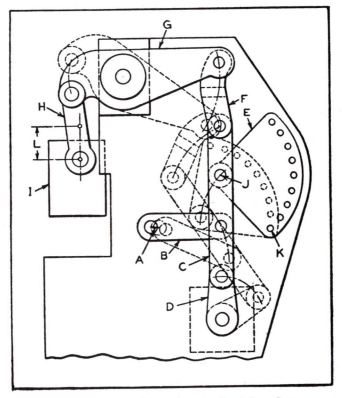

Fig. 8. Variable-stroke Mechanism for Toggle-lever Press.

The three views in Fig. 9 show the mechanism with the bolt K of member E located in three positions which give the maximum, medium, and minimum strokes. Precise adjustment of the stroke can be made before the press is started or while it is running. The crank A rotates through a complete circle. With clockwise rotation, the arc for the downward stroke is larger than for the return stroke; this results in the downward speed being somewhat reduced. Reducing the length of the stroke, in turn, reduces the time required for the downward movement.

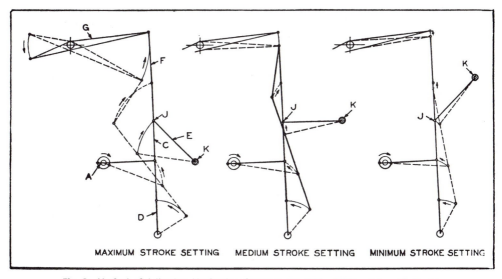

MAXIMUM STROKE SETTING   MEDIUM STROKE SETTING   MINIMUM STROKE SETTING

Fig. 9. Method of Adjusting Mechanism Shown in Fig. 8 for Different Length Strokes.

Variable-Stroke Mechanism with Adjustable Crankpin.—

Fig. 10 shows an interesting design for a variable-stroke mechanism that was devised as part of the development of a feeding arrangement. The device consists of a crankpin wheel A which is provided with an integral shaft supported in a bearing in the plate B. A slot is accurately milled in the face of wheel A to accommodate the movable eccentric crankpin block C.

Block C is provided with rack teeth which mesh with pinion D. The pinion is an integral part of the shaft E which extends through a bore in the shaft of wheel A. A locking disk F is fastened to the shaft of wheel A by two set-

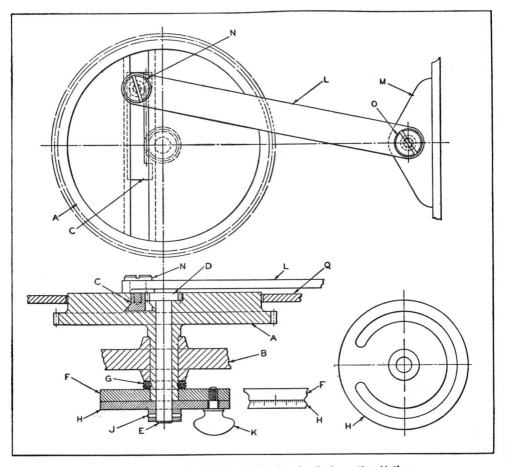

Fig. 10. Mechanism for Varying the Stroke of a Reciprocating Motion.

screws G. The adjusting disk H is fastened to shaft E by pin J. Disks F and H are locked together by screw K.

The eccentric movement is transmitted through link L to the pusher M. Studs N and O serve to connect the ends of link L to their respective members.   The wheel A is arranged in position below a table top Q.

The operation is quite simple.   Power is transmitted through a system of gears, not shown, to the gear that is an integral part of wheel A.   The eccentric location of stud N produces the required reciprocating motion in the pusher M through link L.   To adjust the eccentricity, thumb-screw K is loosened and disk H is turned slightly in the direction required to produce the necessary eccentricity.   The adjustment is transmitted through pinion D to the gear rack on the eccentric block C.

The pinion and rack are so constructed that they mesh with practically no free motion.   Disks F and H are of a large diameter and are provided with graduations.   The adjustment of the disks relative to each other results in a correspondingly small movement of pinion D and block C. This feature, coupled with the elimination of free motion in the linkage and gear teeth, produces a very accurate adjustment of pusher M.

Variable-Stroke Mechanism that Can be Adjusted While Operating.

—The principle of mechanically shifting the crankpin to increase or decrease the length of the stroke, as employed in the variable-stroke mechanism just described, was applied to machine shop shapers sixty or more years ago.   In the shaper applications, a screw was used in place of the gear and rack incorporated in the more recent design for changing the position of the crankpin.

A means of changing the stroke of a crank has been developed so that it can be employed to make adjustments while the mechanism is in operation.   With this improved design, shown in Fig. 11, the stroke can be adjusted to any length within close limits from zero to the maximum length for which the mechanism is designed.

Referring to the illustration, driving gear B is keyed to driving shaft A.   When gear B makes one revolution, it causes bevel gear J, which carries crankpin K, to make one revolution. Driving shaft *A* extends through the differential gear assembly to the bearing *O*, but is reduced in diameter to accommodate the small bevel gears *F*.

The bevel gear *C* is keyed to driving shaft *A*. Worm-gear *M* carries two bevel gears *D* and *E* on studs fastened within its rim. Worm-gear *M* and bevel gears *F* have a free bearing on shaft *A*. Worm *L* engages the worm-wheel, and, unless turned by means of its handwheel, remains stationary, in which case bevel gear *C* transmits its motion through bevel gears *D*, *E*, and *F* to gear *G*. Gear *G* is keyed to shaft *H*, to the upper end of which is keyed the small gear *I* that is in mesh with rack teeth cut in sliding bar *N* which carries crankpin *K*.

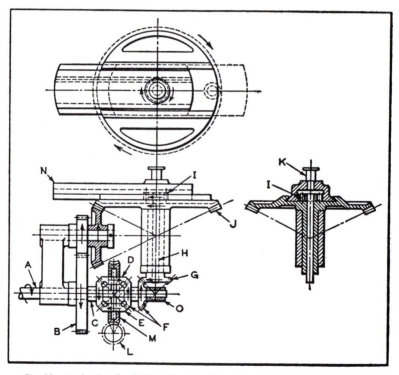

Fig. 11. Mechanism for Driving Crankpin K from Shaft A, which Enables Throw or Distance of Crankpin from Center of Its Shaft to Be Adjusted While Mechanism is in Operation.

When shaft A makes one revolution in the direction indicated by the arrow, gear J also makes one revolution. Simultaneously, the small gear I makes one revolution in the same direction as J, the result being the same as though gears I and J were fastened together. Any movement of the worm-wheel M by means of its handwheel, however, will cause gear I to revolve either slower or faster than gear J, depending on the direction in which the handwheel is turned. This increase or decrease in the speed of gear I will cause bar N to slide in or out, thus changing the radial position of crankpin K relative to the center of gear J and permitting the desired length of stroke to be obtained.

Exact setting for a given length of stroke is facilitated by a graduated dial on the handwheel. If the mechanism is required to operate at a fairly high speed, a specially designed double sliding bar would have to be substituted for slide N in order to keep the revolving unit properly balanced when adjusted to any stroke within its capacity.

Variable-Stroke Toggle-Lever Mechanism for Press.—

The stroke of most presses operated by an eccentric cannot be changed, because the amount of eccentricity is fixed. However, only a slight adjustment of the position of the top bearing of a toggle-lever press is required to obtain a considerable range of adjustment in the length of the stroke. Crank A of the toggle-lever press mechanism shown in Fig. 12 transmits motion to toggle levers C and D by means of connecting-rod B.

The upper lever is located in a bearing F on a slide G which can be moved or adjusted in a horizontal direction at right angles to the direction in which the press ram moves. The adjustment of slide G is effected by a handwheel which operates an adjusting screw. Thus the rigidity of the press body is not impaired by the adjusting device. The positions of the press ram at the ends of its stroke can also be adjusted by means of the screw E and wedge H without changing the length of the stroke. The press - operating

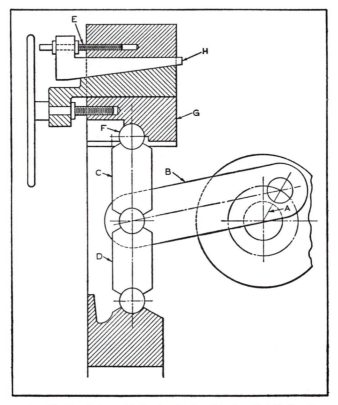

Fig. 12. Toggle-lever Mechanism Designed to Permit Adjustment of Position and Length of Press Ram Stroke.

mechanism described, which is the invention of W. Klocke, is covered by a British patent.

Variable-Stroke Cam-Actuated Mechanism.—The problem of providing a means for obtaining a variably controlled range of action from a single cam arises frequently in the design of various types of machines. The double lever mechanism, shown in Fig. 13, is simple, compact, and well adapted for solving problems of this kind. In its simplest form, it consists of a cam A and two levers B and C, pivoted on fixed centers and interlocked by a movable pivot stud D, which can be adjusted to vary the relative lengths of the active lever arms. One of the levers bears against the cam face while the other imparts the required reciprocating movement to the rotating cylinder E.

Cylinder E is free to slide axially on the main shaft F to which it is keyed. A cylinder cam A is fastened to shaft F. The roller follower G which fits the cam groove is mounted on the bellcrank lever B which swivels about the fulcrum stud H. Roller I, which fits the groove in cylinder E, is simi-

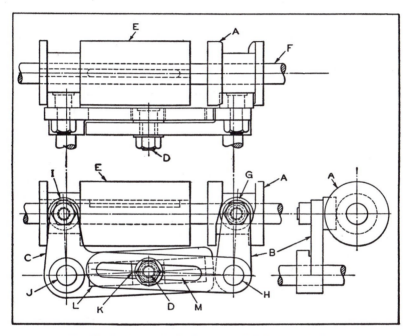

Fig. 13. Mechanism for Reciprocating Cylinder E on Shaft F.

larly mounted on bellcrank C that pivots on the fulcrum shaft J. The stud D carries a rectangular block K which slides in the wide slot L in bellcrank C. Stud D can be clamped in any desired position in slot M in bellcrank B.

When stud D is clamped midway between the two fulcrum points or studs J and H, the lever arms are of equal length, and consequently the axial movement imparted to the cylinder E is equal to the throw of the cam. To increase the movement of the cylinder, stud D is shifted toward fulcrum J. To decrease the movement of the cylinder, the stud is shifted toward fulcrum H. Not only can extremely fine variations of movement be obtained with this mechanism, but any desired range of movement can be obtained by properly proportioning the lengths of the slotted lever arms.

With this mechanism, complex movements are obtainable that might otherwise require elaborate mechanisms by making the fulcrum points adjustable, providing means for varying the position of the sliding block during the operating cycle, or substituting a cam of suitable curvature for the straight groove in cylinder E.

Mechanism for Producing Variable Reciprocating Movement.—A mechanism designed to produce a reciprocating movement of constantly varying length of stroke, constructed for use on a wire weaving machine, is shown in the accompanying illustrations. It was desired to produce an irregular pattern of no definite accuracy in the woven wire product, the main requirement being that the "repeat" patterns be identical. To produce the desired pattern, it was necessary that the length of travel of the reciprocating work-head vary throughout the repeat movement. By means of the mechanism illustrated, the work-head is given a gradual, but not uniform, increase and decrease in the length of stroke.

Referring to the left-hand diagram of Fig. 14, the drive-shaft A is carried in bearing I, attached to a stationary part H of the machine. Gear B is keyed to shaft A, and rotates in the direction indicated by the arrow. The hub on one side of bearing I is machined to support the lever J, which is free to oscillate on the hub. Lever J carries the gear C, which is free on its stud and meshes with gear B. Gear D rotates freely on the stud at the end of lever J, and meshes with gear C.

The connecting-rod F is attached to gear D by the stud L, and to the work-head G at the opposite end. The connecting-

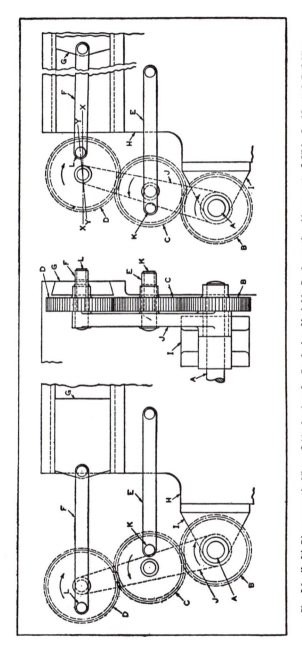

Fig. 14. (Left) Diagrammatic View of Mechanism for Producing Variable Reciprocating Movement of Slide G. (Center) End View of Mechanism. (Right) View Showing Relative Positions of Gears and Slide G after One-half Revolution of Driving Gear B.

rod E is attached to gear C by stud K at one end, and to the stationary part H at the other end. Gears B and C have the same number of teeth, while gear D has one more tooth. On the machine as constructed, gears B and C each have thirty-six teeth, while gear D has thirty-seven teeth, the effect being to maintain a constant definite relationship between gears B and C, and a constantly varying relationship between C and D throughout the cycle.

In the left-hand diagram of Fig. 14, the work-head G is shown at its extreme left-hand position. Gear B, rotating in the direction indicated by the arrow, causes gear C to rotate in the opposite direction. The rotation of gear C causes stud K to change its position so that lever J is moved or pivoted to the right, the extent of the oscillation being determined by the diameter of the circular path of stud K, as measured horizontally on the line of travel of the stud carrying gear C.

The rotation of gear C causes gear D to rotate on its stud in the direction indicated by the arrow, and as stud L rotates around the center of gear D, the radius of rotation is added to the horizontal movement of the upper end of lever J. As gears B and C have the same number of teeth, lever J passes through one cycle of oscillation to each revolution of gear B, but as gear D has one tooth more than gear C, it does not complete a full revolution to each revolution of gear B, and stud L, therefore, has not quite reached its extreme right-hand position, as is shown in right-hand diagram of Fig. 14.

This diagram shows gears B and C revolved one-half turn from the positions shown in the left-hand diagram of Fig. 14, but gear D is one-half tooth short of having completed a half turn. The line $X–X$ drawn radially through the centers of gear D and stud L indicates the position which gear D occupied in the left-hand diagram of Fig. 14. The line $Y–Y$, similarly drawn, indicates the present position after the half-turn of gear B. Although the movement of stud L has been added to the movement of the upper end of lever J, the added movement is slightly, though negligibly, less than would be the case if gear D had completed a half-turn. Thus, with each rotation of gear B, there is one tooth space difference in the relative position of gear D, and a corresponding difference in the position of the work-head G at the ends of its stroke.

The left-hand diagram of Fig. 15 shows gear B after having completed four revolutions. Gear C likewise has completed four revolutions, lever J and rod E occupying the same positions as the left-hand diagram of Fig. 14. Gear D, however, is the equivalent of four tooth spaces short of having completed four revolutions, and the stud L has not reached its extreme left-hand position. The travel of the work-head G to the left, therefore, has been decreased at this point to a distance equal to the horizontal distance between the original and the present position of the stud L, represented by the dimension Z.

In the right-hand diagram of Fig. 15, lever J occupies the same position as in the right-hand diagram of Fig. 14, and the stroke of the slide or work-head G to the right has been shortened by the distance W. Thus, with each revolution of gear B, there is a decrease in the stroke of work-head G, until gear D has completed a half revolution on its stud, when the conditions are reversed and there is a gradual increase in stroke length until gear D has made a full revolution, which completes a cycle or "repeat" movement.

Mechanism for Regulating the Movement or Pressure Applied to a Rod.—The device shown in Fig. 16 is designed to provide adjustable control over the travel or pressure applied to rod A. The normal actuating movement is obtained by having the end B of plunger C rock to a position B_1, there being a spring return (not shown) for rod A which keeps it back in a position to make contact with the end B of the actuating plunger.

Plunger C is supported indirectly on a stud D that also carries a swinging block E, in which the plunger is free to

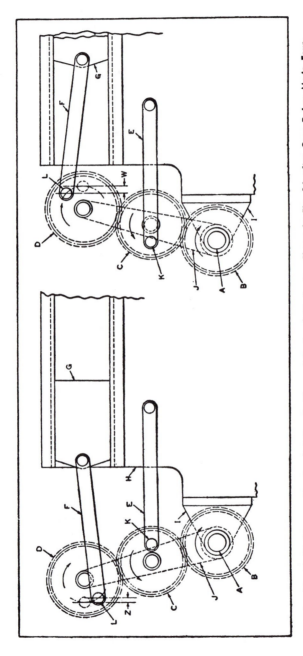

Fig. 15. (Left) Diagram Showing Relative Positions of Parts of Mechanism Shown in Fig. 14 after Gear B has Made Four Complete Revolutions. (Right) Mechanism with Lever J in the Same Position as in Right View of Fig. 14, but with Stroke of Slide G Shortened by Distance W.

slide up or down as required. The upward motion is obtained by raising the lever F through the medium of the link G operated from within the machine. Lever F pivots about the holding pin H and makes contact with a roll J attached to plunger C. There is a tension spring at K which holds the plunger C down against the lever F throughout its range of action.

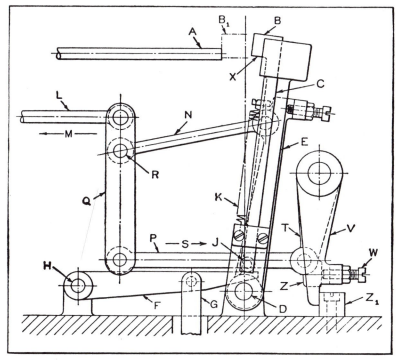

Fig. 16. Mechanism for Controlling Movement Imparted to Rod A by End B of Oscillating Plunger C.

When lever F raises plunger C, the entire swinging unit will rock back and forth without permitting the end B to strike rod A, as that portion X on the plunger will pass over the top of the rod. There is an ingenious arrangement for swinging the arm E back and forth, which is provided with means for adjustably controlling the travel from the right side of the mechanism, where the adjusting means is accessible to the operator of the machine.

Movement imparted to the connecting-rod L from within the machine in the direction of the arrow M pulls rod N in the same direction until the end B of the plunger meets resistance by coming against rod A, the fulcrum point for lever Q being about R as a movable center. This throws the thrust of link P in the direction indicated by arrow S, where it is carried to the lever T, which has an end Z that makes contact with an adjusting screw W in an adjacent lever V. Lever V makes contact with the block Z_1 attached to the machine bed. By adjusting screw W, the relationship between the two levers T and V is changed, thereby controlling within a limited range the action of B against rod A.

Automatic Variable-Lift Cam Mechanism.—A variable reciprocating movement is imparted to the slide of a wire-forming machine by the automatic variable-lift cam shown in Fig. 17. The requirements in designing this machine were that the slide be given four different degrees of movement during the cycle and that the timing of the movements coincide with each revolution of the driving shaft.

Referring to Fig. 17, shaft A, revolving in the direction indicated by the arrow, rotates the gear B, which is keyed to it. Gear E is free to rotate on shaft A, and carries the cam F, which is also free to rotate on the shaft. Gear E is revolved in a direction opposite to that of gear B through the idler gears C and D, which rotate freely on studs attached to a stationary part of the machine. The cam J is keyed to shaft A, and thus is caused to rotate with it.

Cam J consists of a heavy disk, which is grooved to carry the slide H, and a retaining plate, which is screwed to the disk. Slide H is shaped at its upper end to form the lobe of the operating cam. Roller G, which is attached to slide H, passes through a slot in the body of cam J and contacts the periphery of cam F. Slide H is slotted to permit shaft A to pass through. Roller M on slide K follows cam J.

When the mechanism is in the position shown, roller G is in contact with the high section of cam F. Gear E receives its rotary motion from gear B, reduced in the ratio of 1 to 4 by virtue of the relative pitch diameters of the gear train. Cam F, being attached to gear E, also rotates at the reduced rate of one-fourth revolution to one complete revolution of shaft A. As cam F is provided with four sections, each with a different radius, one of the four surfaces will be brought into contact with roller G at each revolution of shaft A. Thus roller G, being attached to slide H, causes slide H to move to one of four positions, depending on the relative position of cam F. The slide K is thereby moved a distance equal to the distance which the end of slide H projects beyond the periphery of the body of cam J. In operation, the outer end of slide H controls the movement of slide K, while the thrust of slide K reacts on cam F.

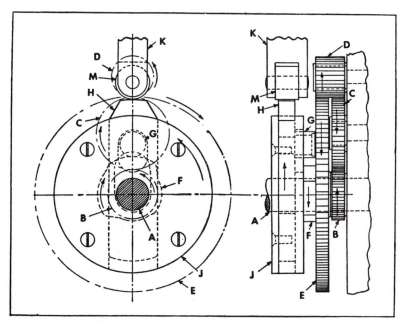

Fig. 17. Automatic Cam Mechanism which Gives a Variable-lift Movement to the Slide K.

Mechanism for Obtaining Variable Intermittent Oscillating or Reciprocating Movement.—The mechanism shown in Fig. 18 was designed to impart a rising and falling movement of magnitude, or height, S to roll G of lever F on a wire-forming machine at each revolution of the driving shaft A. (In the design shown, lever F is given an oscillat-

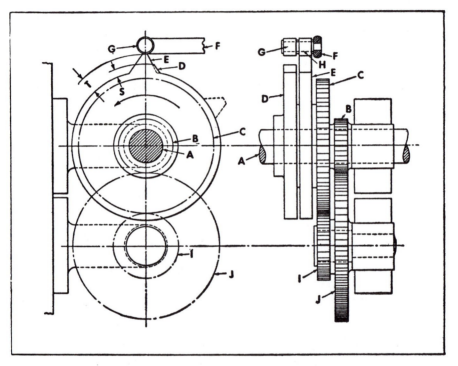

Fig. 18. Automatic Cam Mechanism which Imparts a Rise to Lever F of Magnitude T every Sixth Revolution of Shaft A and of Magnitude S at the Intermediate Revolutions.

ing movement, but if a slide were substituted for lever F, as in Fig. 17, a reciprocating movement would be imparted.) The mechanism is so designed that the magnitude, or height, of every sixth oscillation is approximately twice that of the five preceding oscillations, as indicated by the dimension T.

The required movement is accomplished in a restricted space by means of two cams D and E. The drive-shaft A, rotating in the direction indicated by the arrow, carries the gear B and cam D, both of which are keyed to shaft A. Gear B meshes with gear J, which is keyed to gear I, both rotating freely on a stud. Gear I meshes with gear C, which is keyed to cam E. Both gear C and cam E rotate freely on shaft A. Lever F carries a stud on which the two rollers G and H rotate freely and independently of each other. Cams D and E are of the same size and outline, but it will be noted that cam E provides a lift T which is practically twice the lift S of cam D.

When the mechanism is in operation, cam D rotates with shaft A, transmitting motion to lever F through roller G. Gear B transmits rotative motion to cam E through gears J, I, and C, cam E transmitting motion to lever F through roller H. As the gear train consisting of gears B, J, I, and C is in the ratio of 6:1 with respect to the speed of shaft A, cam E acts on roller H once in six turns of shaft A.

Thus, lever F is given five oscillating movements by cam D followed by an oscillation of greater magnitude imparted by cam E on the sixth turn of shaft A. On the following fifth turn, or revolution, of shaft A, cam E arrives at the position shown by the dotted outline. On completion of the sixth rotation of shaft A, the position of cam E is that shown by the full outline, with the cam roller G raised to its highest position.

CHAPTER 10

Mechanisms Which Provide Oscillating Motion

In the mechanisms described, which produce an oscillating or back and forth rotary motion, there are a number of special features: one mechanism produces rotation that is adjustable for various radii of curvature and will also produce straight line motion or reverse the curvature from convex to concave; several others impart greater or less angular movement to the driven shaft than that of the driving shaft; another produces two oscillations of a shaft for each cycle of a slide; several mechanisms are described for transmitting oscillating motion from one plane to another at right angles to it. A means for providing interrupting control of oscillating movement is also described.

Centerless Oscillating Motion.—Circular motion, such as rotation or oscillation, is usually obtained by means of a guide or constraining member that utilizes as a pivot the center about which the circular motion takes place. This constraining member surrounds the pivot either wholly or partially and causes the moving object to travel in a circular manner about the center. In this case, the center must obviously be accessible.

There are instances, however, where the center does not lie within the workable confines of a given specimen in which true circular motion is desired, and it is necessary for a sliding constraining member to be used that has been specially formed to suit that particular curvature, or one very close to it. Even though such a constraining member does permit some variation in the curvature, the range of differences is comparatively small, because of difficulties in the practical application of the method.

A mechanism that produces circular motion without regard to the center of the curve, that can be adjusted for a wide range of curves, and that produces genuine rotation, though through limited arcs, is shown in Fig. 1. This mechanism is adjustable for various radii of curvature, from a curve of ordinary radius to one of infinity; it will produce straight-line motion or reverse the curvature from convex to concave.

Although the motion is confined to limited arcs, the device produces not merely revolution but rotation; that is, all points in the moving body travel in circular arcs about the same axis or the curve center C. The constraining elements are simple straight-line slides that perform their functions according to the law of geometry pertaining to the location of the vertex of an angle inscribed in a circle.

Referring to the illustration, A is the stationary base of the device, B the oscillating table, on which anything can be mounted for whatever purpose the circular motion may be desired, as for example, machining a curved surface. The particular design represented in the drawing has clarity for its purpose rather than refinement in construction or action, and is therefore somewhat diagrammatic.

For the sake of clarity, the illustration shows the mechanism in that adjustment that places the center of motion C within the confines of the illustration. The top and side views show only half of the entire mechanism, the other half being a duplicate. The following description applies only to the half shown.

From center C a circular arc of radius R may be passed through the four pivots D, E, F, and G. These four pivots will remain on this circular arc in whatever position the table B may be placed. Two of the pivots, D and F, are stationary on the base A, and the other two, E and G, are on the table B. Since the latter two follow the circular path, as will be shown, it follows that the entire table will have a circular rotational movement.

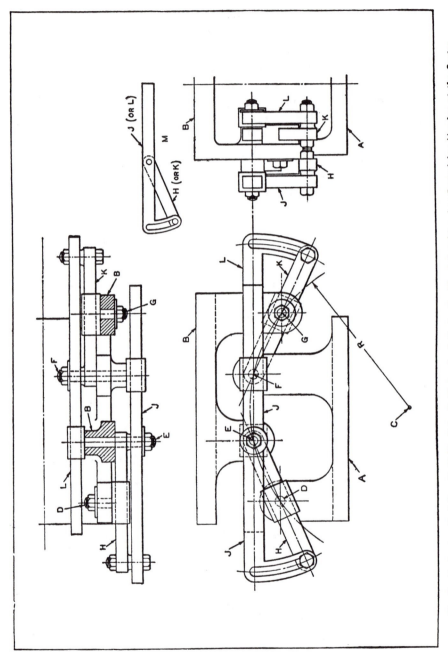

Fig. 1. Mechanism that Permits Table B to Oscillate about Center C without Requiring Any Restraining Member at the Center of Oscillation.

The table B is supported by and attached to the base A by means of two sliding bar sets consisting of bars H and J and K and L, one of which is shown separately at M. Each of these sets consists of two bars H or K and J or L, the shorter one, such as H, being pivoted to the middle of the longer one J. By means of the slotted arc, the shorter bar can be locked at any angle with the longer one within the range of the slot. One of these bar sets, H and J, is pivoted on the table at E, and the other, K and L, is pivoted on the base at F.

The bars of these bar sets slide in swivel guides, two of which are located on pivots E and F in line with but opposite to the pivots mentioned, and two more are located on pivots G and D. These guides are pivoted, respectively, in the table and base, as shown, and are free to turn, so as to assume the positions required by the bars that slide through them. The two bars H and J of the set that is pivoted at E slide in the guides on pivots D and F, respectively. The two bars K and L of the set that is pivoted at F slide in the guides on pivots G and E, respectively.

The action of the mechanism is as follows: The illustration shows the device in the mid-position of its motion. The points D, E, F, and G, are equally spaced. Therefore the angles in the two bar sets must be equal. As the table is moved either to the right or left, the vertex E of the angle in the bar set H and J follows along the circular path in accordance with the geometry of the inscribed angle and its subtended arc. Then the triangle EFG of the other bar set K and L is congruent to the triangle DEF of the first set, and since E and F already lie on the circle, it follows that G also lies on the circle by the same law of geometry. From the fact that the two points E and G of the table move in a circular path around the center C, it follows that the entire table rotates about this center.

The location of the center C and the corresponding radius of curvature depend on the angular adjustment of the bars in the slide bar sets, both of which must be alike—the smaller the acute angle the greater the radius. When the bars are in a straight line, the radius is infinite and the motion is in a straight line. If the angle is changed still further, it becomes negative and the curve of motion is inverted; that is, the curve becomes concave toward the upper side instead of convex. The same concave effect is obtained, however, by inverting the whole device.

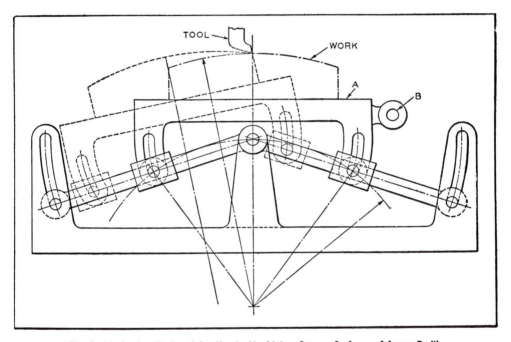

Fig. 2. Mechanism Designed for Use in Machining Convex Surfaces of Large Radii.

Mechanism for Use in Planing Convex Surfaces.—The

mechanism shown diagrammatically in Fig. 2 is designed for use in planing convex surfaces that conform to parts or arcs of true circles having centers that may be located at any distance from the point of the tool. This device is similar to the one just described, although it is much simpler to construct. The diagram shows only one of the two sides of the mechanism, the opposite side being similar to the one shown. The connecting-rod for oscillating the work-table *A* is attached at *B*.

As the motion of the work is not concentric with the convex surface being machined, a somewhat greater heel clearance is required for the tool than for ordinary turning, especially when work of the smallest diameter accommodated by the device is being machined. This clearance can be reduced as the work diameter is increased.

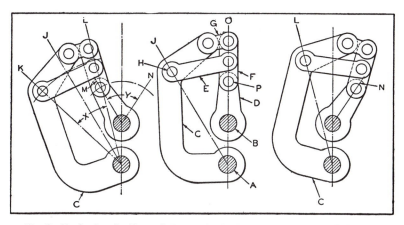

Fig. 3. Mechanism for Transmitting an Oscillating Movement from Shaft A to Shaft B, the Latter Shaft being Oscillated through a Larger Angle than Shaft A.

Crank and Link Mechanisms for Increasing Angular Movement of Shaft.—

In Fig. 3 is shown a simple mechanism designed to transmit an oscillating movement from shaft *A* to shaft *B*. The mechanism is required to give the driven shaft a larger angular movement than that of the driving shaft. A mechanism of this type for multiplying the angular movement of a lever was developed by H. Lindars. It consists simply of the driving lever or crank *C*, the driven crank *D*, links *E*, *F*, and *G*, together with the five pins that connect the cranks and links. The cranks *C* and *D* are securely fastened to their respective shafts *A* and *B* and oscillate with them.

The view at the left shows the mechanism with the driving shaft A and its lever C rotated to their extreme left-hand or counter-clockwise positions, while the view at the right shows the mechanism with the driving shaft and its lever rotated to their extreme right-hand or clockwise positions. The center view shows the crank D and driven shaft B in the vertical position, with pin P on the vertical center line O and pin H of the driving crank C located on the radial line J, which is also indicated as J in the view to the left.

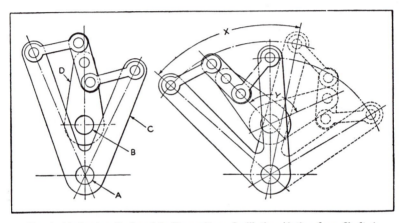

Fig. 4. Mechanism Designed to Transmit an Oscillating Motion from Shaft A to Shaft B with an Increase in the Angular Motion Imparted to Shaft B.

When shaft A rotates to the left so that pin H is in its extreme left-hand position on radial line K, crank D will also have been rotated to the left so that pin P is located on the radial line M. Similarly, when shaft A rotates to the right so that pin H is in its extreme right-hand position on line L, as indicated in the view to the right, crank D will have been rotated to the right so that pin P is located on radial line N. Thus, an oscillating movement of driving shaft A and its crank C through angle X serves to oscillate driven shaft B and its crank D through angle Y, which, as shown in the illustration, is larger than angle X.

It should be noted that, while crank C makes an angular

movement to the left of the central position shown in the center view equal to its angular movement to the right, the angular movement of crank D to the right from the central position is slightly greater than its angular movement to the left.

In the view to the left, Fig. 4, is shown another mechanism that is also designed to transmit an oscillating motion from one shaft to another with an increase in the angular movement imparted to the driven shaft. In this mechanism, which is patented in England under patent No. 465052, the oscillation of shaft A, and member C keyed to it, through angle X, as indicated in the view to the right, causes driven shaft B and its crank D to be oscillated through angle Y. In the view to the right, the mechanism is shown in full lines with its driving and driven shafts in the positions they occupy when they have reached the end of their rotating or oscillating movement in the counter-clockwise direc-

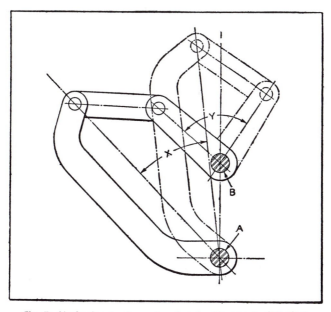

Fig. 5. Mechanism for Increasing Angular Movement of Shaft B as Compared with Shaft A

tion. The dotted lines show the mechanism with its members in their positions at the extreme end of the clockwise movement.

Link Mechanism for Increasing Angular Movement of Shaft.—In Fig. 5 is shown an arrangement of links and levers for transmitting an oscillating movement from shaft A to shaft B. The mechanism is designed to oscillate shaft B through angle Y, which is greater than angle X through which driving shaft A oscillates. This mechanism is intended to accomplish the same purpose as the one just described.

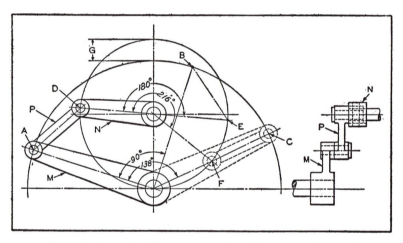

Fig. 6. Crank-and-Link Mechanism for Increasing Angular Movement of Driven Shaft.

Mechanism for Increasing Angular Movement of Shaft.—A crank-and-link type mechanism that produces a somewhat greater angular movement than either of the two designs just described is shown in Fig. 6. When the driving crank M travels from point A to B, through an angle of 90 degrees, the driven crank is rotated through an angle of 180 degrees. This represents a ratio of 1 to 2 in the increase in angular movement. When crank M oscillates 138 degrees, between points A and C, the driven crank will oscil-

late 216 degrees, between points D and F, at a ratio of 1 to
1.56. For this ratio G equals one-sixth the length of crank
M, and the length of link P equals one-half the length of M.

In Fig. 7 is shown an arrangement for oscillating a crank
that has too great an angular motion for practical operation by means of the connecting-rod of a revolving crank.
In this case, crank D must have an oscillating movement of
144 degrees or more. This is accomplished by the introduc-
tion of an auxiliary crank C, which obtains its angular motion of 72 degrees from connecting-rod E driven by a revolving crank (not shown).

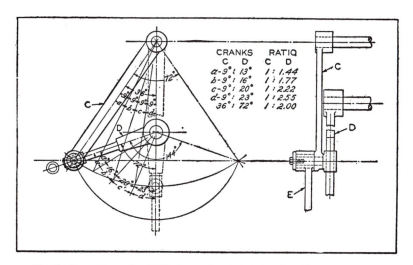

Fig. 7. Arrangement for Oscillating a Crank that has Too Great an Angular
Motion for Practical Operation by Means of Connecting-rod of Revolving Crank.

The connecting-rod of this mechanism could, of course,
be connected at any point on crank C; any position other
than the one shown, however, would necessitate a change
in the length of the stroke, although the 72-degree oscillating motion of crank C would remain unchanged. The motion
of crank D is not uniform, but varies from a ratio of about
1 to 1.44 up to 1 to 2.55, a 1 to 2 ratio of oscillation being
obtained for the half or complete angular motion. The com-

plete angular motion of crank C is 72 degrees, and that of crank D, 144 degrees, giving a ratio of 1 to 2. The like angles a, b, c, and d, representing movements of crank C, produce unlike angular movements of crank D, as shown at a^1, b^1, c^1 and d^1, at ratios indicated in tabular form.

Mechanism for Reducing Oscillating Motion.—On a wire-forming machine, a part of the mechanism is operated by an oscillating shaft which receives its motion, through a system of links, from a crank. Owing to a change in the design of the product, the degree of oscillation of the shaft was required to be reduced. However, as other parts of the mechanism receive their motion from the same source, it was not permissible to change the throw of the crank, nor to change the positions of the connecting levers. Fig. 8 shows the design of a mechanism that successfully met the requirements referred to.

The shaft A, which is required to oscillate, is carried in the bearing B. Lever C is given an oscillating motion by

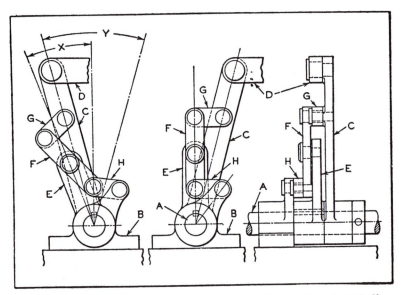

Fig. 8. Mechanism by Means of which Oscillation of Lever C through Angle Y Produces Reduced Oscillation of Lever E through Angle X.

the rod D. Lever C and rod D are of the same dimensions and occupy the same positions as before the change, except that originally, lever C was keyed to shaft A, whereas with the new arrangement it oscillates freely on shaft A. Lever E, which is keyed to shaft A, carries the lever F, which is pivoted at its center. Bearing B carries a fixed arm which is connected to the lower end of lever F by the link H. The upper end of lever F is connected to lever C by the link G. At the left and center of Fig. 8 are end views of the assembly, and at the right is a side view.

In the center diagram of Fig. 8, rod D is shown at its extreme right-hand position. By tracing through the linkage, it will be noted that any motion of rod D is transmitted through lever C to lever F through link G. Lever F transmits the motion further through lever E to shaft A.

In the left-hand diagram, rod D is shown at its extreme left-hand position. It will be noted that, as the fulcrum of lever F is at its lower end, it is acting as a second-class lever, and any motion given to the upper end of lever F, which is the power arm, will be transmitted to lever E in reduced proportion. In the left-hand diagram, the angle Y indicates the movement of lever C, and the angle X indicates the movement of lever E, the difference between the two indicating the reduction in angle of oscillation accomplished through the linkage.

Mechanism that Doubles the Oscillations Imparted to a Shaft by a Slide.

—The fingers of a bag-closing machine are controlled by an oscillating shaft having a variable motion. The motion is imparted to the oscillating shaft from a feed-slide which reciprocates at a constant velocity. In Fig. 9 is shown the mechanism for transmitting the variable oscillating motion to the shaft D from slide A. During each cycle of slide A, shaft D passes through two oscillating cycles. For the first cycle, assuming that point R of link E is at P and that slide A is traveling downward, shaft D will oscillate rapidly through an angle of 190 degrees and

return, at the same velocity, to its starting position. In the
next cycle, the direction of rotation of shaft D is reversed,
after which it oscillates slowly through an angle of 60 de-

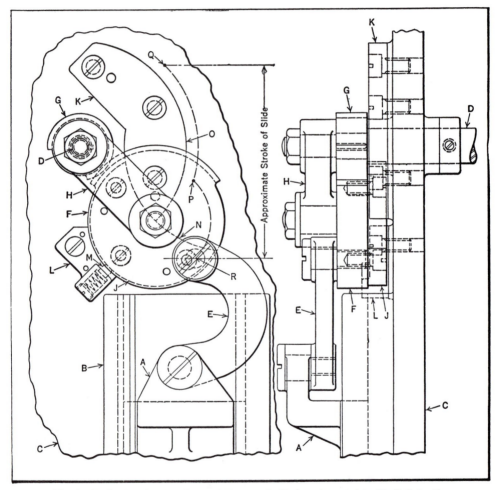

Fig. 9. Mechanism for Doubling and Varying Oscillations Imparted to Shaft D by the
Reciprocating Slide A.

grees and then returns at the same velocity to its starting
position. These oscillations are repeated alternately with
no intermediate dwells.

The feed-slide A operates in the guides B, cast integral with the machine frame C. This slide transmits the required motion to shaft D through the link E, segment gears F and G, and the link H. Link E connects the slide with gear F and is pivoted to these members by shoulder-screws. Gear F is free to turn on a stud secured in link H, and meshes with the gear G, keyed to the driven shaft D. Link H is free to turn at the upper end on the driven shaft D.

On the under side of gear F is fastened a crescent-shaped plate J, which serves to guide this gear in its proper path along the contour of cam K, secured to the machine frame. Block L, containing the spring-actuated pin M, merely acts as a bumper to limit the downward movement of gear F and to absorb the shock of reversal at the low point.

In the position indicated, the slide is at the bottom of its stroke. As it moves upward, the upper screw in link E will be moved to point P. At the same time, gear F will be rotated through an angle of approximately 90 degrees; the lower point of cam K being in contact with the plate J holds link H stationary. This partial rotation of gear F causes gear G and shaft D to rotate 190 degrees in a clockwise direction. At the end of the 90-degree movement of gear F, the curved surface N on plate J coincides with the surface O on cam K, so that further rotation of gear F relative to link H is prevented. However, as surface O is concentric with the center of the upper end of link H and of gear G, continued upward movement of the slide to the end of its stroke will cause both gears and link H to swing upward as one member until the upper pivot screw in link E is at point Q. At the beginning of the latter movement, gear G and shaft D will have reversed their direction of rotation and their angular velocity will be reduced.

As the slide reverses its movement and returns, the slow angular movement of shaft D is also reversed. Shaft D then rotates slowly until the cylindrical part of gear F engages block L. At this point, link H is held stationary and gear F

rotates about the lower point of cam K. In doing so, gear F rotates gear G rapidly in a counter-clockwise direction, causing shaft D to reverse its movement and continue to rotate, but at a rapid rate, until the slide has reached the bottom of its stroke. Incidentally, a slight clearance should be provided between the adjacent surfaces of members K and J at this point in the slide cycle.

Obviously, the velocity of shaft D when the slide is approaching and leaving the bottom of its stroke is the same; and likewise, the velocity of the shaft when the slide is approaching and leaving the top of its stroke is also the same. Consequently, the alternate variation in the point of reversal of shaft D occurs at both ends of the stroke of the slide. By modifying the gear diameters and the stroke of the slide, various angular movements and velocities of shaft D can be obtained.

Lever Mechanism for Transmitting Oscillating Motion Simplified by Means of Universal Joints.—Transmitting a motion or movement developed at some point within a machine to some remote point for the operation of a secondary unit frequently introduces a perplexing problem. Often a combination of levers must be used for this purpose. In some cases, the problem can be simplified by equipping the connecting links or rods with universal joints, as here illustrated.

The mechanism shown, Fig. 10, is designed to transmit to lever X the oscillating motion imparted to lever J by cam C. Cam C, on driving shaft B, imparts a reciprocating motion to yoke E through roll D in the cam groove. Yoke E slides on block F located on shaft B.

It will be noted that both cam C and yoke E are located in a vertical plane and that lever J is in a horizontal plane, while lever X is in a vertical plane. Lever S, which is connected to reciprocating lever J by rod P, is also in a vertical plane. Thus, the problem of transmitting an oscillating motion to lever X becomes one of transmitting motion from a horizontal to a vertical plane. Obviously, this problem is greatly simplified by using universal joints to connect rod P with levers J and S, as indicated by the dot-and-dash lines Z and Y.

Lever S is pinned to shaft T and transmits the required motion to lever X, which is also keyed to shaft T. Lever X, in turn, operates a slide mechanism which is connected to swivel-block W.

Referring to the construction of the mechanism, a portion of the machine frame is shown at A in both views. Bearing brackets on this frame support shaft B on which

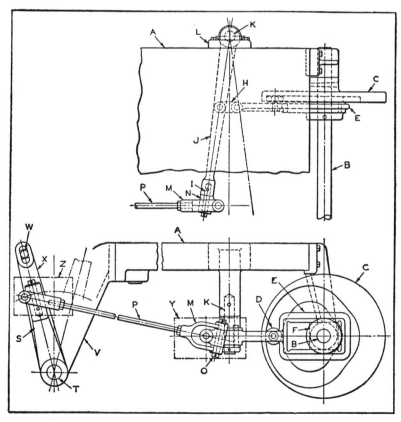

Fig. 10. Mechanism in which Universal Joints are Used to Simplify Transmission of Movement from Horizontal to Vertical Plane.

cam C is mounted. Yoke E is connected by pins to lever J through the medium of swivel-link H. Lever J is free to swivel about stud K as a center, stud K being carried in support L attached to frame A. In the end of lever J is a pin I on which block N is free to swivel. A pin O connects yoke M to block N. This construction provides a universal

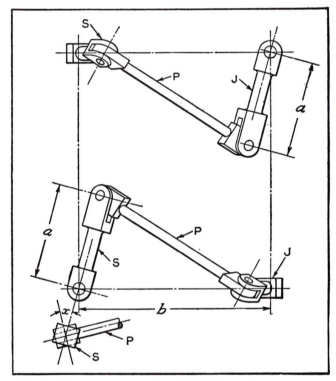

Fig. 11. Simple Mechanism for Converting Horizontal to Vertical Oscillation.

joint which permits yoke M to swivel in any direction. The two trunnion bearings of the two universal joints at the ends of rod P are located at right angles. Two brackets V support bearings for shaft T. With this arrangement, the tendency for the levers and connecting-rods to become cramped or jammed is eliminated.

Converting Horizontal to Vertical Oscillation by Simple Mechanism.

The linkage mechanism just described must have an appreciable amount of play in the pivot joints in order to operate successfully. If a few calculations are made, on the supposition that levers J and S are rigid bars, it will be found that the connecting rod P would be strained in torsion during the oscillating motion. In any practical mechanism, made as described and illustrated in Fig. 10, all distortion for which relief is not provided by the play allowed by the pivots will be taken up by the various parts of the mechanism.

However, the desired conversion from oscillation in the horizontal plane to oscillation in the vertical plane can be accomplished by a linkage mechanism of fewer parts which was devised by G. T. Bennett and described in *Engineering* in 1903. If applied to the mechanism in question, as shown in Fig. 11, it would require levers J and S to be of equal length, and the length of connecting-rod P to be the same as the distance between the axes of rotation of levers J and S. Then each of the universal joints shown would be replaced by a simple pivot joint. The angle between the pivots in rod P would be a right angle, while the sine of the angle x between the pivots in lever J would be equal to the ratio of the length of lever J to the length of rod P, or sine $x = a \div b$. Lever S would, of course, have the same angle x between its pivots as lever J.

The Bennett linkage is not an approximation; it is a mathematically correct method of converting an oscillating motion in one plane into an oscillating motion in another plane. There is no limit to the amount of oscillation; the cranks or levers may execute complete rotations or even continuous rotation if the links are designed to clear each other.

Space Linkage Mechanism for Transmitting Oscillating Crank Motion.

There are cases where obstructions or interferences may make it impossible to join the two links by a single connecting-rod, such as the mechanism just described. The interference can sometimes be avoided by the use of a connecting-rod made in two pieces which are simply

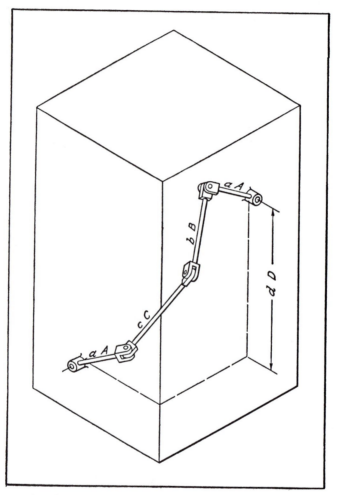

Fig. 12. Space Linkage Mechanism for Transmitting Oscillating Motion.

hinged together. The main frame, the two cranks, and the two parts of the connecting-rod, as shown in Fig. 12, constitute a hinged movable five-bar linkage in space.

A five-bar linkage of this kind must be designed to have certain mathematical relationships between its various members in order to operate. Let the length of a link be defined as the length of the common perpendicular between the hinges in the link. Let the twist of a link be defined as the angle between the hinges in the link. In the diagram and in the formulas below, the small letters refer to the lengths of the links, while the corresponding capital letters refer to the twists of the links.

The five-bar linkage will operate if the following conditions are observed: (1) The two cranks have equal lengths a; (2) the two cranks have equal twists A; (3) the sum of the lengths of the two parts of the connecting-rod is equal to the length of the main frame, $b + c = d$; (4) the sum of the twists of the two parts of the connecting-rod is equal to the twist of the main frame, $B + C = D$;

$$(5) \quad \frac{a}{\sin A} = \frac{b}{\sin B} = \frac{c}{\sin C}$$

To simplify the diagram, the cranks are shown to be oscillating in two perpendicular planes. The twist D of the main frame or link is, therefore, a right angle. Angle D need not, however, be restricted to a right angle; any other angle between the planes can be handled just as readily.

Oscillating Motion Transmitted from Vertical to Horizontal Plane.—The oscillating motion of a part in a vertical plane is transmitted to another part in a horizontal plane by means of the simple mechanism illustrated in Figs. 13 to 15. The magnitude of the oscillations is adjustable to suit various requirements.

A front elevation and plan view of the mechanism in the central or middle position of its cycle are illustrated in Fig. 13. Lever G oscillates in a horizontal plane about pivot rod H. Rod F, the position of which is adjustable in the slot of lever G, is fastened to the lever and operates with it. Part E is free to slide vertically on rod F and horizon-

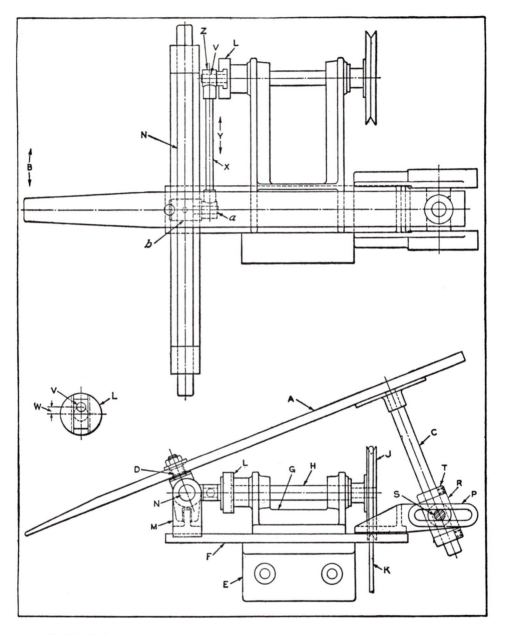

Fig. 16. Mechanism for Imparting Oscillating Motion to Arm A, which can be Adjusted to Various Angular Positions.

Mechanism for Imparting Oscillating Motion to Adjustable Arm.—The mechanism shown in Fig. 16 is designed to impart an oscillating or vibrating motion to the arm A, which can be adjusted to various angular positions. The mechanism is driven by a round belt K from a V-pulley within the machine on which the oscillating motion is employed. The complete mechanism, including arm A, is mounted on a baseplate F, which is secured to the machine by the bracket E.

Arm A pivots about the center of rod C, the oscillating motion indicated by the arrows at B being obtained from the reciprocating motion transmitted to the stud D of the cross-head b. The necessity for adjusting arm A to different angular positions is a factor that makes the design of this mechanism of particular interest. Provision for adjusting the length of the oscillating stroke also lends interest to the design.

The means for obtaining the oscillating motion will be described before considering the adjustment features. The shaft H, driven by belt K, is provided with a crankpin V for the bearing Z on one end of the connecting-rod X. The bearing at the opposite end of the connecting-rod is fitted on a pin a in a sliding block or cross-head b, of which stud D is a part. The reciprocating motion of stud D transmitted by the movement of the connecting-rod X, indicated by the arrows at Y, produces the required oscillation of arm A.

The bracket M supports the shaft N on which the cross-head b reciprocates. There is a slotted arm on the under side of the cross-head which is a sliding fit over the guide rail on the base of bracket M. The guide rail prevents the cross-head from rotating on shaft N. The length of the arc through which arm A oscillates can be changed to suit requirements by adjusting the sliding block in the flange L in which the crankpin V is mounted. Adjusting the block to increase the distance W of the crankpin V from the center of shaft H, as shown in the small insert diagram at the left of Fig. 16, increases the length of the arc through which arm A oscillates.

Stud D is fastened to arm A in such a manner that the arm can be tilted to any desired angle with the horizontal. The block R in which rod C pivots can be swiveled in bearing P and locked in the required angular position by a clamping bolt S. The collars T locate rod C in block R.

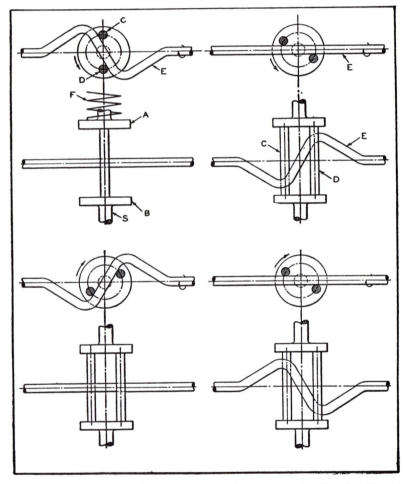

Fig. 17. Four Views of Simple Oscillating Mechanism, Showing How Rotating Shaft E Transmits Oscillating Motion to Shaft S.

Simple Mechanism for Producing an Oscillating Motion in a Shaft.—The mechanism shown in Fig. 17 was developed to produce an oscillating motion in the vertically suspended shaft S through the continuously rotating horizontal shaft E. It consists mainly of two disks A and B, both of which are a part of the vertical shaft assembly; the horizontal shaft E; spring F; and the pins C and D. The horizontal shaft E is bent to the shape indicated in the illustration, and is located between the pins C and D.

The four positions of the horizontal and vertical shafts shown in Fig. 17 indicate that the horizontal shaft serves as a cam, with th pins C and D acting as the cam followers. The shaft S is spring-loaded to keep the pins C and D in contact with the shaft E.

The mechanism can be operated at moderate speeds and, with a reasonable amount of tension in spring F, the oscillations will be imparted to the vertical shaft smoothly and without shock.

By changing the shape of the shaft E, sufficient variation in the oscillations can be obtained to meet different requirements. Some experimental work may be necessary when precise results are desired. However, a fairly accurate lay-out will usually be sufficient to give the desired action.

Double Toggle-Lever Mechanisms for Operating Presses.— The capacity of a straw-baling press designed as shown in Fig. 18 was considerably improved by redesigning the operating mechanism. The machine has an oscillating pressing piston C, actuated by a rod B from a rotating crank A, which is constructed as a gear wheel and rotated by a small gear D. The pressing action takes place when the three links 1, 2, and 3 form practically a straight line, a high pressure then being applied to the straw bale during the interval y_1–y_2 of the crank motion. This interval is so short that the baling action is not quite perfect.

The full exerted pressure loads the mechanism. The

swinging piston is quickly pushed forward and is already on its return stroke when the knot is tied. Therefore the elasticity of the bale stretches the binding cord and hinders the binding mechanism. High friction in the links increases the wear, and the forward and return strokes are performed at an almost identical speed.

In the improved mechanism, shown at the left in Fig. 19, two members E and F are added, and the number of links is thereby increased from four to seven (as the connecting link between the three members B, E, and F must be calculated as two links). The new mechanism has two dead points; therefore, the toggle action shown at the left in Fig. 19 is extended to about three times hat of the older mechanism. The push exerted during the pressing action is taken off the gear and transmitted to the fixed link 5. The speed of the piston is gradually reduced until it is brought to a standstill. It remains in this position until the binding action is completed. The whole drive runs easily and without shock. The wear is consequently reduced to a minimum, and the capacity of the new machine is considerably increased.

Similar mechanisms can be employed in other industries, such as the one shown at the right in Fig. 19, which is used in a press designed for molding plastics. This mechanism

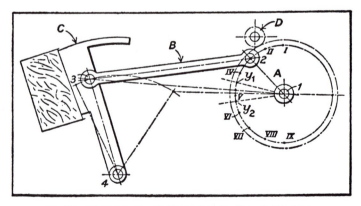

Fig. 18. Original Design of Mechanism for Straw-baling Press.

is the same as in Fig. 18, except that the swinging motion of the piston is replaced by the reciprocating sliding movement of a ram. Owing to its high capacity, better results are obtained with this press than with the older design having a simple crank mechanism.

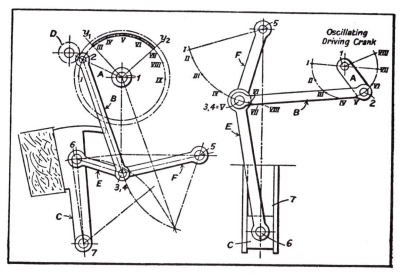

Fig. 19. (Left) Double Toggle-lever Mechanism for Straw-baling Press. (Right) Double Toggle Lever for Plastic Molding Press.

Mechanism for Applying Rotary or Oscillating Motion to a Driven Shaft.

—The mechanism shown in Fig. 20 is designed to give a rotary or oscillating motion to the driven shaft E. The oscillating motion is made possible by the oscillating action of the segment gear M. When this motion is required, the driven shaft changes its direction of rotation every half-revolution and operates at one-eighth the full rotating speed of the motor driving shaft. When a continuous rotating motion is required for shaft E, sliding collar D of the clutch is engaged with clutch member C. Gear F then rides on the cylindrical part of the clutch member D, from which it is disengaged, and shaft E rotates continuously at the motor speed.

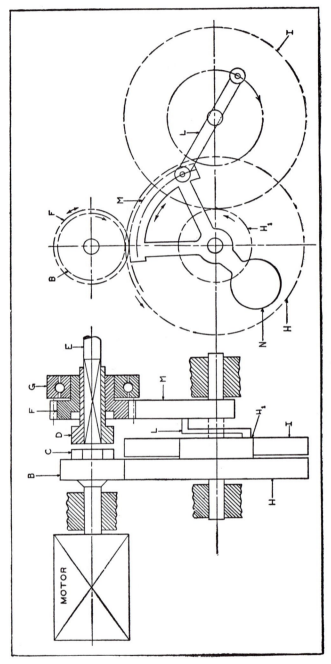

Fig. 20. Mechanism for Imparting Either Oscillating or Continuous Rotary Motion to Driven Shaft E.

The diagram at the left of Fig. 20 shows the mechanism in the neutral position. The gears B, H, and H_1 have a ratio of 1 to 8. The diagram at the right shows the gearing arrangement that makes possible the oscillating motion. Crank L is attached to gear I in an offset position, so that it provides a crank movement for oscillating segment gear M which meshes with gear F. The ratio of the pitch diameters is so selected that gear F is rotated 180 degrees in a forward and return movement; gear F is connected to the driven shaft E, giving it a similar motion when member D is moved to the right. This mechanism can be improved by introducing a second clutch to permit gear B to be disconnected from the motor shaft for the direct drive and also by attaching a balance weight N on the segment gear M, as shown in the right-hand diagram of Fig. 20.

Slight Oscillating Movement with Interrupting Control.– A mechanism for imparting a slight oscillating or vibrating movement to a shaft, which is equipped with an arrangement for interrupting the oscillating movement without

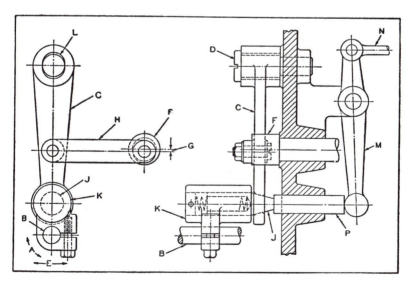

Fig. 21. Mechanism Designed to Impart Oscillating Movement to Shaft B with Provision for Interruption by Lever M.

stopping the driving shaft, is shown in Fig. 21. **This device** rocks shaft B back and forth, as indicated by arrow A, by having lever C, which is pivoted on stud D, rock about the center of the stud with a radial movement, as indicated at E. To operate lever C, an eccentric shaft is provided at F with an offset equal to G. As the shaft F rotates, the connecting link H is moved back and forth, imparting movement to lever C.

Lever C transmits a rocking movement to shaft B through engagement of the spring plunger J mounted in lever K, with a hole in lever C. The hole for the stud about which lever C oscillates is elongated at L to allow the two levers to pivot. Normally, the movement described is continuous, but at certain times it is necessary to stop the movement of lever K and to permit lever C to continue in operation.

This is accomplished by providing an arm at M which is actuated by a connecting-rod N so that a rod P pushes plunger J far enough to the left so that the clearance between the plunger and the tapered hole is great enough to permit lever C to oscillate without transmitting sideways motion to plunger J. This prevents the movement of lever K, which, in turn, stops the rocking movement of shaft B. Shaft B serves to operate other mechanisms within the machine. This plunger type of mechanism is suitable for use where only a slight rocking movement is necessary. The bevel plunger J is raised just a sufficient amount from the beveled hole to permit it to clear the hole at the extreme end of the oscillating stroke of lever C.

Feeding Mechanism Operated by Crank and Cam.—

A very smooth oscillating movement is imparted to the fingers Y of the mechanism shown in Fig. 22 by means of a crank C and the cam B secured to the rotating shaft A. The oscillating movement is approximately as shown by the arrow E.

Arm D is free to slide up and down in a connecting block F, which is pivoted to the stud H. Stud H forms a part of the lever *C*. As shaft *A* revolves, it carries block *F* to the left or right as indicated by the arrow *E*, and up and down through the medium of the connecting stud *H*. However, we are only concerned with the movement to the left and right, as the block merely slides up and down on the arm *D*, accomplishing no work while sliding vertically.

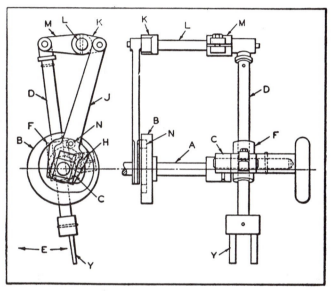

Fig. 22. Mechanism for Imparting Smooth Oscillating Movement to Fingers Y from Rotating Shaft A.

While the oscillating movement is taking place, the cam fork *J* moves up and down, being actuated by the roll *N* which is attached to it and travels in the groove of cam *B*. This causes lever *K* to rock the shaft *L*, thereby causing lever *M* to move arm *D* up and down. The combined up and down movement of arm *D* and left and right movement imparted by the eccentric throw of lever *C* results in a very smooth oscillating movement of the fingers *Y*.

Oscillating Mechanism for Milling Machine Shaping Attachment.—The oscillating mechanism shown in Fig. 23 actuates a slot- or groove-shaping attachment for a milling machine. The grooves G to be cut in the ends of bars W, shown at the upper right of Fig. 23, have their bottom surfaces machined to conform with an arc of a circle. The milling attachment was designed to handle this machining operation, because the bars were too long to be handled in a lathe.

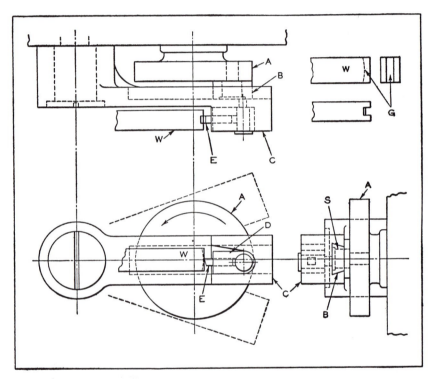

Fig. 23. Mechanism Applied to Milling Machine for Oscillating Shaper Tool Used to Cut Groove G in Work W, Shown at Upper Right.

Referring to the oscillating mechanism, the disk A is driven by the spindle of the milling machine, and carries a block B. Block B is free to swing on its stud S, and slides in a dovetail groove at the rear of lever C. Lever C is free to swing on its stud, which is fastened to the column of the milling machine. The rotation of disk A in the direction shown by the arrow imparts an oscillating motion to lever

C, as shown by dotted lines in the lower left view of Fig. 23, which indicate the two extremes of the oscillating motion.

The block D which carries the tool bit E is fitted into a recess in the front of the lever C, as shown in the lower left view of Fig. 23, and is free to swing on its stud. In this view, the tool bit E is shown in position for cutting, with the block D resting on the lower edge of the recess in lever C. The cutting action takes place on the upward swing of lever C. As the lever swings toward the bottom, the block D turns on its stud to prevent the cutting edge of the tool bit from dragging during the non-cutting half of the oscillating movement. Thus, block D operates in the same manner as the clapper block of a shaper. The work W is supported by clamping it on blocks on the milling machine table, the longitudinal feed being employed to advance the work to the shaping tool.

Oscillating Cam Made Adjustable.— On a recently designed automatic drilling machine, the work-holding fixture is operated by pneumatic cylinders. Compressed air is admitted to or discharged from the cylinders by means of a plunger valve operated by a cam supported on an oscillating shaft. For various types of work and different operations, it was necessary to provide some means of adjusting the timing for the application or release of the air pressure. The cam illustrated was designed for this purpose.

As shown in Fig. 24, plate B is clamped to and oscillates with shaft A. Cam-plate E is free to swing about the common center line of the plate and shaft, since it is loosely mounted on a hub of plate B. Fiber plates C (located between plates B and E) and plate D (on the outside face of plate E) are mounted on studs H. These studs are pressed into holes in plate B and pass through slots in cam-plate E.

Springs located between washers on the studs can be compressed by tightening the nuts. This causes the cam-plate to oscillate with plate B, the amount of oscillation being controlled by adjustable stops G and J.

Roller *F*, which is pinned to the plunger of the pneumatic valve, is shown in contact with the rise of cam *E* in Fig. 24. Movement of shaft *A* in the direction indicated by the arrow will cause the plunger to be depressed, thus permitting air to enter the cylinders.

Referring to Fig. 25, cam-plate *E* has moved to the right and the valve plunger is depressed. The cam-plate has come

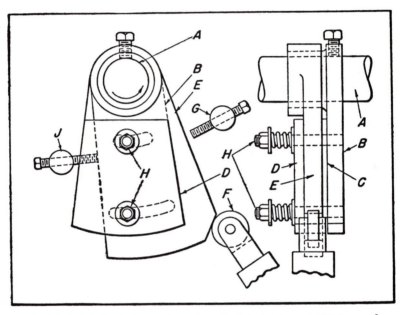

Fig. 24. Adjustable Oscillating Cam which Controls Admission or Discharge of Air to a Pneumatic Cylinder that Actuates Work-holding Fixtures. Amount of Oscillation of Cam-plate E can be Adjusted by Stop-screws G and J.

in contact with stop *G* so that further movement of the plate is prevented, regardless of the extent to which the shaft is oscillated. Then when the movement of the shaft is reversed, plate *E* returns to the position shown in Fig. 24, where further movement of the cam to the left is limited by stop *J*, and the valve is closed.

In this manner, the work-holding clamps are closed at the beginning of the first half of the cycle and released at the

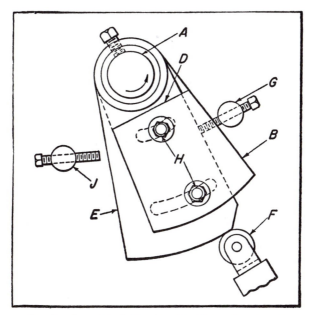

Fig. 25. Cam-plate E Shown in Fig. 24 has Here Moved to Right so that Valve Plunger Supporting Roller F is Depressed and Air is Admitted to Cylinder.

start of the second half of the cycle. Timing for the application or release of air pressure can be varied by adjusting the stop-screws. For example, if it is desired to open the valve later, the screw in stop J is backed off, thus causing a time lag before the valve is open. Similarly, the screw in stop G can be backed off to permit later closing.

CHAPTER 11

Mechanisms Providing Combined Rotary and Linear Motions

Described in this chapter are various mechanisms which impart a combined rotary or oscillating and linear or reciprocating motion to one or more elements. They provide, respectively, a combined rotary and traversing movement for a shaft; a reciprocating and rotary motion for a pneumatic drill; a reciprocating and rotary motion for a "flying needle" used in weaving wire cloth; a reciprocating shaft that rotates during part of its stroke; a ball and socket operation of a sleeve valve which reciprocates and oscillates; a crank which imparts an oscillating motion to one member and a reciprocating motion to another; a means of adjusting the radial position of a member while it is rotating; and a bearing designed for both rotary and reciprocating motions.

Mechanism for Producing Rotating and Traversing Movement.—A simple mechanism designed to impart a combination traversing and rotating movement to a machine member is shown in Fig. 1. Two gears C and D are keyed to the driving spindle K and retained by a flanged nut. The driving shaft L, which serves to rotate and traverse the machine member as required, has a screw thread and a keyway which is a sliding fit for the key P. The sleeve G is arranged to run freely in the bushings F, housed in the web E of the machine. End thrust imparted to this sleeve is taken by the large flange and the flanged nut N. Sleeve G is threaded on the inside to fit shaft L.

The gear B is keyed to the sleeve and the gear A runs freely on it, both gears being retained by the flanged nut M.

The part H is secured to gear A by three screws. Key P is held to part H, as shown, and slides in keyway in shaft L.

To make clear the operation of the mechanism, let it be assumed that gears A and C have forty-eight and twenty-four teeth, and gears B and D, forty-seven and twenty-five teeth, respectively. Let the member to be actuated, and consequently shaft L and gear A, make one revolution. Then gear C and shaft K will make $48 \div 24 = 2$ revolutions.

When gear D makes two revolutions, gear B makes

$$\frac{25}{47} \times 2 = \frac{50}{47} \text{ revolutions.}$$

Thus shaft L rotates relative to sleeve G, and a traverse feed movement is imparted to the member. If the screw thread on shaft L has 10 threads per inch, the distance through which shaft L is fed during one revolution equals

$$\left(\frac{50}{47} - 1\right) \times \frac{1}{10} = 0.0064 \text{ inch.}$$

Different feeds and speeds can be obtained by varying the ratios of A to C and B to D.

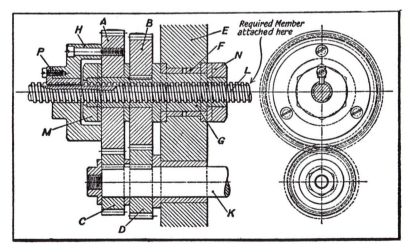

Fig. 1. Mechanism for Imparting a Rotating and Traversing Movement to Shaft L from the Driving Spindle K.

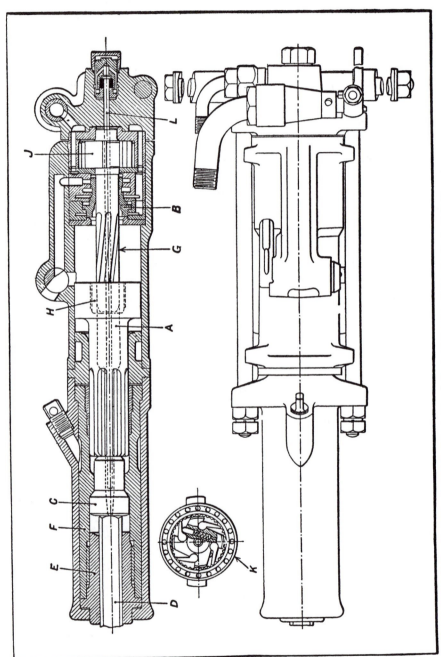

Fig. 2. Anvil-block Type Mechanism of Pneumatic Rock Drill.

Reciprocating and Rotary Motion for Pneumatic Drill.—

Several different types of valve-gear mechanisms are employed for the operation of pneumatic rock drills. One of these types is shown in Fig. 2. This particular mechanism is used for a 3-inch anvil-block type hand hammer drill employed in the mining industry. The movements of the piston A are controlled by a spool valve B. The tool derives its name from the anvil C, which is interposed between the piston and the drill D. The particular drill shown has a shank of hexagonal cross-section. It is guided in the bushing E, which, in turn, is held in the tool-holder F. Anvil C is guided by tool-holder F.

The essential movements for rock drilling are an uninterrupted series of blows in rapid succession on the end of the drill, and a rotary motion of the drill, which must be turned a few degrees between successive blows. The blows, 1600 to 2200 per minute, are imparted by the piston A operating on the anvil block C, the velocity of the piston at the moment of impact being between 20 and 25 feet per second. The turning motion is imparted to the drill between successive blows by the helical-splined rifle bar G. This bar is engaged by a phosphor-bronze nut secured in the piston at H.

The rifle bar is provided with a ratchet mechanism at J, an end view of this mechanism being shown at K at the left of Fig. 2. The ratchet mechanism allows the rifle bar to turn in one direction only. On the return stroke of the piston, the ratchets prevent the rifle bar from turning and, in consequence, the piston turns a few degrees on the rifle bar in accordance with the helix angle of the splines. On the power stroke of the piston, however, the rifle bar is free to turn, and the piston moves forward without rotating, the inertia of the piston and the parts keyed to it causing the ratchet gear to slip. The piston is splined, and the splines are engaged by a nut in the tool-holder F. The turning movement of the piston on its return stroke, therefore, is transmitted to the tool-holder and to the drill.

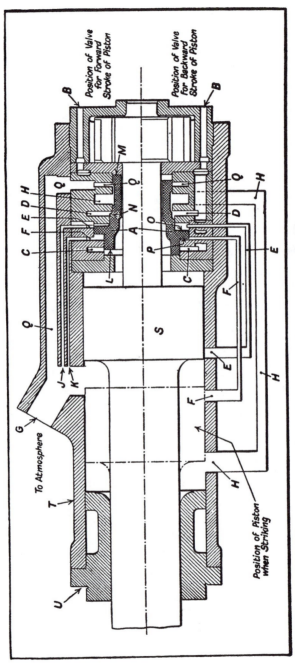

Fig. 3. Spool Valve Mechanism of Pneumatic Drill.

Water or air is almost always required at the drill point to facilitate the removal of the rock fragments or chips. In consequence, the drill is usually hollow, so that air or water can be fed through it to the point from the tube L by way of the hollow anvil block, piston, and rifle bar.

Fig. 3 is a diagrammatic sectional view of the pneumatic drill, illustrating the operation of the spool valve. The top half section shows the valve in position for admitting air to the rear of the cylinder, while the bottom half section shows the valve in position for admitting air to the opposite side of the piston.

The action of the valve is as follows: With the piston S in its rearmost position, air is admitted through the passages B and port C to the cylinder, as seen in the top half section, to drive the piston outward. The valve A is locked in this position by the air pressure acting on its front face L, the area of the front face being larger than that of the rear face M, and both faces being subjected to the same air pressure.

On the outward stroke, when the rear edge of piston S uncovers the port E in the cylinder, air at cylinder pressure passes up through the passage E to a port in the valve box, and acts on the face O of the valve. This moves valve A to the left, closing port C and cutting off the air supply which serves to drive the piston outward. At the same time, the valve opens communication between the live air port D and the port and passages H, permitting air to pass to the front of the cylinder and drive the piston back. Just before piston S delivers its blow, the rear edge of the piston uncovers the main exhaust port G, and the air behind the piston passes out to the atmosphere.

In the live air port D, there is a small face N on the valve, and the air pressure acts on the face, tending always to move the valve to the left. This serves to lock the valve in the position it takes up when air is being supplied to drive the piston back.

On the return stroke of piston S, when the rear edge has passed and closed exhaust port G, the remaining air in the rear of the cylinder is compressed until the pressure is sufficient to move valve A to the right, back to its initial position.

There is a front control port F in the cylinder, which is connected to the port F in the valve box. When the piston uncovers this port on its return stroke, air passes through the passage F and acts on the valve face P. This tends to move the valve to the right, but in ordinary running, the port F in the cylinder is inoperative, the compression pressure in the rear of the cylinder on the return stroke of the piston being the real cause of the movement of the valve. The purpose of port F in the cylinder is to prevent the piston from coming to rest with the main exhaust G slightly open to the front end of the cylinder, which would cause difficulty in restarting. On the return stroke of piston S, the main exhaust port G is uncovered before the piston is brought to rest, and the air in front of the cylinder exhausts to the atmosphere through this port.

As soon as the front end of piston S, in moving outward to deliver its blow to the drill, has closed main exhaust port G, there is a tendency for the air trapped in front of the piston to cushion the force of the blow delivered to the drill. This is, however, taken care of. Referring to the top half section of the illustration, it will be observed that when live air is being supplied to the rear end of the cylinder the front end of the cylinder has an opening to the atmosphere through passage H, the reduced diameter of valve A, and the port and passage Q. The air in front of the cylinder can therefore exhaust to the atmosphere, and cushioning of the piston prior to the blow is avoided.

It will be noted that the front air supply port H is partly covered when the piston is in position for striking the drill. This prevents damage to the piston in case the drill should be run without drilling actually taking place, or in the event that the operator does not hold the tool down to the work.

When the piston passes port *H*, air is trapped between the front face of the piston and the face of the division ring *U*, and the resulting compression prevents damage to the division ring.

Two small exhaust passages *J* and *K* are controlled by the faces *O* and *P* of valve *A*. These passages insure that no residual air pressure will remain in ports *E* and *F* of the

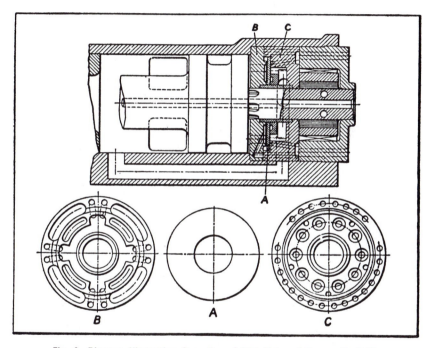

Fig. 4. Diagram Illustrating Operation of Disk Valve of Pneumatic Drill.

valve box after the valve has been moved to either end of its travel. The absence of residual pressure in these ports is essential for free movement of the valve.

Several other types of valve gear are employed. In the design shown in Fig. 4, the valve is of the disk type, the steel disk *A* operating between the valve seats *B* and *C*, and performing the same duties as the spool valve previously described.

Compact Mechanism Provides Combined Reciprocating and Rotating Motion.—A mechanism for producing a combined reciprocating and reversing rotating motion, which is of particular interest because of its simplicity and compactness, is shown in Fig. 5. This mechanism is used to operate a "flying needle" on a machine for producing a woven-wire product.

The shaft A, which is supported in bearings B and C, has a series of modified gear teeth on it which mesh with the teeth of gear E. The base of bearing C carries the roller D which engages a helical groove in shaft A. In operation,

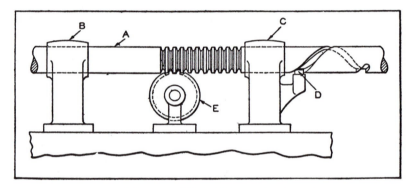

Fig. 5. Mechanism for Reciprocating and Rotating Shaft A.

gear E is given a reversing rotary motion by means of a sliding dog mechanism. As shaft A is reciprocated by gear E, it is also given a reversing rotating motion by the action of roller D in the helical groove. This arrangement provides a smooth uniform motion.

Reciprocating Shaft that Rotates During Part of Stroke.—A somewhat similar arrangement to that just described is shown in Fig. 6. Arm A receives a reciprocating motion from an independent source and transmits this motion to shaft B. When shaft B moves to the left, it is rotated through roller C engaging a helical groove in the shaft. This combined movement continues until the collar D comes in

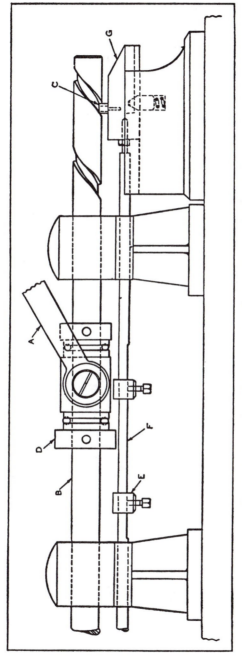

Fig. 6. Mechanism for Obtaining Combined Reciprocating and Rotating Motion.

contact with stop E. Upon further movement of shaft B to the left, rod F and slide G are drawn to the left at the same speed. It can be readily seen that when this simultaneous action is taking place, shaft B has no rotary motion. The position of stop E can be adjusted to vary the rotary motion of shaft B.

Ball-and-Socket Mechanism for Operating a Sleeve Valve.— In Fig. 7 is shown a ball-and-socket mechanism for imparting a combined oscillating and reciprocating motion to a sleeve valve. This mechanism was patented in England.

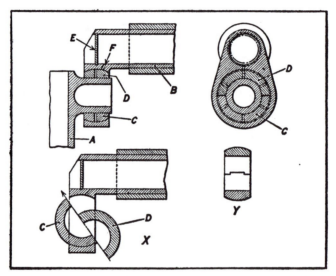

Fig. 7. Ball-and-Socket Mechanism for Imparting a Combined Reciprocating and Oscillating Motion to a Sleeve Valve.

Its most interesting feature is the novel method of assembling the simple but efficient ball-and-socket joint. The sleeve A is driven from the rotating crankshaft B, the socket being formed in the crank and the ball mounted so that it can slide on a pin projecting from the sleeve, as shown at the upper left of Fig. 7.

As shown, the annular ball member is made in halves C and D. To assemble the joint, one half of the ball member, say C, is set in the socket with its axis at right-angles to that of the socket and its face inclined as illustrated in the view at X. The other half D is then slid into position in the direction of the arrow, and the two halves turned together through 90 degrees, so that they are co-axial with the socket. The pin projecting from the sleeve can then be inserted in the bore of the ball member, the two halves thus being maintained in the correct position. The two halves can, if required, be tongued and grooved, as indicated at Y.

In the arrangement illustrated, the hollow crankshaft is blanked off at E, and provision is made for feeding lubricating oil to the joint through the bore of the shaft and the hole F.

Foot-Operated Mechanism for Transmitting Rotary and Linear Movements.—The natural movement of the operator's leg as it swings a pedal forward is utilized in the mechanism shown in Fig. 8 to transmit a rotary motion to lever D, and, at the same time, impart an upward motion to shaft A, in the direction indicated by arrow B. The rotary motion occurs during the first part of the swinging movement of the foot-lever. Foot-lever E used for this purpose pivots on stud F. A stop-screw G locates the foot-lever in its starting position.

The operator, with his foot on the lever at X, kicks it back in the direction indicated by arrow H. At the beginning of the stroke, lever D is caused to move through the medium of connecting-rod J attached by adjustable rod ends K to studs L and M. This movement is transmitted by connecting link N to a unit within the machine on which it is employed. Lever D rocks on shaft P, which is supported in bearings in the end frames of the machine.

When pressure of the foot at X has moved the lever to the position indicated by the dotted lines at Q, lever D will have been rotated to the position indicated by the dotted lines at a, and surface R will be in contact with a roll at S on bellcrank T. The remainder of the foot-lever action, which carries lever E to position O, moves lever D to position b, and is utilized in actuating the toggle lever of which crank T and link V form a part. A vertical movement is imparted to shaft A through toggle connecting-link stud W traveling along path y.

Bellcrank T pivots on and is supported by a fixed stud Z. Link V is pivotally attached at I to a block pinned to shaft A. The effect of this toggle linkage and rotary lever action

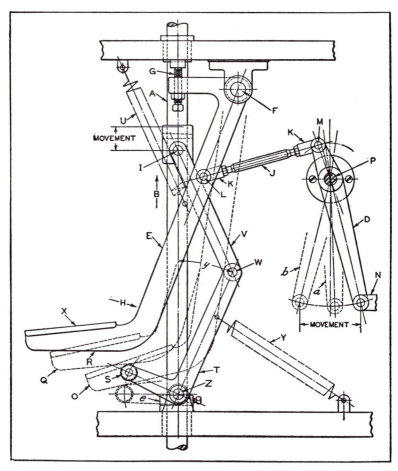

Fig. 8.   Foot-operated Mechanism for Transmitting Rotary Motion to Lever D and Vertical Movement to Shaft A.

is to combine a crosswise movement of connecting link N with a delayed vertical movement of shaft A when the foot-lever is given a full swinging movement from the position shown by full lines to that shown at O. A spring U serves to return the foot-lever to its starting point. Another spring at Y returns bellcrank T and shaft A to their starting points which are determined by collar e.

External Control for Mechanism within Rotating Member.

–Fig. 9 shows a device that provides exterior control of the movements and adjustments of some movable element on a rotating shaft. By such a device, for example, the radial adjustment of the tool on the arm of a boring-bar can be made while the bar is in motion, the adjustment being made through controls that do not revolve with the bar and that can be operated as conveniently as when the bar is stationary; or the blade of a revolving screw propeller can be adjusted for pitch without regard for the motion of the propeller shaft.

The illustration shows a more or less general arrangement of the device, many of its elements being subject to modification to adapt it to special needs and to the ideas of the designer. Also some construction refinements have been slighted for the sake of clarity, thus making the drawing somewhat diagrammatic.

The moving element to be adjusted is represented by the double crankshaft A, mounted in bearings that are carried on and rotate with the shaft B. The cranks are quartered, or set at 90 degrees to each other, and are driven by the connecting-rods C and D. Near the other end of the mechanism is another double-throw crankshaft E with quartered cranks having throws equal to those of crank A. It is to this crankshaft that the initial controlling movement is applied by hand-crank T.

Unlike shaft A, shaft E has no other motion than rotation about its own axis. Through the medium of the two connecting-rods F and G, this shaft imparts a reciprocating

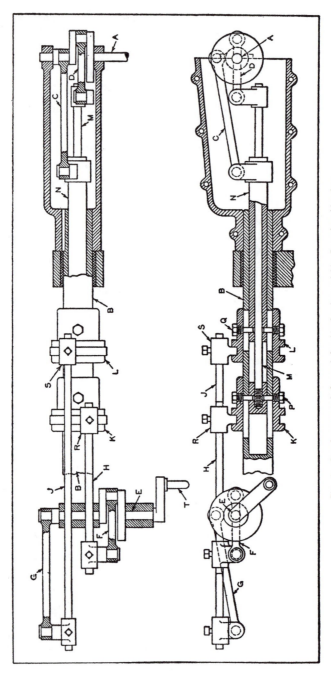

Fig. 9. Crank-operated Control for Mechanism within Rotating Member.

longitudinal motion to the rods H and J. This motion is transmitted to the sliding collars K and L, and thence to the central rod M and the sleeve N, which deliver the motion to the connecting-rods C and D. Any motion of the shaft E is thus duplicated in shaft A.

The two collars K and L rotate with shaft B, but are free to slide longitudinally on it. By means of pins P and Q, passing through slots in shaft B, the sliding motion of the collars is imparted to rod M and sleeve N. The reciprocating motions of rods H and J are transferred to the collars by means of the shoes R and S which ride in the grooves of K and L. It is thus seen that the motion of shaft E is transmitted to and translated on shaft A by longitudinal reciprocating motion at the points R and S, where the external stationary control parts meet the internal revolving control parts. This transmission of motion is thus independent of the rotative position of shafts B, and hence is entirely independent of its rotation.

Ball Reciprocating Bearings.—The term "ball reciprocating bearing" is used by the manufacturers of these bearings to designate a bearing capable of both rotation and axial reciprocation. Obviously, such a bearing cannot have definite ball tracks or grooves. Sometimes rotation only is required, and sometimes axial reciprocation alone is desired. For this reason, the outer raceway is the inside of a hardened and ground cylinder, while the inner race is either the outside of a cylinder or the surface of a hardened and ground shaft. The balls are placed between the two surfaces and are generally retained in the pockets of a metal cylinder serving as retainer.

It is well known that a ball pressed against a flat plate, especially against a surface the curvature of which is opposite to that of the ball, has less load-carrying ability than the same ball placed in a groove similar to that of the conventional Conrad type ball bearing. The load-carrying capacities in the two cases have a ratio of about 1 to 10.

In order, therefore, to secure the greatest load-carrying capacity in a ball reciprocating bearing for a given size, there are but few alternatives. One of these is to provide the greatest possible hardness and uniformity of parts, in view of the fact that all contacts are minute ellipses of very limited area, and the harder the parts the greater the unit pressure that can be applied. Another solution is found in accuracy of manufacture and proper design to distribute the load uniformly over the bearing. A third alternative is by finding a means of increasing the number of balls in the same size bearing so that a proportionate gain in capacity is obtained. Thus, if it is possible to install twice the number of balls, the capacity of the bearing is doubled.

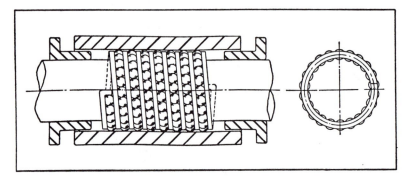

Fig. 10. Bantam Construction Ball Bearing Capable of both Rotation and Axial Reciprocation.

The Bantam Bearings Division of the Torrington Co. has been granted U. S. Patent No. 2,316,468 on means for increasing—in fact, doubling—the capacity of this type of bearing. The usual ball cage is a cylinder of bronze, steel, or aluminum drilled to form ball pockets of the greatest number per inch of cylinder length. The new Bantam construction consists of spirals of balls separated by coils of a specially formed helical spring, as shown in Fig. 10. The section of the spring wire may be triangular, round, square, or rectangular; but by making it in the form of an hourglass not only facilitates loading the balls, but retains them between the races during shipping and installation. This construction is less expensive than the conventional type of retainer, both from the material and weight standpoint, and from a fabrication point of view.

In Fig. 10, it will be noted that the balls are restrained from end exit by the formed ends of the helix. When this bearing is used in rotation only, the helix of the retainer so positions the balls that each has its own special path as the shaft revolves.

In one test bearing, the pitch of the helix was 0.250 inch; each row contained twenty-two 3/16-inch diameter balls. Therefore, allowing no end play, each ball had a path (in rotation only) 0.0113 inch wide. Each path of the stationary race was thus loaded to maximum only once in about two revolutions, whereas in a grooved-race radial bearing with twenty-two balls in line, the maximum load on the stationary race would occur twenty-two times in two revolutions.

Ball reciprocating bearings are often used to operate in an axial direction only, and in such cases a large number of axial paths exist. The most common usage of this type of bearing is, however, in both reciprocation and rotation. When so used, the tracks of the numerous balls make intricate and overlapping designs, often covering every particle of the surfaces of both raceways. In this type of bearing, it is desirable to hold the diametral clearances close; a slight preload is permissible.

As a general guide for the capacity of this type of bearing, it may be said that the static capacity is dependent upon the number of balls, the diameter of the balls, and the radius of curvature of the inner race. Expressed as a formula,

$$C_s = KND^2F$$

in which C_s = static capacity, in pounds; N = number of balls; D = diameter of ball, in inches; F = curvature factor; K = constant = 200.

The constant K is derived from laboratory tests in which balls were loaded on hard flat steel plates up to the point where any greater load would cause "Brinclling" (indentations in the steel plate). The value of K is 200 for plates and balls of 60 Rockwell C hardness.

The value of the curvature factor F is given below. The value is found by determining R, the ratio of the inner race diameter to the ball diameter.

R	4	6	8	10	12	14	16
F	0.820	0.875	0.907	0.930	0.945	0.957	0.965

Values of F not given above may be interpolated.

To obtain the capacity C_r, in rotation only, the values of C_s may be modified by using $K = 100$ and applying speed factors. (These speed factors may be obtained from a manual published by the Bantam Bearings Division of the Torrington Co., South Bend, Ind.)

For both rotation and reciprocation, the capacity is little affected by the rate of reciprocation, as long as it is not excessive. Other factors that influence the capacity of this type of bearing are deflection of parts making up the assembly, accuracy of workmanship, over-travel of load, and lubrication.

Some of the devices in which this type of bearing is used are taper measuring instruments, engine governors, welding machine guides, printing press vibrator shafts, cloth-cutting machines, grinding-wheel dresser shafts, spool-winding machines, propeller-testing equipment, buffing machines, stem grinding and polishing spindles, wire-brushing machines, rotary-tool pipe cut-off spindles, airplane landing gear, inking rolls, sheet-polishing machines, gear grinders and lappers, paper-coating machines, remote-control gear-shift mechanisms, and hydraulic welding machines.

As the use of this type of bearing has occasionally been limited by its relatively low capacity, its field of usefulness should be greatly increased by this improved design which permits, in general, of doubling its load-carrying ability.

CHAPTER 12

Speed-Changing Mechanisms

Providing a fixed or adjustable speed of rotation of a rotating driven member that is different from the speed of rotation of the driving member can be accomplished in many different ways. Mechanisms described in this chapter illustrate the use of gears, ratchets, friction wheels, cams, pulleys and belts in combinations that are noteworthy for some ingenious feature or special function which they perform.

In one of the mechanisms discussed, the driven gear continues to rotate in the same direction at greatly reduced speed when the drive shaft is reversed. Also described are a cone pulley with epicyclic gearing for a high-ratio reduction drive; a friction drive designed for stepless speed variation; a cam-controlled variable-speed drive; a mechanism for controlling the speed of a driven shaft by changing the speeds of two driving motors; a cam-controlled worm wheel drive for controlling speed and direction of shaft rotation; a wabble-gear speed reducer; a device for automatically shifting a back gear in and out as the torsional resistance of the driven shaft varies; and a mechanism for changing cam speeds independently of camshaft speeds.

Other speed-changing mechanisms are described in Chapter 11 of Volume I and Chapter 10 of Volume II of Ingenious Mechanisms.

Mechanism for Producing Speed Change by Reversing Driving Shaft.—The mechanism shown in Fig. 1 is employed on a machine for fabricating a twisted wire product. This machine twists a group of wires together, the pitch or degree of twist being controlled by the rate at which the wire.

is fed into the twisting mechanism. The wire is twisted in one direction for a specified number of turns at a given pitch, and then twisted in the reverse direction for a number of turns at a greater pitch.

To obtain the required twist, the twisting mechanism is reversed while the feeding mechanism continues in the same direction but at an increased speed. The mechanism illustrated simply provides for changing the speed of gear *D*

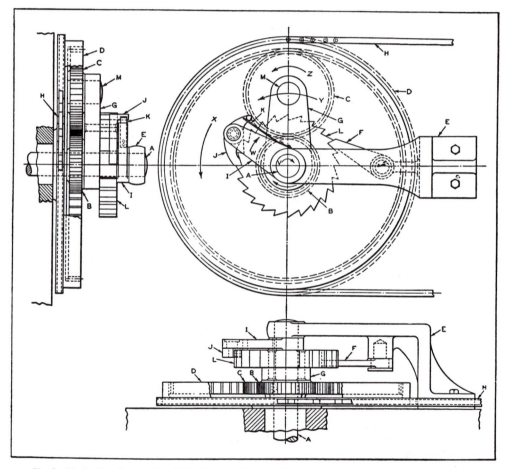

Fig. 1. Mechanism Designed to Drive Sprocket D from Shaft A at a Ratio of 1 to 4 when Shaft A Rotates in Direction Indicated by Arrow V, and to Continue to Drive Sprocket D in Same Direction but at a Ratio of 1 to 1 when Shaft A is Reversed.

when the driving shaft A is reversed without changing the direction in which D rotates.

Referring to Fig. 1, the shaft A rotates in the direction indicated by the arrow V. Gear B, which is keyed to shaft A and rotates with it, meshes with gear C. Gear C meshes with the internal gear D, which is a free running fit on shaft A, causing it to rotate in the direction indicated by arrow X. The ratio between gears B and D of the mechanism, designed as illustrated, is 1 to 4, gear C acting as an idler while gear B drives gear D. Gear D is provided with sprocket teeth for the chain H, which transmits motion to the feeding mechanism at a distant point.

The lever G is free to turn on shaft A, and supports gear C, which rotates freely on stud M. Ratchet wheel L is also free to turn on shaft A, and is riveted to lever G. Lever I is keyed to shaft A, and carries the pawl J, which is held in contact with ratchet wheel L by the spring K. Bracket E supports the outer end of shaft A, which runs freely in its bearing. The pawl F, which engages ratchet wheel L, is also mounted on bracket E. As shaft A operates in a horizontal position, no spring is required to keep pawl F in contact with the teeth of ratchet L. All three views in Fig. 1 show the mechanism at the same point in the operating cycle.

In operation, shaft A, rotating in the direction indicated by arrow V, carries with it gear B and lever I. Gear C, meshing with gear B, rotates in the direction indicated by arrow Z, transmitting its motion to gear D in the direction indicated by arrow X at a reduced speed of rotation in the ratio of 4 to 1. Although lever I and pawl J rotate with shaft A, no motion is transmitted to ratchet wheel L, the teeth of which are designed for engagement with the pawl when rotating in the opposite direction. However, the rotation of gear B, acting on gear C at the point of contact, and the resistance of gear D, acting on gear C at the point of contact on the opposite side, tend to produce a turning movement of lever G in a clockwise direction.

As lever G and ratchet wheel L are riveted together, any motion or reaction affecting one also affects the other. Hence, the reaction on lever G serves to maintain engagement of pawl F with ratchet wheel L, which prevents clockwise rotation of the ratchet wheel. Thus the axes of shaft A and stud M are maintained in fixed positions; the assembly acts as a simple gear train, the power being transmitted from gear B to gear D through idler C in the ratio of their pitch diameters.

When shaft A, Fig. 1, which operates the twisting mechanism, is rotated counter-clockwise, or in the opposite direction to that indicated by arrow V, lever I also rotates in that direction, causing pawl J, which rotates with it, to engage the teeth of ratchet wheel L, so that the ratchet wheel will be rotated in the same direction, the pawl F at this point becoming inactive. As ratchet wheel L and lever G act as a unit, ratchet wheel L, stud M, and gear C rotate in the direction of arrow Y, with the axis of shaft A serving as a center. As the center of gear C is now rotating about the same axis and at the same angular velocity as gear B, there can be no motion of gear C about the center of stud M as an axis. Therefore, gear C can no longer operate as a gear, but merely serves as a connecting link for transmitting motion from shaft A directly to gear D. In this manner, the chain H is given a uni-directional movement at two speeds, as controlled by the direction of rotation of shaft A, which operates at a uniform speed, but is periodically reversed.

Two-Gear Speed-Reduction Mechanism Redesigned to Obtain Uniform Rotation.—Fig. 2 shows how the speed-reducing mechanism of the two-gear type illustrated and described in Volume I of "Ingenious Mechanisms for Designers and Inventors," pages 340-342, was redesigned to obtain uniform rotation. A standard internal gear G and pinion C were modified to operate without tooth interference in this redesigned mechanism.

On the driving shaft A is mounted an eccentric B. The

axis of driving gear C follows the motion of eccentric B, but is kept from rotating about its own axis by pin D, which works in the slot E. Linkage F is actuated by the eccentric B, which constantly maintains slot E in a perpendicular position through the action of the parallel links H, pivoted on studs J. Since the axis of gear C follows the motion of eccentric B and the gear does not rotate about its own axis,

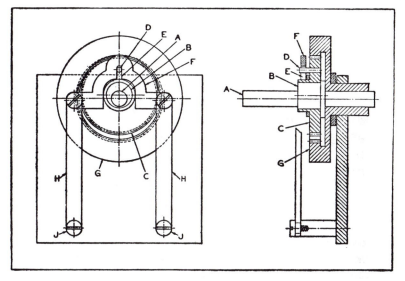

Fig. 2. Internal-external Gear Reduction Mechanism which Provides Uniform Speed.

the motion imparted to the driven gear G will be uniform and equal to

$$(\text{R.P.M. of } A) \times \left(\frac{N - n}{N} \right)$$

, in which N equals the number of teeth in gear G and n equals the number of teeth in gear C.

Cone Pulley with High-Ratio Epicyclic Reduction Gearing.—The cone pulley with epicyclic gearing shown in the cross-section view in Fig. 3 was designed to provide a high-ratio reduction drive for the feed camshafts of several horizontal drilling machines. For this application, the cone pulley P was required to make 338 1/3 revolutions for each revolution of the driven shaft S. The drilling machines were being used on special automobile work at a time when it was impossible to obtain new or improved machines. The problem of designing the high-ratio reduction-gear drive was complicated by the necessity for keeping the size of the unit within certain dimensions, in order to permit it to be assembled in the space available. The design and construction of the reduction gearing unit was still further complicated by the fact that only a set of 16-pitch milling cutters was available for cutting the gears required.

Under these conditions, it was necessary to deviate from standard practice with respect to center distances between gears and the depth of the gear teeth. The special gears made to meet the unusual requirements, while theoretically incorrect, have given years of useful service. Exact center distances and details of the special gears are not given here, since it is unlikely that readers who may wish to build a similar high-ratio gear-reduction unit will be handicapped by the restrictions under which this unit was designed. With modern facilities for cutting any size and pitch of gear desired, it should be comparatively easy to build a similar gear-reduction unit of almost any desired ratio.

Referring to Fig. 3, the cone pulley P is of cast steel, finished all over and accurately balanced. Pulley P is fitted with an Arguto wood bearing and is mounted on the hub of the fixed 47-tooth sun gear A. The integral hub extending from the opposite side of sun gear A is pinned in a fixed position to the stationary housing G, which encloses two of the gears and is fastened to the drilling machine frame. The cone pulley is retained on the hub of stationary gear A by a fiber washer held in place by a dowel-pin and hexagonal head stud as shown.

The follower planet pinion B, having 20 teeth which mesh with the teeth of sun gear A, is pinned to the hub of the

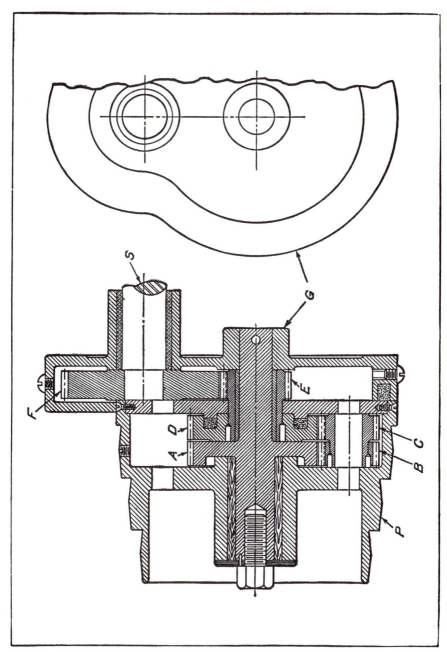

Fig. 3. Cross-section View of Cone Pulley with High-ratio Epicyclic Reduction Gearing and Partial End View of Gear Housing.

shown in Fig. 3, and is used to facilitate making the mathematical computations required in designing the epicyclic gear train. The same reference letters are used in both illustrations to indicate the same parts. This latter diagram has been made to a scale of one-half actual size, the circles in Fig. 4 indicating the pitch circles of the gears and pinions, and arm P representing the pulley P.

In the diagram, the arm P is shown rotating clockwise and carrying the pinions B and C with it. Since the teeth of pinion B are in mesh with the teeth of the stationary sun gear A and planet pinions B and C are keyed together, the latter two pinions revolve together in a clockwise direction as they rotate or are carried around the fixed sun gear A by pulley P. As the teeth of pinion C are in mesh with those of sun gear D, the latter gear will be driven in either a clockwise or counter-clockwise direction, as determined by the epicyclic gearing ratio formula given in the fourteenth edition of MACHINERY'S HANDBOOK, page 844, Fig. 17. Applying this formula, with the correct reference letter for each of the gears and pinions, we have the ratio

$$R = 1 - \frac{A}{B} \times \frac{C}{D} = \text{ratio of gear reduction or fraction of}$$

a turn imparted to sun gear D by one revolution of the pulley or driver arm P in which

$P =$ the pulley or driver arm;
$R =$ ratio of gear reduction;
$A =$ fixed sun gear with 47 teeth;
$B =$ planet pinion with 20 teeth;
$C =$ planet pinion with 21 teeth; and
$D =$ driven sun gear with 49 teeth.

Substituting numerical values (number of teeth in each gear) in the formula

$$R = 1 - \frac{A}{B} \times \frac{C}{D}$$

we have $R = 1 - \dfrac{47}{20} \times \dfrac{21}{49} = 1 - \dfrac{141}{140} = -\dfrac{1}{140}$

Therefore, one revolution of the pulley or driver arm P will turn the driven sun gear D 1/140 of a revolution. The direction of rotation will be counter-clockwise, or opposite that of the driver P, as indicated by the minus or negative sign which precedes the final result. Thus pulley P must make 140 revolutions for one complete revolution of sun gear D.

Now, since pinion E, which has 24 teeth, is keyed to sun gear D and drives gear F, which has 58 teeth, the number of revolutions of pulley P required to obtain one revolution of gear F, or the camshaft S, is obtained by the equation

$$140 \times \frac{58}{24} = 338 \ 1/3$$

The direction of rotation of the camshaft is clockwise, or the same as that of the driving pulley P, as will be seen in Fig. 4.

It is interesting to note that a reduction ratio of 10,000 to 1 can be obtained with only the four gears A, B, C, and D if they are made with 101, 100, 99, and 100 teeth, respectively.

Quick-Change Two-Speed Belt Drive.—The two-speed belt drive shown in Fig. 5 was developed for use on a certain machine. The design provides two speeds without the usual requirement of shifting or moving belts. Two sets of pulleys, A and B, fastened to the driving shaft C and the driven shaft D, respectively, by means of pins, are employed.

Belts E and F are applied to their pulleys rather loosely. The rocker arms G and H, pivoted on the bearings J and K, are provided with two idler wheels L and M. Spindles N and O serve to carry the idler wheels and tie the two rocker arms together into a single unit. The bar P provides a means of connecting the rocker unit to a suitable actuating mechanism.

The operation of the drive is very simple. In the particular application illustrated, a very slow starting speed is required. After the driven pulley has reached the starting speed, it is immediately accelerated by shifting the drive to the next pulley. In other words, the rocker unit is placed in position 1 for starting, and is finally shifted to position 2. In thus operating the drive, the tension is first placed on belt E for slow speed and then shifted to belt F to obtain the high-speed drive.

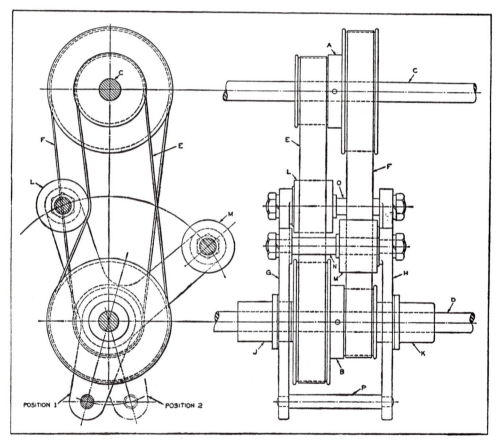

Fig. 5. Two-speed Quick-change Belt Drive.

In addition to providing a very easy means of varying speeds, this drive also permits placing the equipment at a remote place and controlling the speed through suitable linkage to the bar *P*.

Friction-Drive Mechanism Designed for Stepless Speed Variation.—Speed variation without steps can be obtained by sliding a belt along opposed tapering driving and driven drums; by a V-belt, steel ring, or chain driving between pairs of adjustable conical disks which permit varying the effective diameters; or by friction disks at right angles.

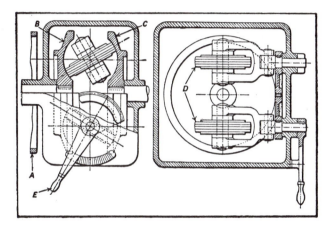

Fig. 6. Friction-drive Mechanism with Stepless Speed Changes Obtained by Moving Hand Lever.

The single-lever control friction drive mechanism shown in Fig. 6 is designed to give stepless speed variation between a minimum and maximum speed.

While heavy power transmission is not to be expected of friction drives, the arrangement shown possesses the advantage of a double drive. Driving pulley *A* is keyed to the duplex friction bowl *B* and transmits motion to member *C* by means of the two intermediate disks *D*. The disks are mounted in swiveling forks attached to gear segments operated by a single lever *E*. With the disks in the horizontal

Cam-Controlled Variable-Speed Drive.— By properly proportioning the four gears and cam that constitute the important parts of the drive shown in Fig. 8, it **can** be caused to give any of a great variety of speed **actions,** the

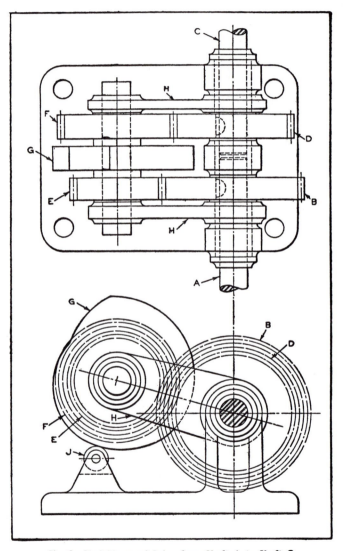

Fig. 8. Variable-speed Drive from Shaft A to Shaft C with Variation in Speed Obtained by Cam G.

simplest of which may be described as an intermittent or stop motion. When designed as a stop motion, it can theoretically perform this action in any one of a variety of ways. The ratio of the period of stop and the period of motion may be of any arbitrarily chosen value.

The number of stops per revolution may be either fractional or integral. To some degree, the stop intervals may have a non-uniform or non-periodic sequence. The action, instead of being the usual total arresting of motion, may also consist of merely a slowing down of the speed or it may be even more than a total stop—that is, it may be a momentary reversal of the motion.

Fig. 8 shows a simple but general arrangement of the device. The drive-shaft is shown at A and the driven shaft at C. Gear B, keyed to shaft A, drives gear D, keyed to shaft C, by means of back-gears E and F. At G is a cam that revolves as a unit with gears E and F. The frame H of the back-gear is a rigid unit, but pivots or is free to oscillate about the common axis of shafts A and C. The frame H, with its gears and cam, rests on the cam-roller J.

The gears B, E, F, and D constitute a train that may have any desired driving ratio within certain limits other than unity; that is, the ratio of B to E must not be the same as D to F. The action of the mechanism is as follows: The cam G rotates with the gears E and F, and as it rides on the roller J, its undulating contour gives a rocking motion to the frame H, which impresses a motion on the rest of the mechanism, in addition to that given by the driving shaft A. Thus if shaft A is held stationary, a movement of H will cause C and D to turn in one direction or the other, since the train value is not unity.

On the other hand, if cam G is lifted free of the roller J, and H is held stationary, the only motion that A can impart to C will be that transmitted by the gear train B, E, F, and D. Thus the motion of C due to that of H can be combined with that due to A, either positively or negatively, with the consequence that the motion of C results at one time from the sum of the two motions of A and H, and, at another time, from the difference of the two motions. By giving the proper value to the gear train B, E, F and D, and the proper contour to the cam G, any one of the actions mentioned can be obtained.

Mechanism for Starting, Stopping, Changing Speed, and Reversing Output Shaft.—A mechanism with two driving motors which permits starting, stopping, changing speed, and reversing the output shaft without stopping the motors is shown in Fig. 9. This control over the driven shaft is obtained by means of differential gearing without the use of friction clutches, gear shifts, or other well-known speed-changing and stopping devices, and is accomplished by simply changing the speeds of the two driving motors. This provides a stepless variation in driven shaft speed.

Each of the four bevel gears A, C, D, and F have the same number of teeth. Gear C is keyed to the motor shaft M, and gear A is keyed to the output shaft B. Both of these shafts extend to the center of the cross-piece G, which is cast integral with the spur gear H. The cross-piece provides a working bearing for gear H, as well as a support for the shafts of bevel gears D and F. Pinion E is keyed to the motor shaft N. The gears E and H are in the ratio of 2 to 1. The arrows on the faces of these gears indicate the direction of rotation.

Four tables of speeds, not shown, are used in operating the mechanism. These tables indicate the speed and direction of rotation of the gears E, C, H and A. The tables show that E, C, and H revolve in one direction continuously. The tables give the speed at which each motor must be operated to give the output shaft B any forward or reverse speed from 0 to 320 revolutions per minute in steps of 20 revolutions per minute, with the motor speeds ranging from 340 to 660 revolutions per minute. Of course these speeds may be increased.

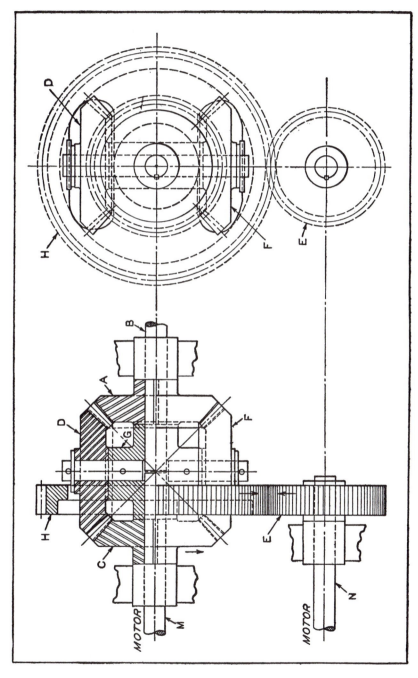

Fig. 9. Mechanism for Controlling Driven Shaft by Changing Speeds of Two Driving Motors.

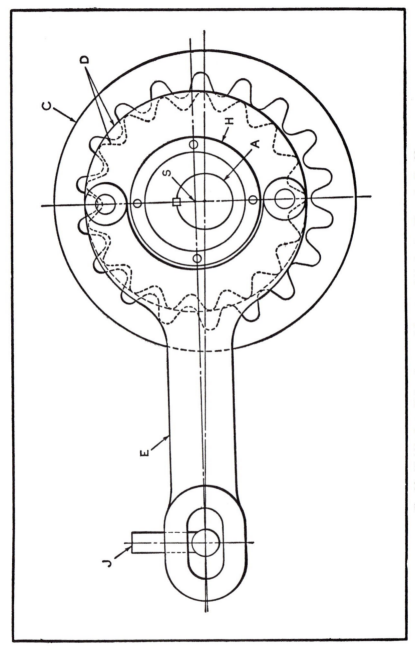

Fig. 11. Wabble Gear, Pinion, Strap and Locator of Mechanism Shown in Fig. 12.

Speed Reducer with Wabble-Gear Mechanism.—A wabble gear, cast in a mold made from a pattern originally designed for use in the production of gearing for boiler grates more than seventy years ago, performs a major function in the rugged, 18-to-1 ratio speed reducer shown in Figs. 11 and 12. Although this mechanism may appear

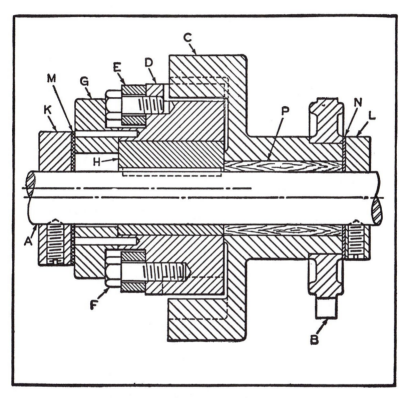

Fig. 12. Wabble-gear Speed-reduction Mechanism Used to Drive Tumbling Barrel.

somewhat crude, it has given excellent service for many years as a drive for tumbling barrels and similar equipment. It has the advantage of being comparatively simple and inexpensive to construct. The jack-shaft *A*, which runs at a speed of 180 R.P.M., drives the link-belt sprocket *B*,

Fig. 12, at a reduced speed of 10 R.P.M., and further speed reduction can be obtained by selecting sprocket *B* and its driven sprocket to give the desired speed, which in some cases may be as low as 2 R.P.M.

The tooth profiles of the wabble gear *C* and the cast wabble pinion *D* developed to operate with it are shown in Fig. 13.

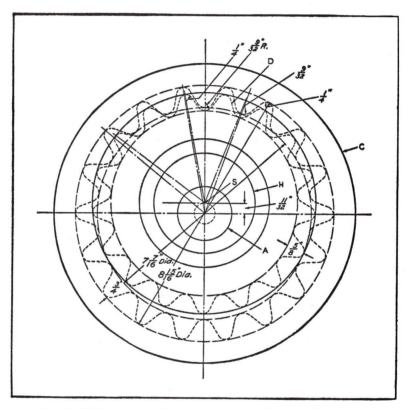

Fig. 13 Wabble-gear and Eccentrically Mounted Pinion of 18-to-1 Ratio Speed-reduction Mechanism

The strap *E* which is secured to pinion *D* by cap-screws *F*, as shown in Fig. 12, is also a casting. The retaining collar *G* serves to keep the eccentric *H* in place and to prevent the screws *F* from working loose. The eccentric *H* is keyed to the jack-shaft *A*. The wabble lever locator *J* is positioned in the slot of wabble lever *E*. As indicated in **Fig. 11, the** end of locator *J* has 1/8 inch clearance or play on both sides of the slot in wabble lever *E* when it is in the **center-line** position.

Jack-shaft *A*, Fig. 12, is 1 15/16 inches in diameter, and the reduction mechanism is mounted as near to one end of the shaft as possible to facilitate assembly. Collars *K* and *L* are made light drive fits on shaft *A* and are tapped to give a tight fit for the retaining set-screws. **Fiber washers** *M* and *N* and the Arguto oilless bearing *P* give good **service** over a long period of time.

Referring to Figs. 11 and 12, it will be obvious that when shaft *A* is revolved, the eccentric *H* keyed to it will **cause** the center *S* of the wabble pinion *D* to follow a circular path about the center of shaft *A*. Since the wabble lever *E* prevents the wabble pinion gear *D* from revolving about its axis, the circular motion imparted to pinion *D* serves to revolve the wabble gear *C* through the equivalent of one tooth space for each revolution of eccentric *H*. As there are eighteen teeth in wabble gear *C*, shaft *A* will make eighteen revolutions to one revolution of the sprocket *B* attached to the wabble gear.

Automatic Back-Gear Shifter.—A device for automatically shifting a back-gear in and out as the torsional resistance of a shaft varies is shown in Fig. 14. This mechanism is a part of an automatic drill-press feeding device. Shaft *Q* is connected to the drill-press feed-spindle through an adjustable slip clutch (not shown), and motor-shaft *A* is the source of power. The automatic shifting of the back-gear permits the use of a small motor to deliver the high thrust necessary at the drill point for drilling large holes, and at the same time permits rapid traverse of the drill to and from the work. The back-gear is automatically disengaged for drilling small holes.

Hub *B* is keyed to input shaft *A*. Cantilever leaf **springs** *C*, attached to hub *B*, are under sufficient **initial strain to** transmit full torque through spherical-ended pins *D* without slippage. Collar *E*, gear *H*, and worm *N* are pinned to shaft *F*. Cluster gear *J*, which is supported in frame *M*, is in constant mesh with gear *H* and idler gear *G*. Worm-gear *P* is pinned to shaft *Q*. The thrust collar *K* floats on shaft *F* and supports the compression spring *L*.

At the start of the cycle, input shaft *A* drives shaft *F* at the same speed through the gripping action of pins *D* on collar *E*. During this time, gear *G*, which is a sliding fit on

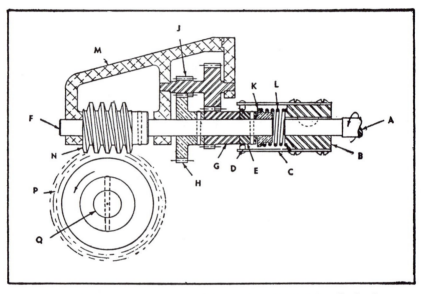

Fig. 14. Back-gear J is Automatically Engaged or Disengaged as Torsional Resistance of Shaft Q Increases or Decreases.

shaft *F*, is idling at high speed. When the reaction of worm *N* to the torsional resistance of shaft *Q* builds up sufficiently to overcome the load exerted by spring *L*, shaft *F* will shift axially to the right. Thus pins *D* will transfer their torque from collar *E* to the hub of idler gear *G*, thereby transferring the drive through the cluster gear *J*, or back-gear, to shaft *F*. The back-gear is automatically disengaged upon the reversal of the input shaft *A*.

On this particular mechanism, the input shaft delivers 1/12 H.P. at 1500 R.P.M. Gears having 24 pitch furnish a 4 1/2 to 1 reduction through the back-gears. Pins *D*, collar *E*, and the hub of gear *G* are hardened. The initial load on spring *L* must be set lower than the minimum worm reaction to prevent pins *D* from dwelling at the parting line of collar *E* and gear *G*.

Mechanism for Changing Cam Speeds Independently of Camshaft Speeds.—A mechanism for reducing the rotary speed of a cam without altering the speed of the shaft on which the cam is mounted is shown in Fig. 15. This

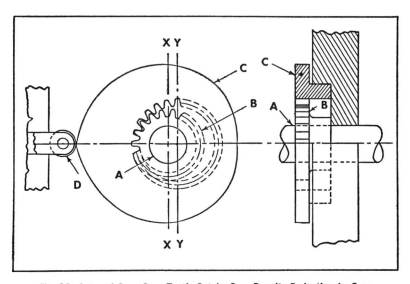

Fig. 15. Internal Spur Gear Teeth Cut in Cam Permits Reduction in Cam Speed without Changing Camshaft Speeds.

mechanism was applied to a machine for winding flat wire on spools. The wire-guiding mechanism was required to be operated by a uniform-motion cam, maintaining a definite rate of travel over a specific distance.

Because of a reduction in the width of the flat wire being spooled, it became necessary to decrease the rate of travel of the guiding mechanism by approximately one-third without altering the distance traversed or the speed of the cam-shaft, which operated other mechanisms.

To accomplish the required difference in rotary speed between the cam and its shaft, gear teeth were machined in the bore of the cam C to engage a spur gear B, which was assembled to the shaft A. The ratio of internal gear teeth to the spur gear teeth was made approximately 3 to 2. Thus the cam, which rotates about center line Y—Y, revolves at two-thirds the speed of the shaft, which rotates about the original center line X—X. In this way, the follower roll D, which actuates the wire-guiding mechanism, is driven at one-third the speed of shaft A.

CHAPTER 13

Speed Regulating Mechanisms

Machines which wind material such as paper, cloth or metal strip on spools or reels or which form or twist wire may require a synchronous rotation of two shafts with or without an occasional momentary acceleration or retardation of one shaft with respect to the other. In other machines the speed of the driven shaft must be maintained within close limits. The mechanisms described in this chapter have been designed to perform such special speed controlling functions. They include a hand-controlled arrangement for maintaining constant speed of a pull-roll unit; a friction-driving device which maintains web tension of rolled material within close limits; an automatic speed control for providing constant cutting speed on a lathe; a hand-operated mechanism for advancing a gear driven shaft while it is operating; a mechanism for insuring synchronous operation of two hand-operated shafts; a mechanism for obtaining a special cycle of speed relationships between two shafts; a mechanism for varying the pitch and twist in wire twisting machines; a mechanism which is used with a fluid power variable speed drive to maintain speed within 1/4 per cent of desired values.

Constant-Speed Pull-Roll for Winding Metal Strips.— Certain classes of metal strip material, after being put through various cleaning processes, are wound on steel spools mounted on a revolving shaft. The diameter of the spool hub, on which the metal is wound, is usually 6 inches. The spools were originally driven at a speed of 100 revolutions per minute. With this arrangement, the surface speed at the beginning of the winding operation was about 150 feet per minute. As the coil of metal became larger **diametrically**, the surface speed increased proportionately. The result was that by the time the full length of metal had been

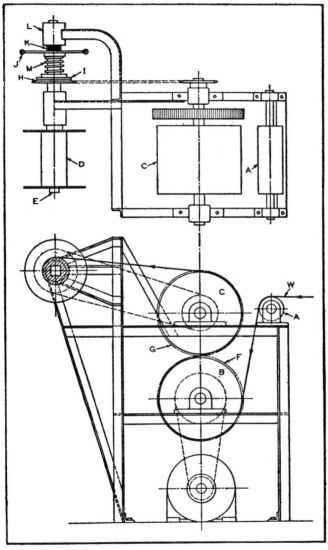

Fig. 1. Constant-speed Rolls for Pulling Metal Strip through Processing Machine and Winding it on Metal Spool.

wound around the spool, the surface speed had been increased about 100 per cent. This increase in speed resulted in scratched stock, which frequently had to be rejected.

As the processing operations required certain speeds for best results, it was necessary to maintain a constant winding speed. To meet these requirements and to overcome previous difficulties, the constant-speed pull-roll unit shown in Fig. 1 was built. This unit consists primarily of two large rolls B and C, driven at a constant speed and mounted one above the other on suitable bearings in an all-welded frame of standard structural steel. This unit also contains a square-ended revolving shaft E, mounted in bearings located on the outside of the main frame. The steel winding spool D slips over shaft E, the square end of which drives the spool.

As the metal strip W comes from the processing machine, it passes over the guide roll A. From roll A it passes under and part way around the pull-roll B, thence upward and half way around the pull-roll C to the winding spool D on the shaft E. From the illustration it will be noted that there is a small space between the pull-rolls. This space is necessary to allow the joints of the material to pass between the pull-rolls.

The unit is driven by a chain from the motor located below the pull-rolls, spur gears F and G transmitting power from one roll to the other. From pull-roll C a chain drives the winding spool D through friction disks H and I. More or less friction is obtained by turning handwheel J on the threaded bronze sleeve K which fits over the main shaft and is fastened to the rear bearing L.

As handwheel J is moved either clockwise or counterclockwise, the compression spring M increases or decreases the pressure on the friction disk I, so that the desired tension is maintained on the metal being wound. Thus the speed at which the metal is wound on the spool remains constant no matter what the diameter of the coil may be, the slipping that occurs between the friction disks compensating for the difference in speed. The rolls *B* and *C* do all the pulling of the metal through the processing machine and they also regulate the speed of the machine and govern the winding spool speed.

Friction Driving Mechanism for Rewinding Roll.—A friction driving mechanism that has been found practical as a means for driving a roll used in rewinding a paper or cloth web is shown in the accompanying illustration. This mechanism was designed specifically for rewinding cloth on a dyeing pad. Modern printing presses and paper-coating machines are, of course, equipped with an elaborately designed unit for automatically controlling the web tension and compensating for the constantly increasing diameter of the roll of rewind material. For a small proving press or laboratory model, however, the use of a simple and flexible friction driving unit is desirable.

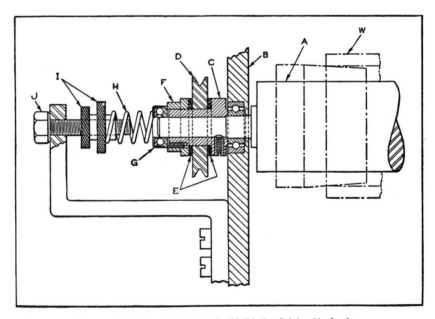

Fig. 2. Rewinding Roll Equipped with Friction Driving Mechanism.

In this case, the rewind unit is not required to pull the web through the operating rolls. It is only necessary for the unit to keep the material under tension within limits that will prevent wrinkling. Owing to the light amount of power needed, the friction adjustment must be very sensitive.

Referring to Fig. 2, the speed of the driven pulley D must be so adjusted that it is approximately 2 per cent higher than the minimum web speed when starting at the small diameter of the rewind core W. Pulley D is a running fit over bronze hub C, which is secured to the extension shank of the rewind roll A by set-screws or dowels. Two leather washers E on each side of the pulley provide the necessary friction for driving the roll when sufficient pressure of spring H is applied to collar F. All bearings are of the anti-friction type, including the roll-supporting bearing in the frame B.

A feature of this rewind mechanism is the spring pressure adjustment provided by stud J and nuts I, which can be easily changed while the machine is in operation. The thrust ball bearing G is preferred to an ordinary thrust washer because the face of the latter, revolving against the end of spring H, would reduce the sensitivity of the mechanism.

Automatic Variable-Speed Control Maintains Constant Cutting Speed when Facing Disks.—An automatic speed control applied to a Gisholt Simplimatic lathe in the machine shop of the Reeves Pulley Co., Columbus, Ind., makes it possible to maintain a constant cutting speed when facing disks. In other words, the number of revolutions per minute of the work decreases as the tool moves toward the circumference of the disk.

The machine is used for turning and facing disks in sizes up to 20 inches in diameter. The drive consists of a motorized mechanical variable-speed transmission with mechanical automatic speed control, arranged so as to provide a

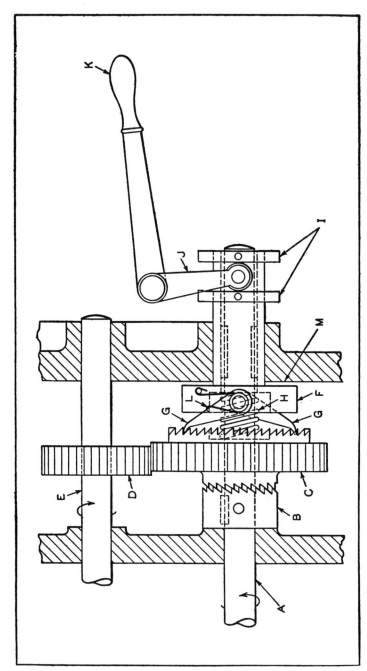

Fig. 3. Mechanism which Permits Driven Shaft E to be Advanced Ahead of Driving Shaft A by Means of Hand-lever K.

uniform cutting speed as the tool moves from the smallest diameter at the hub to the largest diameter at the circumference.

The speed control device is connected by a cable to the cross-feed of the lathe. As the cutting tool moves outward from the hub of the disk being faced, its movement is transmitted through the cable to the indicating lever of the mechanical variable-speed device, which, by means of a chain drive, transmits power to the speed control unit. The latter, in turn, is connected through a reducer and coupling to the drive-shaft of the lathe. Thus, as the cutting tool moves out, the motorized variable-speed transmission drive unit is automatically regulated to reduce the number of revolutions per minute of the disk, thereby maintaining a uniform cutting speed.

Lever-Operated Mechanism for Advancing Gear-Driven Shaft.—In operating a wire-forming machine, it is necessary to maintain a certain amount of slack in the wire as it passes through the machine. At times, the slack is taken up as a result of uncontrollable conditions and must be increased while the machine is in operation. Fig. 3 shows the construction of a mechanism by means of which the amount of slack is manually controlled.

The shaft A, rotating in the direction indicated by the arrow, normally drives shaft E through the gears C and D, the ratchet teeth cut on the side of the collar B, which is keyed to shaft A, being engaged with similar teeth cut on the hub of gear C. Spring H serves to maintain contact between the ratchet teeth, so that gear C is positively driven. Fig. 3, however, shows the ratchet teeth partly disengaged.

The sleeve F is splined to shaft A and rotates with it, but is capable of axial movement, being moved to the left, as shown, when pressure is applied to the left-hand collar I by the handle K through the fork-shaped lever J. The spring H normally holds sleeve F in contact with the frame at M. Two pawls G are carried on sleeve F and are held in contact with the ratchet teeth on the right-hand side of gear C by springs L.

As sleeve F is splined to shaft A, it normally rotates with it, the ratchet teeth in collar B and gear C being engaged and the pawls G merely resting on the ratchet teeth on gear C. In this position, the pawls G do not in any way aid in transmitting motion to gear C. Thus the entire assembly on shaft A rotates as a unit.

When it is necessary to increase the slack in the wire, the handle K is depressed, causing lever J to move sleeve F to the left, as shown, and the pawls G to spread or flatten out similar to toggle levers. As these pawls spread, they exert a turning effect on gear C, advancing it relative to collar B and causing the ratchet teeth on the hub of gear C to ride over those on collar B. When gear C has been thus advanced a distance equal to one tooth space, spring H causes it to slide to the left and re-engage the teeth on collar B.

Mechanism for Insuring Synchronous Motion.—On a machine for fabricating a wire product, two shafts, operated by means of hand-levers, actuate a tension mechanism. The shafts are required to operate practically in unison, a narrow range of latitude being allowed. Gearing the shafts together appeared to be the solution, but this method proved unsatisfactory, as, at times, one shaft was driven by the other. resulting in torsional stresses, which were objectionable. In order to insure equal power application and synchronous movement of the two shafts, the arrangement shown in Fig. 4 was devised.

The two gears A and B are carried on the shafts and mesh with the double rack C, which floats between them. Rack C carries a serrated slot at its lower end. A serrated pin D, fastened in a stationary part of the machine, is located in the slot of rack C. Pin D is given a minimum amount of clearance in the slot, the clearance being shown somewhat exaggerated. The plate E, which serves as a guide for rack C, is broken away to show the pin D.

In operation, if either of the handles is moved ahead of the other, the action of the gear on that shaft causes the rack C to swing on the teeth of the other gear as a fulcrum, so that the serrations in the slot of the rack engage the serrations on pin D, thus preventing further movement until the other handle is given a corresponding movement, thus disengaging the slot serrations from pin D.

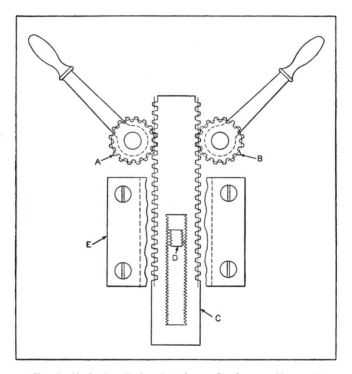

Fig. 4. Mechanism Designed to Insure Synchronous Movement of Two Levers under Equal Torque.

Mechanism for Obtaining Irregular Rotating Movement.

A machine for fabricating a wire product has two spindles that perform twisting operations in synchronism through approximately half of the operating cycle. During the remainder of the cycle, it is necessary for one of the spindles to accelerate its speed of rotation or to advance

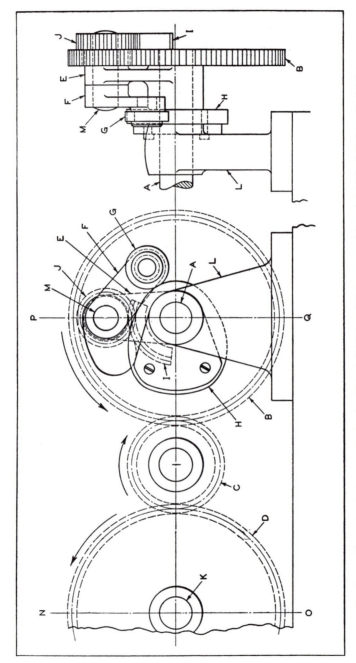

Fig. 5. Mechanism for Producing Irregular Rotation of Shaft A and Constant-speed Rotation of Shaft K from the Driving Gear C.

ahead of the other, returning to synchronous operation after a specified point in the cycle has been passed. The mechanism developed to obtain the necessary movement is shown in Figs. 5 and 6.

Referring to Fig. 5, gears *B* and *D* are driven by pinion *C*, rotating in the direction indicated by the arrows. Gear *D* is keyed to shaft *K*, while gear *B* is free on shaft *A*. Shaft *A* is supported in bearing *L*, which has a flange on one side on which cam *H* is mounted. Lever *E* is keyed to shaft *A*, and carries at its upper end the shaft *M* to which lever *F* and pinion *J* are keyed.

Lever *F* carries roller *G*, which rolls on cam *H*. Pinion *J* meshes with gear segment *I*, which is carried on, and rotates with, gear *B*. The boss on the upper end of lever *E* is extended to pass through the slot in gear *B*, as shown.

In operation, the rotation of pinion *C* is transmitted to gears *B* and *D*, which rotate in synchronism at a uniform rate. As gear *D* is keyed to shaft *K*, the latter rotates with it. Gear *B*, however, not being keyed to shaft *A*, does not transmit motion directly to shaft *A*. As shown in Fig. 5, the extended boss of lever *E* is in contact with one end of the slot in gear *B*; thus the motion of gear *B* is transmitted to shaft *A* through lever *E* in the direction indicated by the arrow, the effect being the same as though shaft *A* were driven directly by gear *B*.

Referring now to Fig. 6, in which bearing *L* is cut away to expose lever *F*, the rotation of gear *B* has caused roller *G* to rise to the high point of cam *H* and begin descending the other side of the cam. When roller *G* begins to ascend the rise of cam *H*, a rotative motion is imparted to pinion *J* through lever *F* and shaft *M*. This rotative motion of pinion *J*, which is in mesh with gear segment *I*, causes a slow rotative motion to be imparted to lever *E* in the same direction as the driving motion transmitted to gear *B*. As lever *E* is keyed to shaft *A*, any movement of *E* causes a change in the relative positions of gear *B* and shaft *A*.

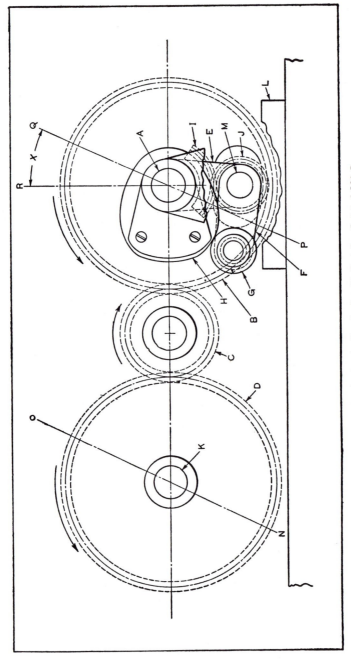

Fig. 6. Diagram Showing Mechanism in Fig. 5 after Shaft A has Rotated 180 Degrees.

When roller G reaches the high point of cam H, the entire assembly again rotates as a unit, their relative positions remaining unchanged until roller G begins to descend to the low point of cam H. In effect, shaft A is first given an accelerated movement, because the motion imparted by lever E is added to that imparted by gear B. The accelerated movement of shaft A is followed by a decelerated movement due to the reverse motion of gear J when lever E is returning to its original position.

The frictional resistance of the assembly, as used on the machine, combined with the resistance of the twisting operation, is usually sufficient to maintain contact of roller G with cam H; at high speeds, however, it may be necessary to attach a spring to lever F to maintain contact of the roller and cam.

In Fig. 5, the perpendicular radial center lines NO and PQ indicate the relative positions of gears B and D, and of lever E. In Fig. 6, center lines NO and PQ indicate that gears B and D have rotated approximately 150 degrees in synchronism. Center line R, through lever E indicates a rotation of 180 degrees of lever E, the angular advance of shaft A relative to gear B being indicated by angle X.

Wire-Twisting Mechanism Designed to Vary Pitch of Twist.

— The purpose of the mechanism shown in Fig. 7 is to twist two lengths of wire A and B together, the pitch of the twist being varied to suit certain requirements. The two wires are fed at a uniform rate of speed through the twisting spindle C, the rotating speed of the spindle being automatically varied, so that a definite number of twists of uniform pitch are produced, followed by twists of constantly varying pitch. The wires twisted together in this manner are later cut to length. The two views in Fig. 7 show the interesting mechanism which was designed to produce the required variations in the speed of rotation of the twisting spindle.

As shown in Fig. 7, the bearing I supports the driving

shaft J from which the mechanism receives its motion. Lever K is keyed to shaft J and carries the roller L at its outer end. Gear E is keyed to shaft M, which is carried on one end of the lever N. The roller L fits in a groove of gear

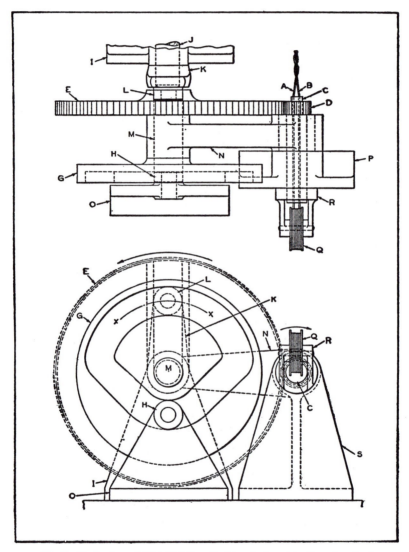

Fig. 7. Mechanism of Wire Twisting Machine Designed to Vary Speed of Gear E and Spindle C.

E and is used to transmit the rotary motion of shaft *J* to gear *E*. The internal cam *G* is keyed to the shaft *M*, and therefore rotates in unison with gear *E*.

The roller *H*, carried on stationary bracket *O*, travels in the groove in cam *G*. Lever *N* is supported freely on the hub of bearing *P* within which the spindle *C* rotates. Spindle *C* is provided with two holes through which the wires *A* and *B* pass, and receives its rotary motion from the gear *E* through the pinion *D*. The wire reel *Q* is carried on the hub *R*, which rotates with spindle *C*.

Referring to Fig. 7, the gear *E* is rotated in the direction indicated by the arrow, receiving its motion from shaft *J* through lever *K*. The spindle *C* is rotated in the direction indicated by the arrow, receiving its motion from gear *E* through pinion *D*, the ratio in this case being 8 to 1. The relative positions of shafts *J* and *M* are controlled by the position of the roller *H* in the groove in cam *G*.

In the position which is shown in the view at the top of Fig. 7, the axes of the shafts *J* and *M* coincide, and the arc of travel of roller *L*, indicated by the line *XX* in the lower view, is concentric with the axis of shaft *M*. Therefore, at this point, there is no movement of roller *L* in the groove on the back of gear *E* and the speed of rotation of gear *E* is uniform and at the same rate as that of shaft *J*. As the wires *A* and *B*, which are wound double on reel *Q*, are fed through spindle *C* at a uniform rate of speed, a twist of uniform pitch is formed as long as the roller *H* remains in the lower portion of the groove in cam *G*.

Referring to Fig. 8, continued rotation of gear *E* has brought the high point of cam *G* into operation on roller *H*, causing cam *G* and gear *E* to be raised by the lever *N*, swiveling on bearing *S*. In this position, the axes of shafts *J* and *M* no longer coincide, being separated by the distance *T*. Owing to the change in the position of gear *E*, it will be noted that the path of roller *L*, indicated by the line *YY*, is now closer to the pitch line of gear *E* than shown in Fig. 7.

The effect of this change is twofold. The peripheral speed at the pitch circle of gear E, when the gear is in the position shown in the lower view in Fig. 7, is greater than at the path followed by the roller L, in the ratio of the difference of their radii. In the view shown in Fig. 8, the path of roller L is closer to the pitch line of gear E, and although gear E

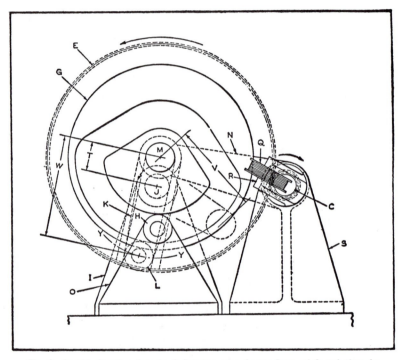

Fig. 8. In this Position Spindle C Rotates at a Lower Rate of Speed than in Position in Fig. 7, and Gear E is Given a Variable Speed Rotation.

and lever K are still rotating on the axes of their respective shafts, the change in the relative positions of the axes of shafts J and M produce the same effect at this point as would occur had the length of lever K been increased a corresponding amount.

As the speed of rotation of spindle C is governed by the peripheral speed of the pitch circle of gear E, spindle C is rotating at a lower rate of speed when in the position shown in Fig. 8 than when in the position shown in Fig. 7, the effect being to give a greater pitch twist to the wires A and B.

The second effect of separating the axes of shafts J and M is to produce a variable rotation of gear E. Regardless of the relative positions of the two axes, the effective length of the lever arm in transmitting motion from shaft J to shaft M is equivalent to the distance between the axes of these two shafts at any given point in the cycle, as indicated by the distance W. This distance will be greatest, and the rotation of gear E will be slowest, when the axes of shafts M and J and roller L are on the same straight line. As shown in Fig. 8, this point has not quite been reached.

When this point is passed, the length of the lever arm is gradually reduced, as indicated by the dimension V. During this period, spindle C is being rotated at a variable speed, thereby producing a twist of varying pitch in the wires. When the lower portion of cam G is again brought into action on roller H, the axes of shafts J and M again coincide, and spindle C rotates at uniform speed.

Precision Variable-Speed Mechanism Employed in Oil-gear Drive.—The mechanism to be described is used in conjunction with fluid-power variable-speed drives designed to give precisely the driving speed desired, regardless of fluctuations in the speed of the drive to the pump end of the unit resulting from variations in the load, oil temperature, running fits, or power-line currents.

These precision variable-speed drives, developed by the Oilgear Co., Milwaukee, Wis., comprise a standard Oilgear fluid-power transmission with a Micro Servo-Motor stroke control cylinder which adjusts the pump stroke to give exactly the required hydraulic motor speed. Oil is admitted to the control cylinder by a pilot valve, actuated by a small differential unit which continuously compares the hydraulic motor speed with an accurate time-measuring unit or with the speed of any desired master unit.

One of the Oilgear units, when driven by an ordinary induction motor or lineshaft, will drive its load at any desired speed, either constant, adjustable, or continuously varying between zero and maximum.

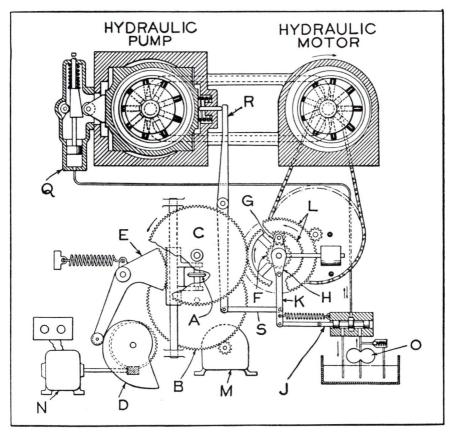

Fig. 9. Fluid Power Drive with Precision Speed Control Employing Synchronous or Selsyn Motor M as Master Speed Unit.

The mechanism, which maintains the speeds within 1/4 per cent of the desired values, is shown in Fig. 9 equipped with a small synchronous or Selsyn motor M, which serves as the accurate time-measuring unit. In cases where the frequency error in the power current supply available for motor M is as much as 0.3 per cent and a closer control of speed is desired, the pendulum type precision control shown in Fig. 10 may be used. In this mechanism, pendulum P takes the place of motor M shown in Fig. 9.

The essential principle of the differential time precision control, as shown in Figs. 9 and 10, is that the hydraulic motor, driven by the flow of power from its variable-stroke pump, transmits the resulting speed to a differential comparator. This differential also receives a standard or master speed time control from the small synchronous motor M (Fig. 9) or the pendulum P (Fig. 10) which serves as time control units, or from some other unit such as a roll stand or float roll measuring control. In any case, the differential continuously compares the actual speed with the master speed and translates every discrepancy into exactly the increase or decrease of pump discharge necessary to correct the error.

It should be noted that while the use of a pendulum eliminates frequency error, it also limits the speed at which the comparator acts, and thus results in a slower response. When the master speed disk is driven by a motor, the action of the disk type control is so quick that fluctuations of speed due to ordinary load changes are caught and corrected within about one-tenth of a second. With moderate load fluctuations, such as usually exist in a continuous processing plant, the momentary governing errors are held within 1/4 of 1 per cent, plus or minus, while the integrated error over a period of time—say, ten seconds or more—is too small to be measurable by a stop-watch.

The differential time disk type control is shown in its usual form in Fig. 9, with a small synchronous or Selsyn motor M furnishing the master speed. If M is a synchronous motor, the control is by "time," and the speed of the hydraulic motor is given in revolutions per minute. If M is a Selsyn motor, it receives its current from a Selsyn generator driven by its master unit, and follows the speed of that master through its entire range from the standstill or stationary position to maximum speed.

The master speed is transmitted to the differential mechanism through the friction-disk ratio-changer, comprising a movable idler disk A that is spring-pressed between the driving or master disk B and the driven disk C. Disk A is mounted in a slide-block, and is adjustable by push-button control motor N or by a manual or float-roll control through cam D and segment E.

The master speed, as modified by the ratio-changer, is transmitted by gear teeth in the edge of disk C to a gear integral with the sun pinion F, which is the first leg of the differential comparator already mentioned. Planet gear G is mounted on a radial valve actuating crank H (the third

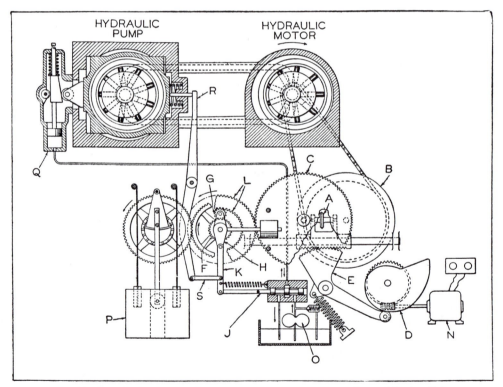

Fig. 10. Drive Similar to That Shown in Fig. 9 but with Different Type of Timing Unit.

leg of the differential) which, in turn, actuates pilot valve J through floating lever K. Planet gear G also meshes with an internal ring gear L which is exactly twice the diameter of sun pinion F.

Any movement of the pilot valve resulting from movement of the axis of the planet gear and crank H changes the pump stroke and hydraulic motor speed. To hold pilot valve J stationary, ring gear L (the second leg of the differential) must be driven by the hydraulic motor at exactly one-half the standard speed of gear F. Any discrepancy will result in a slight movement of valve J, permitting oil from gear pump O to enter the pump-stroke changing cylinder Q, and thus increase the pump discharge, or permitting oil to escape from Q, which will reduce the pump discharge. When the change in pump discharge has corrected the motor speed, so that it again balances the speed of the differential gears, the pilot valve will resume its neutral position.

An essential part of the mechanism is the follow-up lever R which responds to the movement of the pump slide-block, and by means of link S, closes the pilot valve as soon as the required correction has been made, thus preventing the so-called "hunting" action. As the disks A, B, and C are made of hardened steel, are accurately ground, and transmit no appreciable power, they show little or no wear after long service. This is true even when the speed is not changed and idler disk A runs in one location continuously.

A characteristic feature of the drive described is that it cannot be overhauled by a decelerating inertia load, so that winders and unwinders for paper machines, super-calenders, printing presses, textiles, strip and foil, etc., driven by the differential time controls will automatically start and stop heavy rolls of material without sag or added strain in the web. By means of the precise speed adjustment of the disk control, two or more rolls of paper or foil can be unwound from rolls of different diameter at identical surface speeds and at uniform tensions, and laid together.

CHAPTER 14

Feed Regulating, Shifting and Stopping Mechanisms

In all machines which perform operations on parts or on material, means must be provided for regulating, shifting and stopping the feed of either a tool or the work. Such mechanisms may be utilized to obtain coarse or fine feeding movements; to adjust the amount of automatic feed; to proceed through a complete cycle of advancement, cutting, withdrawal and replacement of workpieces; to properly position work for drilling or cutting; to stop the rotation of work in a particular angular position; to shift from one work-performing mechanism to another; and to stop operation of a machine when no more material or work is available for feeding. These and other operations are performed by the mechanisms described in this chapter.

Other mechanisms which perform similar functions are described in Chapter 16, Vol. I and Chapter 14, Vol. II of Ingenious Mechanisms.

Mechanism for Obtaining Coarse and Fine Feeding Movements.—It was in conjunction with the building of a small milling machine type of grinder for cutting slots 1/16 inch wide by 1/8 inch deep in some square, hardened pieces that the screw and worm-wheel mechanism shown in Fig. 1 was developed for the purpose of obtaining a quick approach to the grinding wheel and a fine feed for grinding the slot. A rubber cutting wheel was used for the slot-grinding operation.

The screw or worm A is a close sliding fit in the bore B and engages the worm-wheel C as shown. The worm-wheel is pinned to the stud D which has a bearing in the cast-iron bracket E. The stud D is made square at F to accommodate the lever G, and is threaded to receive the knurled nut H.

In operation, the handle K of lever G is moved in either direction. This causes the screw A to slide in or out with much the same action as a rack operated by a pinion. After using the lever to bring the work into the grinding position, the knurled nut H is tightened to hold the lever stationary. A fine feeding movement can then be obtained by turning the knurled nut J, worm-wheel C acting as a fixed nut for the screw A.

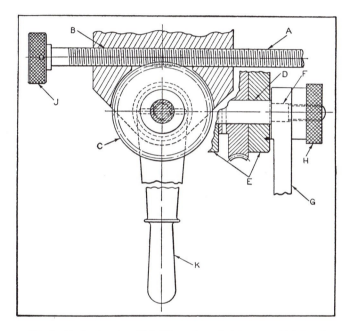

Fig. 1. Mechanism in which Worm Gear C Acts as a Pinion and Worm A Serves as a Rack.

Fine Feeding Mechanism Adjustable from 1 to 50 Microns per Revolution.—The fine feed of carriage U, Fig. 2, adjustable in increments of 1 micron from 1 to 50 microns (1 micron = one millionth of a meter or 0.000039 inch) per revolution of drive-shaft C, was obtained by means of the mechanism illustrated in Figs. 2 to 6. This design eliminates the reciprocating parts commonly encountered in feeding

ratchet teeth. This action is eliminated in the mechanism described.

The feeding mechanism consists essentially of a base *A*, on which is mounted an upright casting *B* that supports drive-shaft *C*, as shown in Figs. 2 and 3. Gear *D*, which is fastened to one end of this shaft, drives gear *E*. The latter gear, which is secured to pawl-carrier *F*, is mounted on

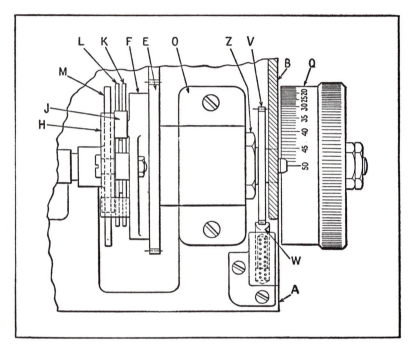

Fig. 4. Partial Plan View of Feeding Mechanism. Index Setting is Held by Plunger W Engaging Gear V.

sleeve bearing *G*. Pawl *H* carries a roller *J* on its driving end, and is held to carrier *F* by a stud on which the pawl pivots. As the pawl-carrier is rotated by driven gear *E*, the pawl is carried above the ratchet wheel *M* for a portion of each revolution when roller *J* rides on cams *K* and *L*. When roller *J* is carried past the cam surfaces, the pawl engages and drives the ratchet wheel through spring action.

Ratchet wheel *M* is pinned to shaft *N*, which is mounted on bearings *P*. Worm *S*, which engages worm-wheel *R*, is also pinned to this shaft. The worm-wheel is keyed to the feed-screw *T*. The 30 to 1 ratio of the worm to the worm-wheel is such that an advance of one tooth of the ratchet wheel rotates the feed-screw 1 micron. The feed-screw moves carriage *U* back and forth on ways which are not shown in the accompanying illustrations.

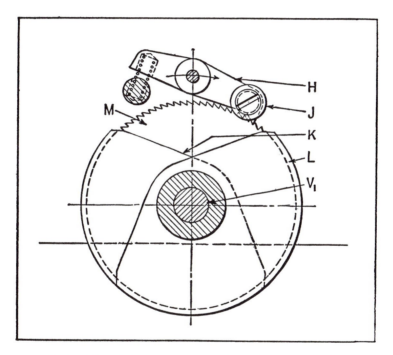

Fig. 5. Section Z-Z of Fig. 3 Showing Positions of Cams K and L for a 25-Micron Feed.

The relative position of cams *K* and *L* shown in Fig. 2 occurs when the index-wheel is set at 50. This setting and cam position cause the pawl to be lifted out of engagement with the ratchet wheel for half of the cycle—180 degrees— of the pawl-carrier. During the remaining half of the cycle, the pawl engages the teeth on the ratchet wheel, and needle valve H by way of suitable passages, at a rate determined by the setting of the needle valve. After completion of an operation, the oil returns through a valve to the other side of the piston under the influence of the return spring. Screws K act as filling plugs.

When slide D comes into contact with stop-screw E, it will meet the central rod F, and its rate of progress will be limited by the rate at which oil or other fluid can escape

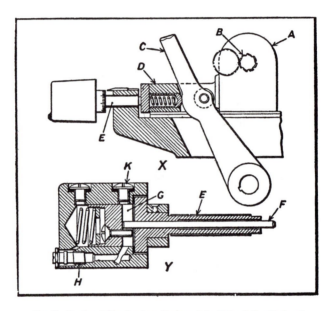

Fig. 7. Dashpot Mechanism Designed to Retard the Rate of Movement of Milling Machine In-Feed.

through the needle valve. The spring and plunger shown in the view at X provides a slight additional movement of lever C after contact with screw E, as required, to permit the operation of a trip motion.

Variable Feed Arrangement for Automatic Wheel-Dressing Device.—The actuating screw of an automatic wheel-dressing device employed on a grinding machine was designed to move the dressing diamond a given amount each time it functioned, but there was no provision for varying the amount of this movement. Variations in the hardness and bonding of a grinding wheel, however, make it desirable that the amount of feed or movement of the truing diamond be adjusted to suit individual grinding wheels and the conditions under which they are used. Fig. 8 shows an

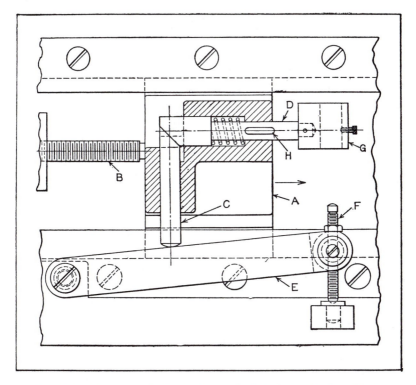

Fig. 8   Feed for Automatic Pressing Device Used for Truing Grinding Wheel.

arrangement designed to permit such adjustment. With this device, the feed can be varied any desired amount by adjusting the bar E.

Referring to Fig. 8, slide A is moved a fixed amount in the direction indicated by the arrow each time the actuating screw B functions. Mounted on slide A at right angles to each other are two plungers C and D. Plunger D is kept in contact with plunger C by means of the coil spring. Plunger C, in turn, is kept in contact with the adjustable bar E. The bar is also held in a fixed position by the pressure exerted by the spring.

The angle of the bar is adjusted by means of screw F which can be clamped in place by a locknut. The diamond-holder G is kept from rotating by means of the key H, which slides in a keyway cut in slide A. The amount of movement imparted to holder G by a given movement of screw B is determined by the angular setting of the sine bar E.

Automatic Feeding Mechanism for Drill Press.—The attachment shown in Fig. 9 has been applied to a drill press to provide automatic power feed and, in addition, to feed the drill into the work, disengage the feed when the drill has reached the required depth, return the spindle to the starting point, and then re-engage the feed. The work is removed and replaced by a new piece during the return movement of the spindle. This arrangement resulted in a considerable increase in production, with greatly reduced operator fatigue.

A front view of the attachment is shown at the lower right of Fig. 9, and a plan view at the upper right of Fig. 9. The spindle A carried in quill B is fed into the work by the rack D and pinion C. The rack and pinion are part of the original equipment; a handle attached to shaft S which carries pinion C has been removed. A worm-gear P is carried on shaft S in place of the handle. The bracket I pivots on stud J, which is carried on a bracket attached to the head of the press. This bracket is not shown, however, as it must be made to suit the individual application.

Bracket I supports the shaft M, which carries the worm K and sheave N. Worm K is keyed to shaft M, and is provided with a hub at one end which supports the compression spring L, shown partly compressed in Fig. 9. Shaft M is rotated in the direction indicated by the arrow by the

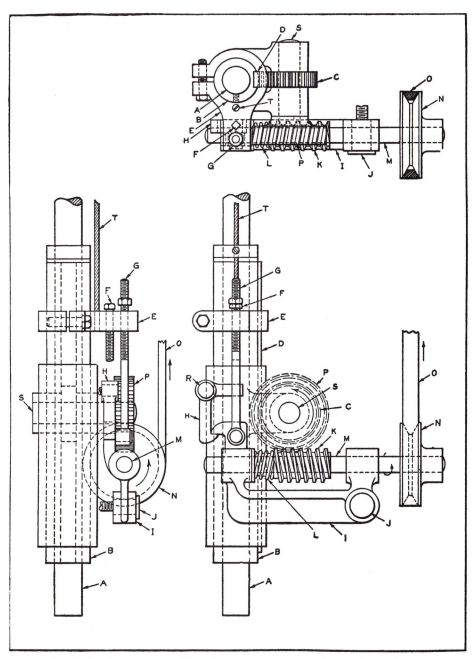

Fig. 9. Side, Front, and Top Views of Automatic Feeding and Reversing Mechanism Applied to Drill Press.

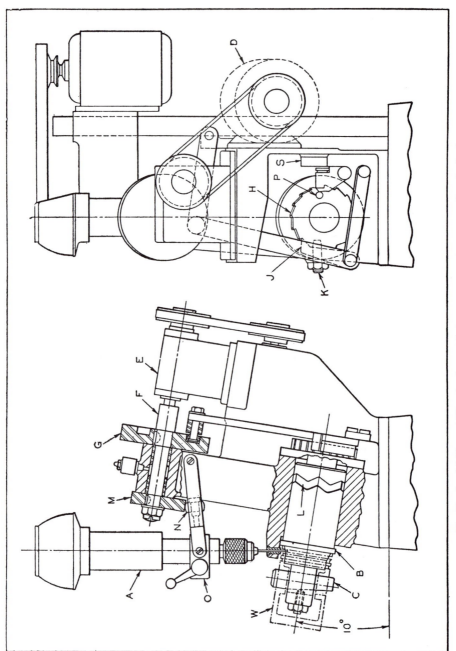

Fig. 10. Special Piston-drilling Machine with Automatic Indexing and Feeding Mechanism.

required for the angular spacing of the holes. A fixed plug K, engaging the cam groove L milled on the periphery of the work-arbor, causes the latter member to move in and out as required for drilling the two rows of equally spaced holes. The cam M, actuating the roller N, imparts the required feeding movement to the drill. The collar O can be adjusted to compensate for drill wear. A pin P, which is located on the ratchet wheel, comes in contact with a micro-switch S, thus stopping the machine automatically after the last hole in the piston has been drilled.

Spindle-Control and Collet-Operating Mechanism for Screw Machines.—The feeding of irregular-shaped automobile parts into the work-holding collets of a special multiple-spindle screw machine necessitated the provision of some means for automatically stopping the rotation of the spindles in exactly the same angular position at the loading station, opening the collet to receive the work, closing the collet on the work, and releasing the spindle-driving clutch. The mechanism developed to perform these functions consists primarily of a three-ball type spindle stopping and starting clutch and a collet opening and closing device, arranged as shown in Fig. 11. Details of the principal members are shown in Figs. 12, 13, and 14.

The spindles of the machine are all housed in a rotor A, Fig. 11, operated by a Geneva motion indexing mechanism. A center gear (not shown) which engages the pinions E serves to drive all the spindles. The collets are closed by the action of a stiff helical spring S.

This arrangement left little space for the spindle clutch, which was required to be of very compact design. Referring to the assembly view of the spindle, Fig. 11, the rocker arm B opens the collet by moving rod C to the right, causing spring S to be compressed. The spindle D is driven by the pinion E, which is bored out to house clutch ring F. The clutch ring is a drive fit in pinion E and is pinned to it, so that members E and F comprise an assembled unit. This

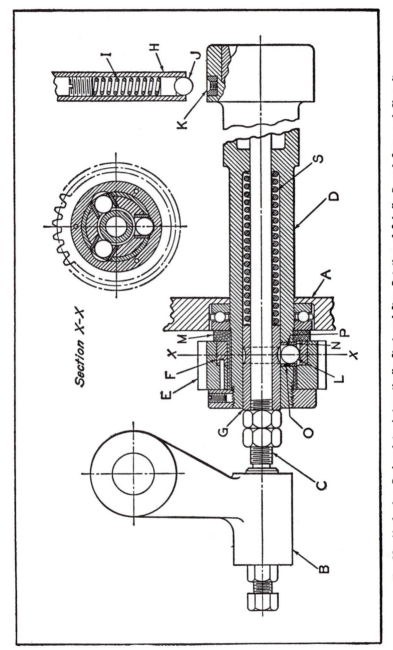

Fig. 11. Mechanism Designed to Automatically Start and Stop Rotation of Spindle D and Open and Close Its Work-holding Collet.

Section X-X

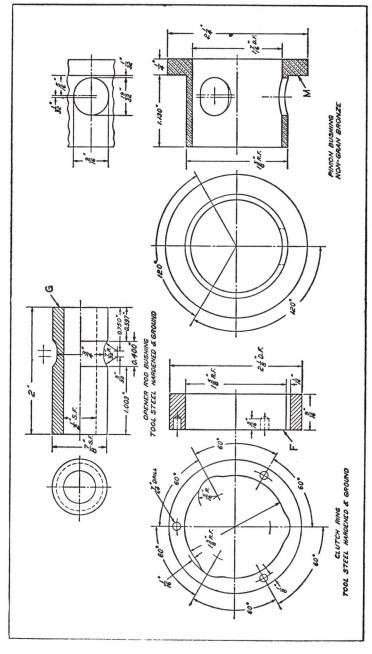

Fig. 12. Details of Collet-opening Rod Bushing G, Clutch Ring F and Pinion Bushing M of Mechanism Shown in Fig. 11.

unit runs continuously at a speed of 900 R.P.M. during the
operation of the machine.

Three holes are drilled and bored through pinion bushing
M to receive the bushings N which house the steel balls L of
the clutch engaging and disengaging mechanism. These
bushings are spaced 120 degrees apart, and are a drive fit

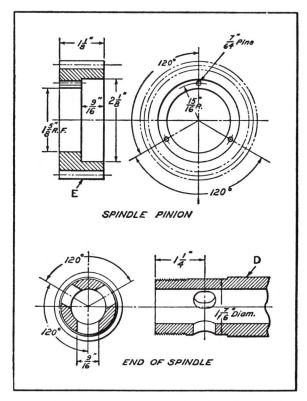

Fig. 13. Details of Spindle Pinion E and the Driven End
of Spindle D in Fig. 11.

in the bored holes. Bushings N have a slot for holding
spring O (see Fig. 14). A 5/64-inch hole P is also drilled
in the bushing to form a retaining seat for the ball L.

The bushing G, Fig. 11, which is employed to guide the
collet-opening rod C, is also used for operating the spindle

clutch. The tube H at the front end of the spindle rotor carries a spring I and ball J. The end of tube H is machined to prevent ball J from falling out of the tube. The hardened cup K in the spindle D is threaded to facilitate its removal when necessary. Tube H is moved downward by a cam device when the spindle clutch is released, so that the ball J is forced into the cup K to stop the spindle. Thus the

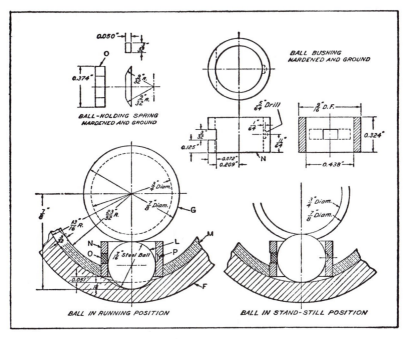

Fig. 14. Enlarged Scale Views of Details N and O in Fig. 11 and Broken Section Views Showing Spindle Clutch Balls L in Running and Idle Positions.

spindle is always stopped in the same position with respect to the rotor A, so that the work, in turn, can be fed into the collet in the correct position. This requirement presented a rather difficult problem, since the spindle runs at a speed of 900 R.P.M. A clutch made up for testing purposes was operated successfully at a speed of 1200 R.P.M. before the design was incorporated in the mechanism described.

In the assembly shown in Fig. 11, the 7/16-inch ball L at the bottom position is shown resting in the 5/16-inch radius cut-out in clutch ring F. The ball remains in this position only momentarily, however, since rotation of clutch ring F brings the cylindrical 1 5/8-inch diameter section of the latter member into contact with the three balls, forcing them inward toward the center of the spindle, where they are held by spring O and depression P.

The two enlarged views in Fig. 14 show exactly how the ball is located both in the running and in the stand-still positions. When arm B, Fig. 11, swings to the left, the draw-rod C closes the collet, and the three 7/16-inch balls are forced outward, so that they come in contact with the bottom of the 5/16-inch radius cut-outs in the clutch ring F and thus start the spindle rotating. When arm B swings to the right, the channel in the bushing G clears the three balls, causing the spindle to stop and the collet to open.

Device for Automatic Shifting between Two Cam-Operated Packing Mechanisms.— In designing a device for disposing of sheets from the delivery end of a paper finishing machine, it was desirable to provide a means of automatically and periodically shifting from one cam-operated packing mechanism to another. The shifting device here described permits five hundred sheets of paper to be placed on one tray while a second full tray adjacent to it is being replaced with an empty one.

Referring to Figs. 15 and 16, arms A and B, which are actuated by cams C and D, are secured to and alternately rock their respective pivot shafts A_1 and B_1. Each pivot shaft is connected to a separate packing mechanism (not shown).

One arm rocks its pivot shaft five hundred times, whereupon it stops at the top of its stroke and the other arm immediately begins its cycle of five hundred strokes. The working stroke occurs when the follower of the arm rolls into the hollow of its respective cam. The cams are secured to shaft E, which rotates continuously at one revolution per cycle of the finishing machine.

A plate F, equal in diameter to the cams, is located between them and slides back and forth on a key in shaft E. When this plate is in contact with either cam, the follower is prevented from rolling into the hollow of that cam and is held at the top of its travel. Meanwhile, the follower on the other arm is free to roll into the hollow of its cam and

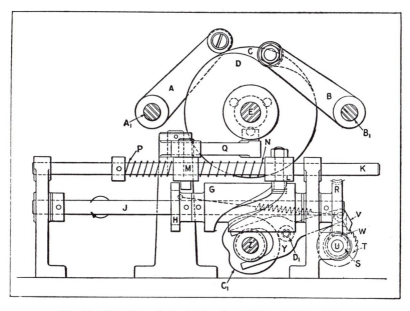

Fig. 15. Side View of Mechanism for Shifting Packing Action of Mechanisms on Shafts A_1 and B_1.

thus rock its pivot shaft. Upon completion of the cycle of five hundred strokes, plate F is shifted from one cam into contact with the other, thus shifting the packing cycle from arm A to arm B, or vice versa.

This shifting is accomplished by means of drum cam G and notched disk H, which are both pinned to shaft J. The follower carrier L for the drum cam is pinned to sliding shaft K, while block M and compression springs N and P

are free to move on this shaft. The follower carrier and block M are both free to slide on the parallel shaft K_1, which is secured to the frame of the mechanism.

Two notches, 180 degrees apart, in hardened disk H provide a sliding fit for the lower portion of block M, which is square in cross-section and also hardened. As the drum cam revolves in the direction indicated by the arrow, with the parts of the mechanism in the relative positions shown,

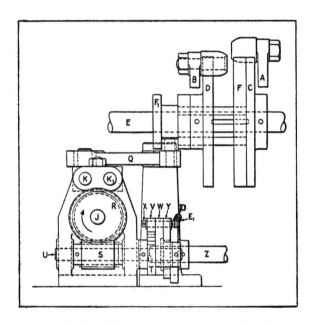

Fig. 16. End View of Mechanism Shown in Fig. 15.

the follower will be forced to move to the left. This rotation will prevent block M from moving to the left and entering the notch in the top of disk H. Thus, compression spring N will become loaded during a half revolution of the drum cam.

At this point, the second notch, shown at the bottom of disk H, becomes aligned with the block. Spring N then unloads, moving the block through the notch and thus rotating bellcrank Q, one arm of which is pinned to the upper portion of the block. The other arm of this bellcrank is provided with a follower located between the flanges of spool F_1. The spool is connected to plate F by three rods, which run through cam D. Thus, the rotation of the bellcrank causes plate F to be shifted from one cam to the other.

Further rotation of the drum cam brings block M into contact with the opposite side of disk H, thus loading spring P. When the block reaches alignment with the notch shown at the top of the disk, it will pass through the disk, thus turning bellcrank Q in the opposite direction and again shifting plate F.

The purpose of the gearing at the right-hand end of shaft J is to turn this shaft one revolution for every thousand revolutions of shaft E. One revolution of shaft J, as previously explained, imparts two shifts to plate F, thus resulting in the desired packing of five hundred sheets in each tray.

Worm-wheel R, which is pinned to shaft J, meshes with worm S. This worm and ratchet T are pinned to shaft U. The ratchet is driven by pawl V through arm W, which turns freely on shaft U. In the outer end of arm W is securely fixed a pin X, which serves as a pivot for the pawl and cam-follower yoke Y.

Cam C_1 is secured to shaft Z, which rotates at the same speed as shaft E. Roller D_1, which is held in contact with this cam by spring E_1, is fastened to yoke Y. One tooth of ratchet T is thus indexed per revolution of shaft Z or E. The number of teeth on ratchet T and worm-wheel R is such that shaft J will make one complete revolution for every thousand revolutions of shaft Z or E.

Hand Control for Reciprocating Slide.—To facilitate setting up one type of wire crimping machine, the reciprocating slide that feeds the wire stock to the machine is equipped with a hand-lever which, when shifted, not only stops the movement of the slide at any point in its stroke, but

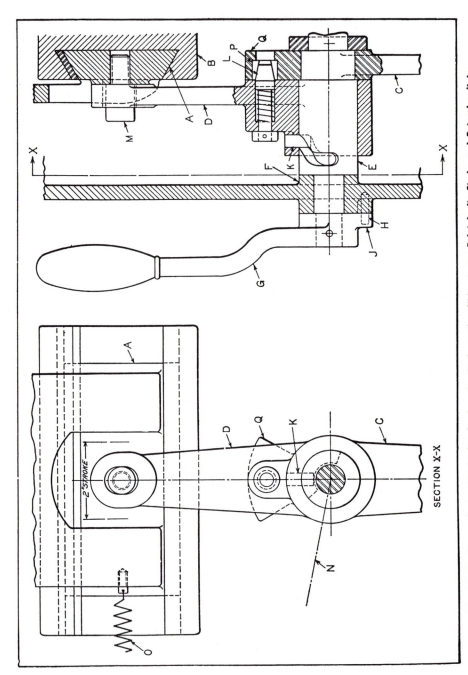

Fig. 17. Hand-lever Control Mechanism for Stopping Reciprocating Slide at any Point in its Stroke and Returning it to the Starting Position.

SECTION X-X

instantly returns the slide to its starting position. The arrangement for obtaining this result is shown in Fig. 17.

The feed-slide is indicated at A. It operates in the dovetail guide B, which is cast integral with the machine base. This slide receives its motion from the oscillating lever C through arm D, connected to the slide by stud M. Both arm D and lever C are fulcrumed on the shaft E. A bearing F on the machine casing supports one end of this shaft, and the other end is supported in another bearing (not shown) at the right. On the hub of hand-lever G, pinned to the end of the shaft, is cast a lug J, located between two pins H in the casing. With this arrangement, the angular movement of lever G is limited to about 80 degrees.

There is a cam slot in the body of shaft E which is engaged by a pin K in the hub of arm D. Also, in the hub of this arm is a spring-actuated plunger L, which locks the arm to lever C, causing these members to oscillate together.

However, if the operator wishes to stop the slide at, say, the center of its stroke toward the right and cause it to return to its starting position at the left, the hand-lever G, which at this time is in a vertical position, is swung toward the left until it coincides with the center line N. As this lever is shifted, shaft E turns with it, causing the cam slot to carry pin K with arm D toward the left (see right-hand view) until plunger L is entirely disengaged from lever C. At this point, coil spring O carries the slide back to its starting position at the left. Arm D remains connected to the slide, as pin M is long enough to compensate for the axial movement of the arm.

Now, if the operator wishes to start the slide again, the hand-lever is shifted back to its vertical position. This causes the cam groove to force arm D back into the position shown. Plunger L is held flush with the arm hub by the segment Q on lever C until the bushing and plunger are in alignment. The plunger then enters the bushing and locks the arm and lever, causing the slide to reciprocate.

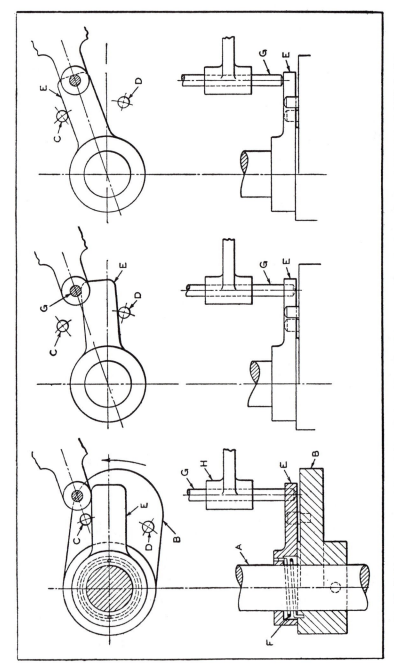

Fig. 18. (Left) Members B and E In Rotating Position With Rod G Lowered to Stop Rotation. (Center) Members B and E in Stationary Position. (Right) Release Rod G Raised to Permit Rotation of Members B and E.

It is important to note that the section of the cam groove engaging the pin K when the lever is in its vertical position, is at right angles to the axis of the shaft. This is essential, since at this time, the pin oscillates in the groove, and a helical path, however slight, would impart an undesirable axial movement to the arm for each cycle of lever C.

Stop with Safety Catch for Controlling Rotation of Shaft.—A machine that was developed for a special process is equipped with a device for feeding certain ingredients. This feeding device is controlled by a mechanism designed to proportion the amount or quantity of the ingredient fed, one complete revolution of the mechanism serving to feed the correct amount. If the supply of the ingredient has been exhausted or there is an insufficient quantity available in the bin, the mechanism will cease to function.

The feeding device (not shown) is attached to shaft A, Fig. 18, left-hand diagram. It is permitted to make one revolution only at a time. When the mechanism is once released for rotation, however, an arrangement is provided whereby the system is held open for operation at any time, regardless of whether or not it is prepared for rotation.

Segment B of the device is fastened to shaft A. The segment is provided with two pins C and D. The arm E is located between the pins and is attached to segment B by means of a small coil spring F. Spring F is arranged to force arm E against pin C. The release rod G slides vertically in the bearing H. The direction in which segment B rotates is indicated by the arrow.

The positions of the elements when the device is rotating are shown in Fig. 18, left-hand diagram. It will be noted that arm E rests against pin C under the action of spring F. Releasing rod G is in position to come in contact with arm E as rotation continues. When the device is stationary, the elements are positioned as shown in Fig. 18, center diagram, with arm E resting against pin D. In this case, arm E is forced away from pin C by the action of rod *G*, combined with the tendency of segment *B* to continue rotation.

Fig. 18, right-hand diagram, shows the positions of the various elements at the moment when release rod *G* is raised and the device is free to rotate. It will be noted that arm *E* has moved away from pin *D* under the action of spring *F* and is now resting against pin *C*. Any attempt to stop the rotation is now prevented by arm *E*, which is located under releasing rod *G*. Thus rotation of segment *B* and arm *E* will be certain to take place as soon as shaft *A* begins to rotate. The operator cannot stop the rotation or prevent its taking place unless he interferes with the action of arm *E*.

The operator can tell at a glance whether or not any of the ingredient has been fed into the machine by noting the position of the elements of the device. For instance, if they are set in the positions indicated in Fig. 18, center diagram, he knows immediately that rotation took place after he raised rod *G* because the device is now locked. If the device is as shown in Fig. 18, right-hand diagram, he is certain that rotation has not taken place and that no ingredient has been fed into the machine.

CHAPTER 15

Automatic Work Feeding and Transfer Mechanisms

This chapter deals with the automatic delivery of work-pieces in the proper position for the operation to be performed on them. The mechanisms described include an automatic magazine-feed attachment for a centerless grinder; an automatic hopper feed for small cylindrical parts; an arrangement for inserting and heading tubular rivets; devices for filling containers and applying covers; an intermittent feeding mechanism designed to operate two slides from one cam; a rapid-motion, short-stroke wire-feeding mechanism; a novel intermittent feeding mechanism; and several transfer mechanisms.

Other automatic feeding mechanisms are described in Chapter 16, Vol. I and Chapter 14, Vol. II of Ingenious Mechanisms.

Automatic Magazine-Feed Attachment for Centerless Grinder.—The automatic magazine-feed attachment shown in Fig. 1 was designed for grinding the part or component shown in the detail view at *W*. This part, because of the shoulder, must be fed downward in the correct lateral position for the shoulders to pass into the clearance groove provided in both wheels. To accomplish this, a mechanical attachment is necessary. The illustration shows the attachment in position on the grinding machine, mounted on the side of the wheel-truing device. It can be easily adjusted to suit the position of the grinding and control wheels, both longitudinally and crosswise. The vertical position is permanent, and needs no further adjustment after the initial setting.

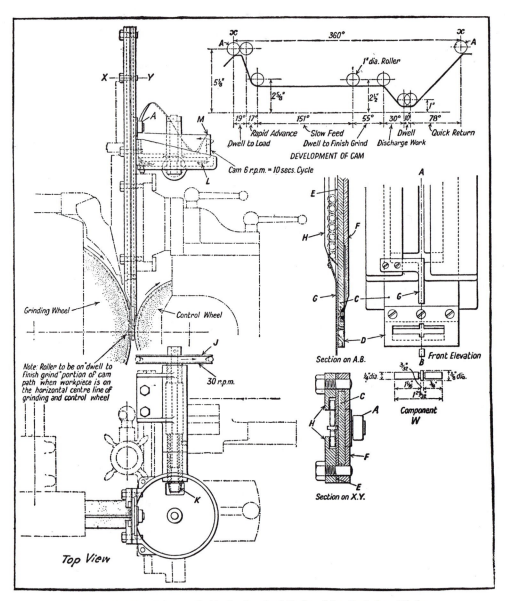

Fig. 1. Details of Automatic Mechanism Feed Designed for Centerless Grinder.

The section on the center line A—B shows an enlarged partial view of the sliding member C, the work-carrier D, which is made of hardened steel; the spacer plates E; the sliding member guide F; the spring finger G, which keeps the work-piece from falling forward when approaching the grinding wheel; and the strips H, which form the magazine into which the work-pieces are loaded.

The section X—Y shows an enlarged plan view of the magazine and its component parts. A piece to be ground is shown by dot-and-dash lines in the feeding position. The enlarged partial view of the front elevation (section A—B) shows the hardened-steel work-holder or carrier D which is attached to the sliding member C and the slot where the work-piece is retained and guided. Also shown is the position of the spring finger G in relation to the work-holder. The action of the whole mechanism is more readily understood if the cam development diagram at the top of the illustration is carefully inspected. This diagram shows clearly all positions of the sliding member C. It is necessary that the weight of the sliding member be sufficient to keep roller A in contact with the face of cam M.

Pulley J is driven by a belt running from the control-wheel shaft. The bevel pinion K is integral with the pulley shaft and engages bevel gear ring L attached to the under side of cam M, thus causing it to rotate in the direction and at the speed indicated.

The operator drops the work into the magazine at the open end; and when the sliding member C is in the top position, the slot in the work-holder D coincides with the curved angular path at the bottom end of the magazine strip H. The piece at the bottom of the magazine then rolls forward into holder D, which is carried downward on the rapid advance portion of the stroke actuated by cam M until it reaches the grinding position, where it is finished to size in the predetermined time period indicated on the cam lay-out.

Proceeding from the "dwell to finish-grind" portion on

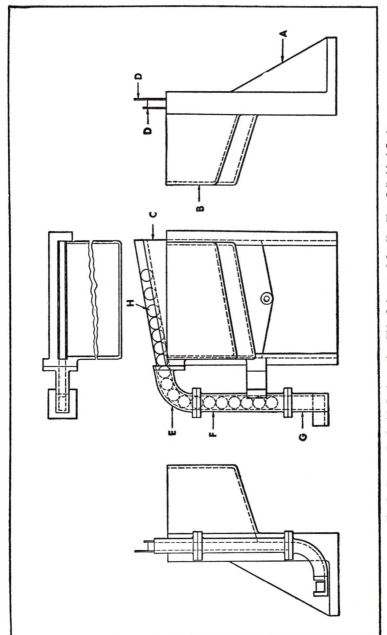

Fig. 2. Hopper Feed with Reciprocating Slide C Arranged for Handling Cylindrical Parts.

cam M, the sliding member C passes to the lowest position in its travel. During this period, the work is carried clear of the grinding and control wheels, and is free to drop out at the control wheel side of the work-holder into a receiving box. Sliding member C is now returned to the highest point of the cam path, which completes the time cycle per piece.

Automatic Hopper Feeds for Small Cylindrical Parts.— Rings or cylindrical parts can be automatically fed from a hopper by means of the mechanism shown in Fig. 2. The hopper shown in Fig. 2 is used to feed cylindrical pieces 1 1/4 inches in diameter by 1 inch high at the rate of forty per minute. This hopper consists essentially of a base A; a bowl B, in which the parts are placed; and a vertical reciprocating slide C. Slide C has two steel strips D attached to its sides. The strip nearest the base of the hopper is made higher than the one adjacent to the bowl. The slide has a vertical reciprocating movement of five strokes per minute. At the bottom of the stroke which aligns the slide with the bottom of the bowl, the work-pieces fall into the groove formed by strips D on slide C.

Slide C may be reciprocated vertically by means of a crank and slotted cross-head, or a Scotch yoke type of mechanism, which will permit the slide to dwell slightly at the lower and upper positions of its stroke, thus allowing the parts to fall into or roll out of the slide. When the slide reaches the upper end of the stroke, the pieces roll into the covered chute E, down the tube F, and into elbow G. The elbow changes the path of the pieces 90 degrees, and delivers each piece resting on a flat face. The weight of the parts following keeps the leading part against a stop. This part is fed by means of a pusher mechanism similar to the one shown in Fig. 3.

The hopper shown in Fig. 3, which is used for feeding brass rings, is similar to the one just described. However, the work-pieces are fed vertically instead of horizontally.

Details of the feeding mechanism are shown in the illustration. When slide C reaches the upper end of the stroke, the parts roll into the vertical covered chute J. As the slide moves down, roll K comes in contact with lever L. This action causes link M to move plunger P up, pushing one or more parts above the spring S. The spring supports the column of rings, which is advanced by each stroke of the slide.

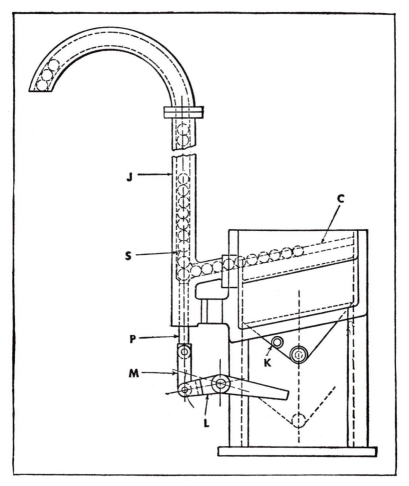

Fig. 3. Arrangement Designed for Feeding Rings Vertically from a Hopper.

Mechanism for Inserting and Heading Tubular Rivets.–
The operating mechanism of the Chobert magazine riveting
machine employed in a British bomber manufacturing plant
is shown in Fig. 4. The sectional views at X show the successive stages in the riveting operation. The machine employs a pull-rod with a dolly or conical expanding tool F
at one end, the diameter of which is slightly larger than
that of the main bore through the tubular rivet. As the
dolly is pulled through the rivet, it forms a head at the
inner end and expands the shank so that it is a tight fit
in the hole.

The machine has a hand-grip at A and a housing B for
the crank-operated gear and cam mechanism whereby a
sliding movement is imparted to sleeve C. Shroud D is
attached to the end of this sleeve and moves with it to
operate the jaws of chuck E.

The pull-rod or dolly F extends through the barrel of
the machine and is held fast by jaws at G. Pressure on the
jaws is applied by screw H, the arrangement permitting the
rod to be removed when it is necessary to load the rivets,
which are slipped over or threaded on the rod with the
flanged ends toward the rear of the machine, as shown. A
shouldered check-spring K is next threaded on the rod, the
small end of which is inserted in the barrel, passed through
the center of feed unit L, and gripped by jaws G. With
chuck E closed, the first rivet is exposed with its flange
seated against the chuck end.

The first rivet is inserted in one of the drilled holes in
the gusset and tubular member, the dolly and end portion
of the rivet projecting into the tube. Three turns of crank
M are required to complete the riveting operation and reset
the machine.

As mentioned, the cam mechanism in housing B imparts
a forward movement to sleeve C, and with it, to shroud D
and chuck E, the rod and rivets remaining stationary. Any
movement of crank M when the rivet and chuck are pressed

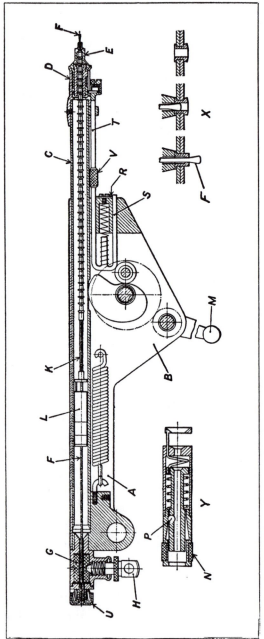

Fig. 4. Cross-Section of Chobert Riveting Machine Showing Essential Features of Operating Mechanism.

against the work will cause the body of the machine, and with it the rod, to retire. Thus, the dolly is withdrawn, completing the riveting operation as explained. This occurs during the first turn of the crank, after which the riveting machine is withdrawn slightly from the work.

During the second and third turns of the crank, the chuck jaws advance, permitting the passage of the next rivet, and close behind it. The machine is then ready for operation again. When operated by an experienced worker, the machine will close 1200 rivets per hour.

The feed unit of the machine is shown to a larger scale in the cross-section view at Y. A split ring N is a good sliding fit in sleeve C. Ball P rides to the large end of the taper bore shown when rod F is in position and is moved to the left during the riveting stroke. On the return or loading stroke, ball P is forced down the taper so that it grips rod F, the movement being assisted by the coil spring, and the entire unit is carried forward through a distance corresponding to the length of one rivet.

It is necessary to reload the rod when the magazine is empty, and to do this, jaws G are released by loosening screw H. Crank M is then given 2 3/4 turns to open the chuck jaws at E. With the jaws open, stop R is moved to engage a slot in the portion S of control rod T. Thus, the jaws are locked in the open position to permit the withdrawal of the empty rod and the insertion of another rod loaded with rivets. With the loaded rod in position, stop R is released, allowing the jaws to close. There should be a clearance of from 0.6 to 1 millimeter between the head of the exposed rivet and the end of the chuck jaws; if this is not the case, adjustment can be effected through the medium of cap-screw U, the threaded portion of which engages the end of rod F.

Mechanisms Designed for Filling Containers and Applying Covers.—The mechanisms here illustrated are employed in an automatic machine that places twelve pencil leads in a box and pushes the box cap in place. The box, or container, is molded by the injection process, the dimensions being indicated at the left in Fig. 5. The selecting mechanism, shown at the right in Fig. 5, takes the boxes from a hopper and presents them to the machine open end forward. The revolving disk carries hooks, so spaced as to allow a single box to be fed by each. If a box is in the wrong position, the hook enters its open end and retains it until the rubber-faced rotating roller *A* returns it to the

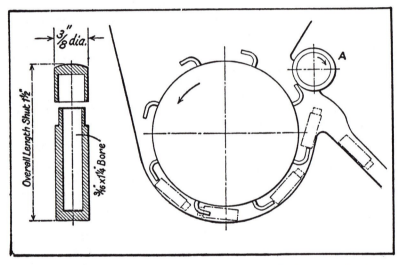

Fig. 5. (Left) Molded Box for Holding Pencil Leads (Right) Mechanism for Feeding Boxes Open End Forward.

hopper. Boxes fed correctly tip over into the delivery feed and slide toward the loading machine.

The continuous belt conveyor, shown in Fig. 6, has successive stations to which boxes, leads, and caps are brought in correct synchronism and quantities. Falling from the selector shown at the right in Fig. 5, the boxes drop into a single-column, vertical stack, the lower end of which is open at the sides. This opening faces one side of the belt conveyor, Fig. 6, which carries properly spaced finger clips at regular intervals. An intermittent feed of the conveyor belt is obtained by means of a Geneva mechanism. The driver of the Geneva motion carries a face-cam that is set to impart a sharp blow to the lowest box in the stack during the stationary period of the belt, propelling the box into the spring clip on the belt. The belt then transports the box to the lead-filling station.

The lead-filling station, shown in Fig. 7, includes a counting and feeding mechanism. A circular shutter valve is

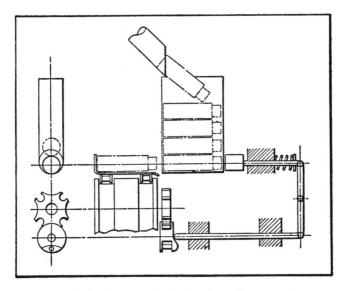

Fig. 6. Arrangement for Feeding Boxes Intermittently
on Belt Conveyor.

employed to take twelve leads from the hopper. This valve contains only two rows between its shutters, which, in this case, are open-ended hollow cylinders provided with suitable ports. Only a sector of the hollow cylinders is required for the counting operation, but complete cylinders are employed for the sake of simplicity and rigidity of construction. These cylinders are oscillated by the lever arm, the lower shutter opening as the upper one closes. No overlap is shown in the illustration, but a certain amount of overlap is necessary, and this is provided by increasing the stroke of the operating lever.

From the valve, a dozen leads fall into the guide chamber, which is bored rather smaller in diameter than the boxes to be fed. The boxes are carried along the belt until they register accurately with the guide chamber. When in this position, a constant accelerating drum type cam moves a plunger forward and backward, feeding a supply of leads into the box without shock.

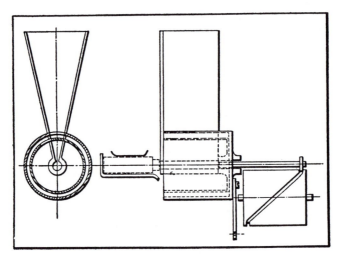

Fig. 7. Mechanism Employed to Select Correct Number
of Pencil Leads and Feed them into Box.

The belt is next traversed to bring the box to the cap or cover station, where the cover is fed into position by a selector similar to the one used for the boxes. The "hand" of the selector must, however, be reversed, as the open end of the cover must be presented in a position opposite to that in which the boxes are fed to the conveyor belt. This station is shown in Fig. 9. The caps are fed down a tube, rolling out though an open incline into the single stack container which has open sides. The feeding mechanism is similar to the one used for the boxes, although a more gradually accelerated thrust is applied by the face-cam. The conveyor chain belt is kept at the proper tension to insure accurate register at the different stations. The belt passes a deflector which ejects the filled boxes.

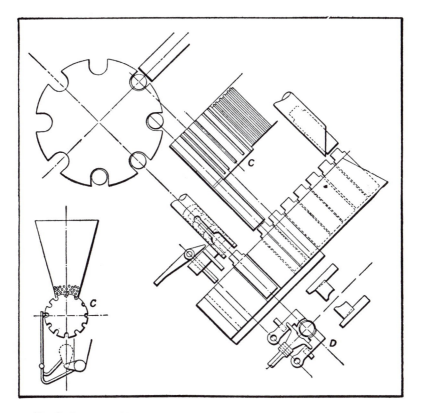

Fig. 8. Drum-type Mechanism Designed for Loading Pencil Leads into Boxes.

Before describing the alternative loading mechanism shown in Fig. 8, attention is called to a feature, without which much loss of time might be experienced in the high-speed automatic operation of the mechanisms described; namely, evidence that the feeding mechanism has functioned properly in placing the required number of leads in the container. The simplest way to detect empty boxes is by weight; loaded holders or boxes, on rolling down a flexible shelf, fall sooner than the empty ones. A dividing line, correctly positioned, is therefore used to separate the filled and unfilled boxes, thus providing a simple means for this inspection.

In the alternative mechanism, shown in Fig. 8, the belt is replaced by an inclined revolving drum. The selected boxes fall into a single-column chute. When a slot in the drum reaches a position opposite the chute, one box drops

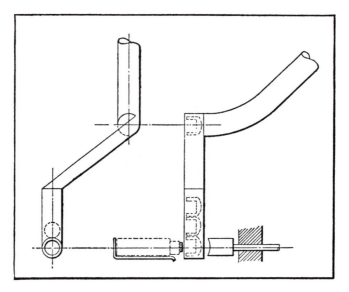

Fig. 9. Mechanism for Placing Covers on Filled Boxes.

into place and is carried around to the lead-filling station. The leads are also counted out by a revolving slotted drum C, two of the slots of which are extended beyond the others to operate the shutter movement that closes the inclined trough into which the leads fall from the drum.

At the capping station, selected caps slide down the supply tube, which has a spring arrangement designed to catch the lowest cap just before it reaches the exit. At the correct instant, the capping lever is rocked about its bearing, closing two jaws on the sides of the lowest cap while carrying it forward. A plan view of this mechanism is shown at *D* below the drum, cam-slides being provided to close the jaws during the forward movement. As the main drum revolves to the next position, the loaded boxes fall from their slots and again pass a mechanism that is arranged to detect unfilled boxes in the manner previously described.

Mechanism for Feeding Wooden Pegs into Magazine, Large Ends Foremost.—The purpose of the device shown in Fig. 10 is to feed round wooden pegs *A* into the magazine *B* with their large ends downward to facilitate assembling them in the product. The wooden pegs are 1 1/4 inches long and 1/4 inch in diameter for a length of 1 inch, and 3/16 inch in diameter for a length of 1/4 inch. These pegs are fed along a chute *C* into the selector cavity *D*, as shown in the plan view. A constant pressure is maintained against the line of pegs in chute *C*, so that when the first peg in the line is removed another moves forward to replace it.

The plate *E* is moved backward and forward, as indicated by the arrows at *F*, being actuated by means of a cam and return spring, not shown. Plate *E* carries a block *G* and has a rectangular slot or hole *H* cut through it which opens into the chute leading to the magazine *B*. Block *G* pushes the first peg in line to the opposite side of the cavity, and at the same time, temporarily prevents any other pegs from entering the cavity. The rectangular slot serves to open and close the opening in the floor of the cavity above the magazine chute. It will be noted that the slot in the slide is slightly wider than the 1/4-inch diameter of the wooden peg and also that it is shorter than the peg, being about 1 1/16 inches long.

In the right-hand wall of the selector cavity *D* is a notch *J*, as may be seen in the plan view. This notch is slightly wider than the 3/16-inch diameter of the peg but not as wide as the 1/4-inch diameter and is somewhat deeper than the 1/4-inch shoulder on the small end of the peg. Opposite this notch and on the same center line is a similar notch *K*, cut in the end of a movable block *L*, which is made to extend into and be withdrawn from the cavity of the selector by means of a suitable cam and return spring, not shown. The motion of this block is timed relative to that of the plate *E* in such a manner as to produce the operating cycle to be described.

Let it be assumed that all of the parts of the selector are in the position indicated by heavy lines in the illustration, and that the first peg has entered the cavity with its small end pointing backward along the chute *C*. The block *L* is next withdrawn and the slot *H* opens the floor of the cavity into the upper part of the chute leading to the magazine. Now the plate *E* moves forward, carrying the block *G* to the

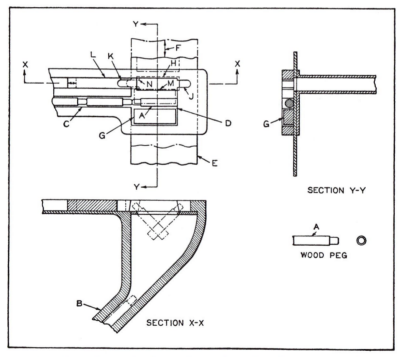

Fig. 10. Mechanism for Feeding Pegs A from Chute C to Mechanism B with Large Ends Foremost.

position indicated by the dot-and-dash lines at *M*. This pushes the first peg across and in line with the two notches *J* and *K* and, at the same time, cuts off the surplus flow of pegs into the cavity.

Next the block *L* moves to the right a distance of 3/16 inch, as indicated by the dot-and-dash lines at *N*, and the notch *K* engages the small end of the peg. The plate *E* then moves back to its original position, the peg being prevented from following the block by the confining notch. As soon as the slot *H* returns to its position above the magazine chute *B*, the peg, being unsupported at all points except under the small end, drops downward, the large end falling into the chute first so that the peg slides into the magazine with the 1/4-inch end to the left. The block *L* then returns to its original position.

The second peg, which enters the cavity small end first, will be considered next. The block *G* also pushes this peg across to the position between the notches, but when block *L* moves into the cavity this time, as the notch *K* is too small to engage the 1/4-inch diameter, the peg is pushed to the right, causing the small end to enter the notch *J*. When the slot *H* returns, it again leaves the peg unsupported at all points except under the small end, so that the large end falls downward and the peg slides along the chute into the magazine with its large end to the left.

This cycle of cam-actuated movements of slides *E* and *L* continuously feeds the pegs into the magazine *B* with their large ends downward. The magazine on which this selector is used has a capacity for handling about fifty units per minute. At this speed, the feeding device functions perfectly, and there appears to be no reason why it could not be operated at a much higher speed.

Intermittent Feeding Mechanism Designed to Operate Two Slides from One Cam.—The mechanism shown in Figs. 11 and 12 was designed to feed a continuous strip of corrugated flat wire stock *W* through a machine for further

fabrication. The rapid, positive, intermittent feeding movement required is obtained through the operation of feed-
slide *A* and work-gripping slide *B* by a single cam *C*. It is
interesting to note that both slides *A* and *B* are operated
simultaneously by cam *C* and that slide *B* is mounted in a
dovetail groove in slide *A*.

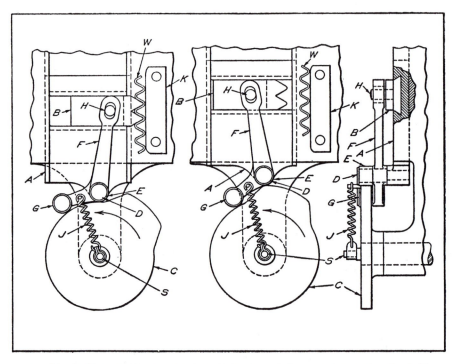

Fig. 11. (Left) Mechanism for Intermittent Feeding of Corrugated Flat Wire Stock W
at End of Dwell Period in which Stock is Held Clamped against Block K by Slide B.
(Center) Slide B Withdrawn from Work W and Slide A Moved One Corrugation to
Rear through Action of Cam C. (Right) Side View of Feeding Mechanism.

Referring to Fig. 11, left-hand diagram, shaft *S*, carrying
cam *C*, rotates in the direction indicated by the arrow. Stud
D on slide *A* carries the cam roller *E* and the bellcrank lever
F, which is free to oscillate. Lever *F* carries on its short
arm the roller *G*, which is free to rotate on its stud. Pin *H*
is fixed in slide *B*, and passes through the slot in the upper

end of lever *F*. Spring *J* serves to hold rollers *G* and *E* in contact with cam *C*. Work *W* is passed through the machine in contact with the guide strip indicated at *K*.

In Fig. 11, left-hand diagram, both rollers *G* and *E* are shown in contact with the low portion of cam *C*, which holds the slides in a fixed position during the rest period of the

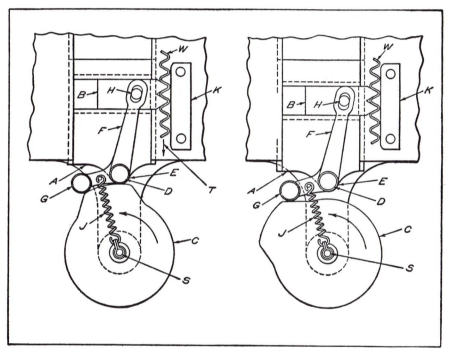

Fig. 12. (Left) Slide B Advanced to Engage Work W Preparatory to Feeding it Forward in Direction Indicated by Arrow T. (Right) Mechanism Engaged in Feeding Work W Forward with Cam C Approaching Dwell Position.

cycle. It is during this rest period that the fabricating operations are performed on the work. The formed end of slide *B* engages the corrugations in the work *W* as shown during this dwell period, holding it in contact with guide strip *K*.

Fig. 11, center diagram, shows the rise or lobe of cam *C* engaging roller *E* and causing slide *A* to move to the rear. Roller *G* has not yet been affected by the action of cam *C*, but the position of lever F has been changed so that, in addition to the rearward movement of slide A, the action of lever F on pin H has caused slide B to be withdrawn from contact with work W. Since the rearward movement of slide A is equal to the pitch or center-to-center distance of the corrugations of the work, no rearward movement is transmitted to work W. The slide B, however, has moved to the rear with slide A, and its forward end is in position to engage another set of corrugations in work W.

Continued rotation of cam C, as shown in Fig. 12, left-hand diagram, causes roller G to rise to the high portion of cam C. As roller E is in the same relative position as shown in Fig. 11, center diagram, there has been no movement of slide A. The change in the position of roller G, however, has caused lever F to pivot on stud D, actuating slide B and causing its forward end to again engage the corrugations in work W. Comparison of the left-hand diagram in Fig. 12 with that in Fig. 11 shows clearly the changes in the position of slide B relative to work W.

Fig. 12, right-hand diagram, shows the mechanism after cam C has passed from under roller E. The pressure of slide B on work W against guide strip K at this stage of the operating cycle prevents any movement of lever F and also prevents roller E from following the contour of cam C. In the position shown in Fig. 12, right-hand diagram, roller G still rests on the "drop" side of the cam lobe. The action of spring J serves to hold roller G in contact with cam C, causing slide A to be drawn forward. This action, in turn, causes slide B to feed work W forward. The forward motion of slide A continues until both rollers G and E come in contact with the low portion of the lobe on cam C, as shown in Fig. 11, left-hand diagram. Until the lobe of cam C again comes in contact with roller E, work W is held firmly in position. With this mechanism, work W is given a rapid, short, intermittent feed with a long rest period between the feeding movements.

Rapid-Motion, Short-Stroke Wire-Feeding Mechanism.— The mechanism shown in Fig. 13 is designed for feeding a strand of wire rapidly through a wire fabricating machine in a series of short strokes, with equal rest periods between strokes. Owing to the high speed of the machine, it was necessary that the mechanism provide a reciprocating move-

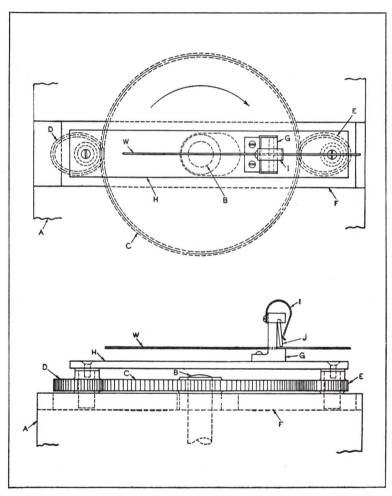

Fig. 13. (Top) Wire Feeding Mechanism Designed for Rapid Short Stroke Motion. (Bottom) End View of Mechanism in which Slide F is at End of Feeding Movement to Right.

ment with a minimum of vibration, and that a positive gripping arrangement be used for feeding the wire.

Referring to Fig. 13, upper diagram, the shaft B rotates in the stationary part of the machine A, in the direction indicated by the arrow, and carries the gear C that is keyed to it. The slide F is carried in the part A, and is slotted, as shown, to clear the hub of gear C. Slide F carries two studs on which the elliptical gears D and E revolve freely, receiving their motion from the gear C. In this view, the teeth on

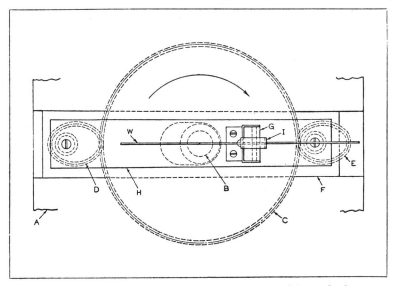

Fig. 14. Wire Feeding Mechanism with Slide F at End of Return Stroke.

the shorter side of the major axis of gear D are in mesh with the teeth of gear C, while the teeth on the longer side of the major axis of gear E mesh with the teeth on the opposite side of gear C. In this position, the center-to-center distance between the axes of gears C and E is greater than that between gears C and D. As the studs that carry the gears D and E are fixed in slide F, the latter is thrown off center, to the right, with relation to shaft B, the slide F being at its extreme right-hand position. The bar H carries the gripper mechanism G, which feeds wire W, as will be explained.

In Fig. 14, gear C has made a partial revolution, resulting in a half revolution of gears D and E and reversing the condition shown in Fig. 13; at this point the slide F has been moved to its extreme left-hand position, the distance traveled by the slide being equal to the eccentricity of gears D and E.

The details of the gripper mechanism are shown in Fig. 13, lower diagram, which is an end view. The bar H is carried on the studs that carry gears D and E, and therefore moves with slide F. The wire W passes through the part G, which is mounted on bar H. The wire W rests on the ledge of part G, against which it is held by the wedging action of plate J, the flat spring I acting on plate J to insure a positive wedging action. As the bar H moves to the right, the plate J wedges between its seat in part G and the wire W, gripping the latter firmly, and feeding it into the machine. As the bar H moves to the left, the wedging action is destroyed and plate J slides over the wire W, without transmitting any motion.

Novel Intermittent Feeding Mechanism.—The mechanism shown in Fig. 15 is designed to intermittently index or feed a block chain or film E longitudinally from left to right. Each feeding movement advances the chain or film a distance L. The indexing is accomplished by a pin D which enters equally spaced holes or slots H in the chain, rising vertically into one of the slots and moving to the right until the chain has been moved a distance L. The pin then moves downward out of contact with the chain. With the simplest arrangement of this mechanism, using only one arm B, and without the block-rotating device M described later, the chain is fed a distance L for each revolution of shaft S, the feeding movement taking place during about one-fourth revolution of shaft S.

Pin D is carried in a square block C, which has a pin G that guides the movement of block C in a rectangular path determined by cam groove J. Pin G also enters a slot in the driving crank B attached to rotating shaft S. Block C is further guided in its rectangular path by rim K on the body A of the mechanism, which also has the cam groove J. Rim K prevents block C from turning about pin G, so that pin D will rise vertically into the slot in chain E for the feeding movement.

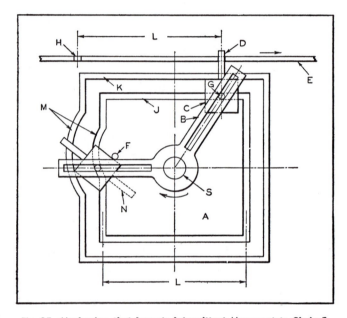

Fig. 15. Mechanism that Imparts Intermittent Movement to Chain E.

By modifying the forms of groove J and rim K as shown at M, and providing a stationary pin F in body A, it is possible to cause block C to rotate through an angle of 90 degrees. With this arrangement and only one block C and one driving crank B, pin D would be brought into the indexing position every fourth revolution of shaft S. By adding another pin N to block C, the chain would be indexed every second revolution of shaft S. Irregularity of the timing of the indexing movements or a variation in the number of indexing movements per revolution of shaft S can be obtained by employing more than one slotted crank B.

A variety of indexing movements can be obtained by varying the design. For example, block C can be fitted with one, two, three, or four pins D, and block-rotating arrangements M can be provided on the bottom and right-hand side of the mechanism.

Feeding Mechanism for Box-Nailing Machine.—The purpose of the device shown in Fig. 16 is to feed wooden boards, two at a time, in a vertical position into a box-nailing machine. The main feature of this feeding mechanism is its positive action in turning the boards from the flat position, in which they are stacked in the hoppers, to the upright position, in which they must be fed to the nailing machine.

The boards, which are 7 inches wide by 9 inches long and from 3/8 to 1/2 inch thick, are fed simultaneously in pairs at the rate of thirty-six pairs a minute. They are first stacked up by hand in the two opposite hoppers A, the two lowest boards being simultaneously carried into intermittently revolving cross-shaped members C by means of lateral feeding mechanisms or "kickers" B. These kickers are operated by the oscillating shaft D through levers E.

Members C, each of which is built up of four L-shaped plates so arranged as to form two passages at right angles to each other, revolve in troughs F, which have apertures G at the bottom for discharging the boards into the chutes H. End pivot shafts I of members C are journaled in an end plate J, and the opposite pivot shafts are journaled in a similar end plate at the opposite end of troughs F. Spur gears K are mounted on shafts I and mesh with the central spur gear L, the latter being attached to the four-notched wheel M, which freely rotates on pivot stud N. With this gearing, by turning wheel M, members C are made to rotate simultaneously.

Wheel M is operated as an ordinary ratchet wheel by pawl P, pivoted at the upper end of rocker Q, which, in

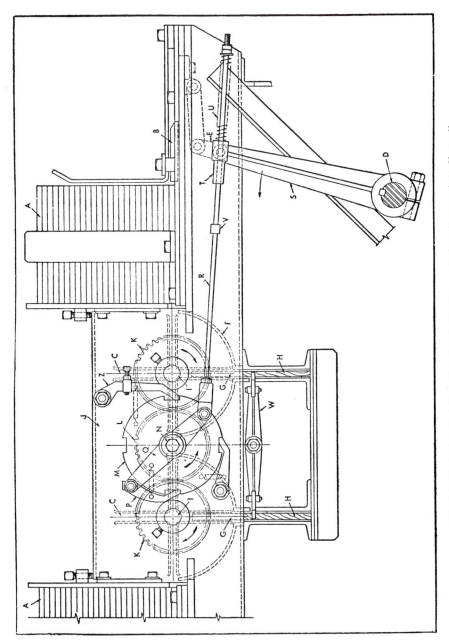

Fig. 16. Mechanism that Feeds Two Boards Simultaneously from Hoppers A to Chutes H.

turn, oscillates on pivot N under the action of the connecting-rod R attached to the lower end of rocker Q. The rod R is attached to the end of lever S, which is mounted also on the oscillating shaft D, by means of block T, through which the connecting-rod is free to slide. A precharged spring U constrains the motion of the connecting-rod in one direction, while a stop V, rigidly attached to the connecting-rod, constrains the motion in the other direction.

As levers E and S move together in the direction shown by the arrow, the lowest wooden boards in the hoppers are introduced into the momentarily stationary members C, while block T approaches stop V, finally reaching it and carrying it forward, together with the connecting-rod. This causes rocker Q to bring pawl P to the point where it will engage the upper notch of wheel M, and at the same time lifts stationary pawl Z, thus setting M free to rotate in the direction of the arrow.

As levers E and S return together to the starting position, kickers B will withdraw from under the stacks of boards, letting them drop under their own weight to the bottom of the hoppers. During this interval, lever S, pressing against spring U, will pull the connecting-rod to the right, causing wheel M to rotate until pawl Z drops into the next notch, thus preventing wheel M from advancing beyond one-quarter of a revolution.

As wheel M stops, lever S will continue to compress spring U until the end of the return stroke. Thus, there is a dwelling period between the completion of the quarter of a revolution of wheel M and the end of the return stroke of arm S, intended to give the wooden boards ample time to drop under their own weight to the bottom of chutes H, through which they are carried to the nailing-machine table by the cross-head W, which moves in synchronism with kickers B.

Mechanism Equipped with Suction Cup for Picking up and Feeding Thin Plates.

—Three distinct phases in the sequence of movements imparted to suction cup W of the

plate-feeding mechanism shown in Fig. 17 are produced
and controlled by a single cam which transmits a sliding
movement to the rack or actuating rod R. The three phases
of the movement required to pick up plate A and carry it to
the position shown at D, as indicated, consist of a vertical
movement of suction cup W which raises plate A to the first
position clear of the pile; a circular motion which carries
the plate to the second or vertical position; and the final

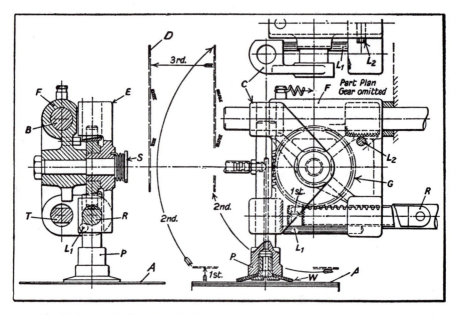

Fig. 17. Mechanism Designed to Pick Up Plate A and Feed it to the Position Indicated at D.

horizontal movement which carries the plate to the third
position at D.

The light frame F of the mechanism slides on bar B and
a flatted rod T, both of which are firmly fixed in the machine body. The triangular-ribbed frame F has its upper
portion formed into a long boss. A central boss carries a
horizontal shaft S on which a carrier C is so mounted that
it can rotate freely through an angle of 90 degrees. The

maximum counter-clockwise position of the carrier, as shown by the full lines, is determined by a limit pin or stop L_1. A second stop-pin L_2 restricts the maximum clockwise rotation.

A gear G having a pitch diameter of 3 1/2 inches, freely mounted on the hub of carrier C, engages rack teeth on a hollow plunger P and the actuating rod R. Plunger P is provided with a suction cup W at the lower end. The required suction is maintained in cup W by connection with an exhaust pump and valve system.

A spiral spring, exerting a force of 2 to 3 pounds at the rack on rod R, tends to rotate carrier C in a counter-clockwise direction. Another tension spring, exerting a force of 4 to 5 pounds, tends to move slider frame F to the right. The weight of plunger P, which is approximately 1 pound, tends to push this member downward.

The relative strength of the springs and the weight equivalent of the plunger are important, as the movements of the several elements depend on the fact that these springs must collapse in a predetermined order. In operation, horizontal rack-rod R moves to the left to rotate gear G and raise plunger P, with its plate, until the shoulder on the plunger strikes the lower boss face of carrier C, actuating rod R being in a position that will not interfere with carrier C.

Continued movement of rod R to the left rotates gear G and carrier C through a quarter-turn against the action of a spiral spring until it is stopped by limit pin L_2. Any further movement of rod R now causes the slide assembly to move to the left against the pull of the spring. A reduction in the height of the pile of plates is automatically compensated for by a longer return movement of rod R by its actuating spring, which causes the plunger to fall to the pick-up position, regardless of the height of the pile of plates. Assuming that the plates are 6 1/2 inches square, the third or horizontal motion must be 2 3/4 inches to obtain adequate swing clearance.

Reversing Transfer Mechanism.–The mechanism shown in Fig. 18, is used on a wire fabricating machine for transferring a flat wire W from its original working position, shown in the upper view, to the position indicated by dotted lines at W_1 in the lower view. It will be noted that this transfer movement turns the flat wire upside down.

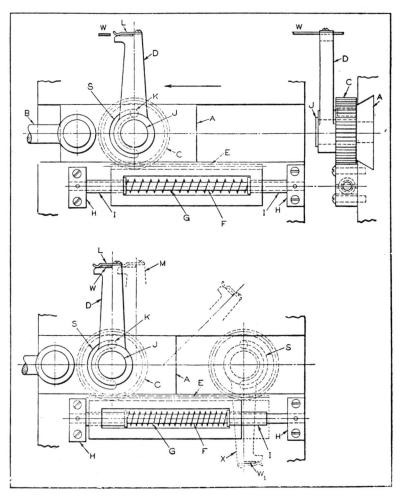

Fig. 18. (Top) Mechanism Designed to Pick Up Wire W and Transfer it to the Position Indicated by Dotted Lines at W₁ in Lower View. (Bottom) Diagram Illustrating Movement of Transfer Lever D.

Two synchronized assemblies or mechanisms like the one shown are used to support the wire at both ends.

Referring to Fig. 18, upper view, the reciprocating movement is transmitted to slide A by a cam-actuated connecting-rod B. Gear C and lever D, which are keyed together, are mounted on stud J which is fixed on slide A. Rotation of lever D is stopped when the ends of a semicircular slot S in gear C come in contact with a pin K in slide A. Gear C meshes with a rack on the upper edge of plate E, which carries the two flanged bushings I and spring G. The bushings I are free to slide in plate E on the shaft F, which is supported by the two blocks H, attached to a stationary part of the machine. The length of the bushings I is such that when in position in plate E, the length of the assembly is exactly the same as the distance between blocks H. Thus any movement of plate E in either direction is resisted by spring G, which serves to equalize the position of the plate between blocks H.

In Fig. 18, upper view, slide A and lever D are shown approaching the end of their travel to the left. It will be noted that the pin K is in contact with the end of slot S in gear C. The movement of slide A, up to this point, gives to gear C and lever D, to which it is attached, a rotative motion. Meanwhile, plate E is held stationary by spring G. In the position shown in Fig. 18, upper view, the upper end of lever D is advancing to receive the work W, which slides under the spring L. When the end of slot S comes in contact with pin K, the rotative movement of gear C is stopped and the continued movement of slide A is transmitted through gear C to slide E, which is then moved against the resistance offered by spring G. In this manner, a straight line movement of lever D is produced near the end of its travel to the left.

In Fig. 18, lower view, slide A is shown in the position it occupies at the end of its movement to the left, the work W being gripped under spring L on the end of lever D. On the return stroke of slide A, lever D returns in a straight-line movement to the position shown by the dotted line at M, which coincides with the position shown in Fig. 18, upper view. As plate E is returned to its central position, continued movement of slide A to the right causes gear C to be rotated on stud J until pin K is in contact with the opposite end of slot S in gear C, at which time lever D is in the position X indicated by dotted lines. A straight-line movement to the right is then transmitted to lever D until work W arrives at the point at which it is to be discharged. As the slot in gear C permits a rotative movement of 180 degrees, the work is given a full reversing or turning movement while being transferred from one position to the other.

Dial Transfer Mechanism for Chain Making Machine.– The dial transfer mechanism shown in Figs. 19 and 20 was developed for use in a chain making machine. The function of the mechanism consists of picking up a piece of work at M, Fig. 19, and transferring it to the position indicated at N. Since the limited space available made it impossible to employ an ordinary cam arrangement for transferring the work, it was necessary to develop the special mechanism here shown.

The problem of transferring the work from M to N, which was too long a distance to permit using a single cam movement, was solved by employing an indexing dial transfer movement, effected by a Geneva motion in combination with a comparatively short reciprocating motion obtained by a barrel cam.

Referring to Fig. 19, the transfer of a piece of work from M to N is accomplished by a combination of two indexing movements of the Geneva actuated dial C which carries the piece from M to O and a traversing movement of the dial C through a distance T by the action of the barrel cam J, Fig. 20. The latter movement carries the piece from O to N.

Assuming that dial C is indexed in a clockwise direction, the piece picked up at M will be indexed to position P, where it remains idle while the preceding piece at O is carried to

N. Then, on the second indexing movement of the dial, it is carried to position *O*, from which it is transferred by the movement of dial *C* through distance *T* to position *N* during the idle period of the dial.

The jaws *H* are spring-operated. The mechanism is driven by shaft *S* on which the barrel cam *J* is mounted. The entire

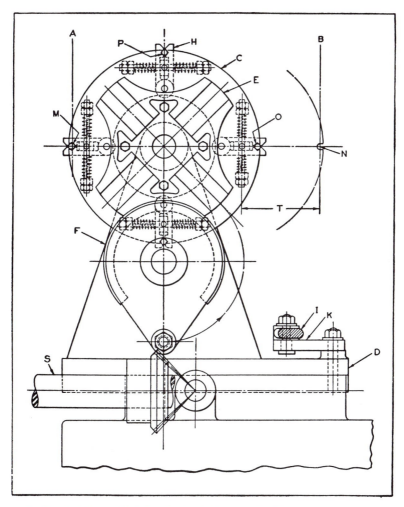

Fig. 19. End View of Dial Transfer Mechanism Used in Chain Making Machine.

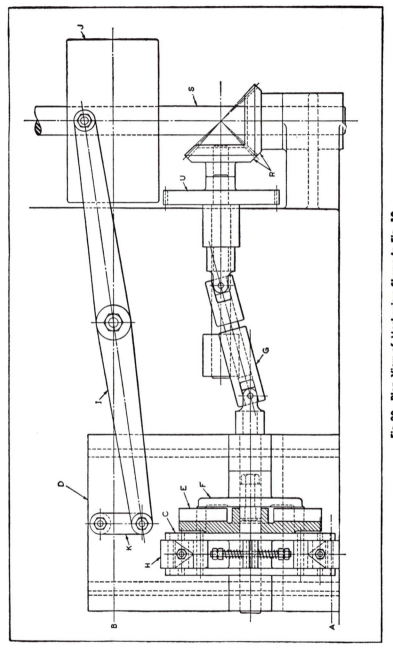

Fig 20. Plan View of Mechanism Shown in Fig. 19.

Geneva mechanism is mounted on a slide D, which is given the required intermittent reciprocating motion by barrel cam J through lever I and connecting link K.

As shown in Fig. 20, the driver F of the Geneva mechanism is driven from shaft S through miter gears R, spur gears U, and universal-joint shaft G, which permits the slide D to move on the base without interfering with the transmission of rotary motion to driver F.

Quarter-Turn Mechanism for Transferring Sheets from Press to Oven.—Tin sheets for making container parts are received from the conveyor belt of a printing or decorating press, swung through an angle of 90 degrees in a horizontal plane, and fed onto the conveyor of a drying oven by means of the mechanism illustrated. The work performed on the press necessitates that the tin plate be delivered in the position shown at A_1 in Fig. 21. However, the long drying oven is narrow and will only accommodate the sheets when they have been turned to the lengthwise position shown at A_2. As it would have been costly to rebuild the oven to handle the greater widths, the quarter-turn mechanism was designed and located between the conveyors of the press and the drying oven.

Sheet A is delivered from the printing press and received on conveyor belts E in the position indicated at A_1. The surface speed of conveyor belts E, which rotate on pulleys mounted on shafts C and D, is greater than that of the press conveyor belts. The front left corner of the tin sheet comes in contact with the conical-shaped, friction disks F and F_1, and a portion of the sheet slips between the faces of the disks. These disks are free to rotate on stud R, and are pressed together by spring G. The tin sheet is thus gripped between the disks, and its forward motion is temporarily halted.

The sheet is swung about the disks, which act as a pivot, by means of the rubber-covered pulley H, which is in contact with the lower surface of the sheet. This pulley, which

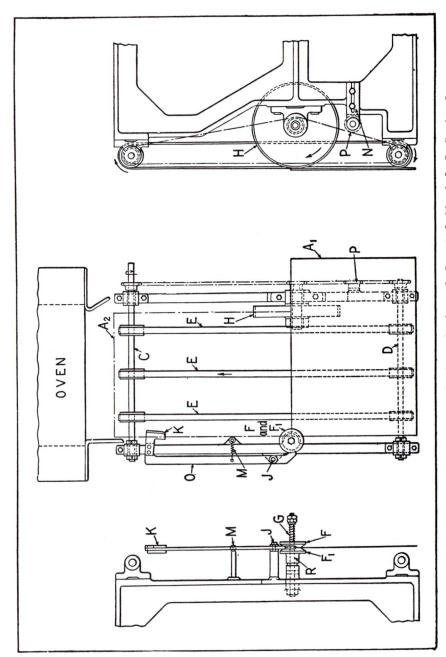

Fig. 21. Metal Sheets are Turned from Position A_1, as they Leave the Press, to Position A_2 for Feeding into Oven.

has a surface speed greater than that of conveyor belts E, turns the sheet to position A_2. As the sheet nears the desired position, its edge comes in contact with the face of the leather-covered bumper K. The force with which the sheet hits the bumper is sufficient to pivot lever O about stud J and cause the bevel-shaped end of the lever to separate friction disks F and F_1, and compress spring G. The sheet is thereby released and allowed to continue its forward motion on conveyor belts E and onto the conveyor of the oven.

Lever O is returned to its original position as soon as the sheet passes bumper K by means of spring M. The tension of this spring must be carefully controlled, so that it will not cause the bumper to apply too much braking action on the edge of the moving sheet. Shaft C is turned by a motor-driven chain. Shaft D and pulley H are rotated by means of sprockets and a single chain driven by shaft C. Idler sprocket P mounted on bracket N is used to adjust the tension of the chain.

Mechanism for Reversing Work and Transferring it to Another Spindle.—The purpose of the mechanism shown in Fig. 22 is to transfer the work W from spindle No. 1 of an automatic machine to spindle No. 2, and at the same time, reverse the position of the work, so that the chamfered end S is inserted in the chuck of the second spindle. The particular set-up shown was used in machining drawn shells such as shown at A to the form shown at B.

After the work has been chamfered, as shown at S, the housing T of the transfer arm is swung or rotated on its supporting bearing X to bring the jaws D and E in line with spindle No. 1. Rod C then pushes the work into place between jaws D and E. The housing T is next rotated or indexed on its bearing X to bring the work into line with the chuck on spindle No. 2. During this transfer movement, the shaft K is rotated 180 degrees, so that the chamfered end S is in the required position to enter the chuck on spindle No. 2 when the work is pushed from between the jaws D and E.

Referring to Fig. 22, spring F exerts sufficient pressure on jaws D and E to cause them to hold the work securely while it is being transferred from one position to the other. The housing T is indexed about bearing X by means of link G which moves forward or backward at the proper time.

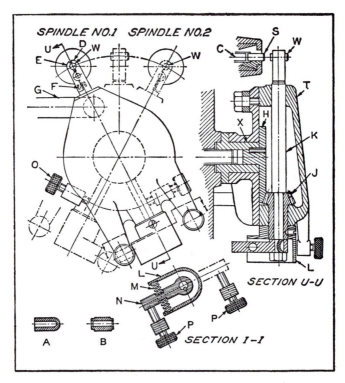

Fig. 22. Mechanism for Automatically Reversing Position of Work W and Transferring it from One Spindle to Another.

The rotation of shaft K for reversing the position of the work is accomplished by means of the stationary segment bevel gear H, which is in mesh with bevel pinion J. To pinion J is secured a member L which transmits the required turning movement to shaft K through springs M and the arm N secured to the end of shaft K. The rotating movement of shaft K is stopped at each spindle position by stops

P, which can be so adjusted that the work is revolved exactly 180 degrees while being transferred from one spindle position to the other. In the stopped position of arm *N*, the spring *M* which transmitted the motion to the arm is under sufficient pressure to keep the arm in contact with the stop. Stops *O* which limit the indexing movement of housing *T* are adjusted to bring the work into accurate alignment with the spindle.

Mechanism for Operating Two Slides Alternately.–

In one automatic wrapping machine, two parallel slides, operating alternately, control the movement of each package through the machine. Both slides, indicated at *A* and *B* in Fig. 23, are given the required reciprocating movement by the cross-head *C* through a combination ratchet and cam mechanism.

The function of this mechanism is merely to lock the slides *A* and *B* alternately to the cross-head. The mechanism consists of locking slide *D* (see also Fig. 24) ; star cam *G* on shaft *H* engaging hardened pins *L* in the locking slide; ratchet wheel *K* keyed to the shaft and operated by the pawl *J* which is pivoted to the machine base; and locking plungers *E* and *F*, backed up by coil springs in the locking slide.

As shown in Fig. 23, slide *B* is stationary and slide *A* is locked to the cross-head *C* by plunger *E*. Locked thus, slide *A* has completed a working stroke and has nearly reached the end of its return stroke. At this time, pawl *J* engages ratchet wheel *K*, and, as the slide *A* completes its return stroke, rotates the ratchet wheel and cam *G* one-tenth of a revolution. This results in the cam forcing slide *D* toward the right, unlocking slide *A* and cross-head *C*, and locking the latter to slide *B*.

With slide *A* now stationary, slide *B* and cross-head *C* are locked and travel together through a working and return stroke. Near the end of the return stroke, pawl *J* once more engages the ratchet wheel and causes the locking slide *D* to move this time toward the left. This movement of slide *D* unlocks slide *B* and cross-head *C* and locks the cross-head to slide *A*, which is then carried through a working and return stroke while slide *B* remains stationary.

To prevent over-run of the ratchet wheel, the ends of the cam lobes were slightly grooved to fit the hardened pins *L*. Both slides *A* and *B* operate at a relatively slow speed. It is

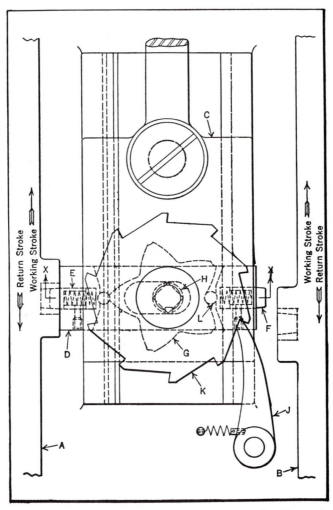

Fig. 23. Combination Ratchet and Cam Mechanism for Alternately Locking Two Parallel Slides to a Continuously Reciprocating Cross-head.

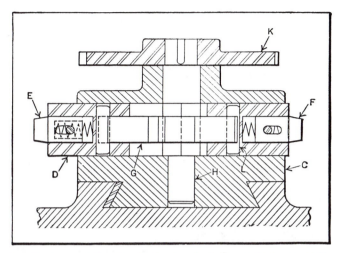

Fig. 24. Section X-X of Fig. 23 Showing the Action of the Locking Slide.

advisable, however, when this mechanism is applied to rapidly moving slides, to provide some sort of a friction stop or brake, so that the slides will stop in the same position at the end of each return stroke.

CHAPTER 16

Feeding and Ejecting Mechanisms for Power Presses

The use of a properly designed feeding and ejecting mechanism is an important factor in power press operation for maintaining a low percentage of spoiled work and a relatively high production rate. The mechanisms described in this chapter were designed for a wide variety of press functions, such as placing a drawn shell in a punching die; feeding strip material at various rates; carrying a printing type frame under the ram of a hydraulic press; operating an indexing type of dial feed; transferring work from one punch press to the proper feeding position in another; ejecting a formed part from a dovetail-shaped punch; retarding an automatic feed device; and stopping a press when stock fails to feed.

Other similar feeding and ejecting mechanisms will be found in Chapter 16, Vol. I and Chapter 15, Vol. II of Ingenious Mechanisms.

Automatic Feed for Placing Drawn Shell in Punching Die.—The die shown in Fig. 1 was designed for punching the bottom out of a drawn shell of the shape shown by the cross-section view at S. The arrangement provided for automatically feeding the shells S to the die resulted in a reduction of spoiled work and trouble from shearing of the punch due to improper location of the work in the die. At the same time, the use of the automatic feed served to increase the production of the die from 1300 to 2400 pieces per hour.

Formerly, the work was pushed down a chute into a locating nest with a stick. Often the work was not in the proper position when the punch came down, with the result that the punch was sheared or dulled. Part of the difficulty was

420

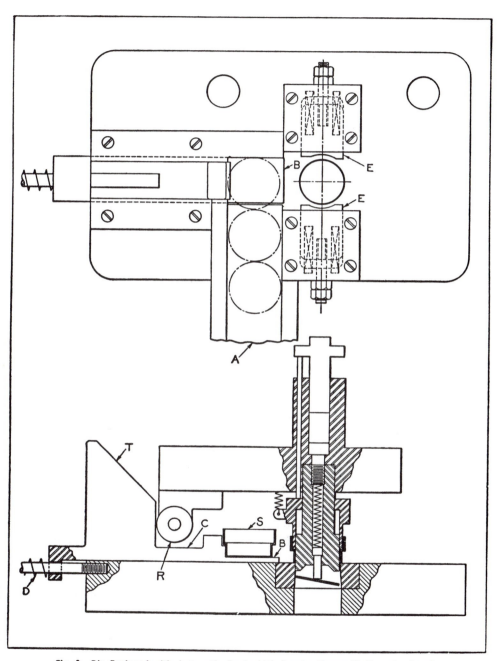

Fig. 1. Die Equipped with Automatic Feed which Locates Drawn Shells under Punch.

due to the fact that the work would not slide over the hole in the die, which was almost as large as the work, but would come to rest against the side of the hole in a tilted position.

With the automatic feed shown diagrammatically in the illustration, the work is fed into the inclined chute A, from which it slides down on a shelf on slide C, which pushes it between two spring jaws E located over the blanking hole in the die. The slide C, actuated by spring D, carries the shell S under the punch on the upward stroke of the press ram. On the downward stroke, the roller R acting on the cam T moves the slide C outward to the position shown in the illustration, and the punch centers the work, pushes it down onto the die, and punches the hole as shown in the lower view in Fig. 1. The round blank punched out of the bottom of the shell drops through the opening in the die. The work is stripped off the punch on the upward stroke of the ram, and ejected from the die by a blast of air.

The spring action used to push the slide forward prevents jamming in case the work is only half way in the slide when this member starts its forward movement. This happens occasionally when the operator fails to keep the chute filled with shells. The principle on which this lateral feeding mechanism operates can be applied, with certain modifications, to other work. For example, the slide shelf may be omitted when it is unnecessary to carry the work over the die opening.

Obviously, the automatic feed applied to this die makes it much safer to operate. Also, the die can be operated continuously, with less danger of damage to the die or work. The shells can be fed by hand into the chute or a hopper feed can be used. They can also be fed by chute directly from the drawing press.

Mechanism for Feeding Washer-Shaped Blanks to Die with Concave Sides Up.—Washer-shaped blanks of slightly concave form, as indicated at a in Fig. 2, are fed by gravity from a hopper through a chute L to the mechanism to be described. The mechanism, in turn, automatically feeds the blanks, one at a time and concave side up, through chute T to the die of a punch press.

The part a is produced by piercing and blanking from mild strip steel. This operation leaves slight shearing burrs around the edge of the blank, as indicated, and gives it a concave or dished effect that varies from 0.010 to 0.040 inch. It is the function of the mechanism shown at b and c to insure feeding the blanks to the die through chute T with the concave side up toward the forming punch.

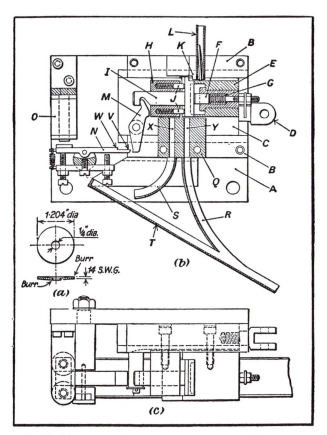

Fig. 2. (a) Views of Blank, Showing Distortion and Location of Burrs. (b) Sectional Side View of Selecting Device. (c) Plan View, with Chute L and Stop K Removed.

Referring to sectional side view *b*, baseplate *A* carries guides *B*, in which the sliding piece *C* is moved horizontally by the pin joint *D*. The pin joint is operated through a linkage connected to the press connecting-rod. A friction slipping device is incorporated to enable the sliding piece to be stopped at certain positions during the cycle. The contact body *E* is fixed to the sliding piece and is bored out, as shown, to receive an insulating bushing which carries the spring contact plunger *F*, this being adjusted to bring the end face level with the face of the contact body, where it is held by spring *G*.

Block *H* is also fixed to the sliding piece and carries the movable clamp *I*, which is forced toward the contact body by two spring pegs *J*. The upper side of the clamping head is milled to form a stop, so that when the slide is moved to the right, the clamp is arrested by a fixed stop *K* and the slide is brought to rest when block *H* comes up against the back of the clamping head.

In this position, a blank in chute *L* may fall into the gap between the contact body and the clamping head. A notch is milled in the clamp stem to engage the upper end of the lever *M* as shown, the lever being freely pivoted on a stud screwed into the sliding piece. The free end of the lever may engage either of the stops *V* or *W* when the sliding piece is moved to the left. The stop engaged depends on whether pivoted stop *N* is up or down, the position of *N* being controlled by an electromagnet *O* connected in series with the contact body *E* and a suitable source of direct current.

The two positions in which the clamp may be withdrawn and the blank released are indicated by the delivery slots *X* and *Y* in block *Q*. The arrangement of chutes *R* and *S* is such that a blank sliding down chute *R* will be turned 60 degrees in a counter-clockwise direction, and a blank sliding down chute *S* will be turned 120 degrees in a clockwise direction before passing down chute *T* to the press tool and its feeding mechanism.

Assuming that a blank is in chute L in the position shown in Fig. 2, and that slide C is moving to the right, the action of the device is as follows: Clamp I is arrested by stop K, and the contact body E moves on, allowing the blank to fall into the opening produced. The slide is then moved to the left, permitting pegs J to clamp the blank against the contact body, which results in the electromagnet O being energized. Stop V is moved out of the path of lever M, and the blank is carried along until lever M comes in contact with stop W, when the clamp is withdrawn and the blank is discharged down slot X. As soon as the blank leaves the contact face, the circuit is broken and the magnet de-energized, so that, on the return of the slide to the starting point, the stop N returns to its original position when cleared by lever M.

If the blank is located in chute L the opposite way round to that shown, then the central part of the disk will not touch the contact plunger and the magnet will not be energized; consequently, stop V will cause the blank to be discharged down slot Y. It will be apparent, therefore, that all blanks sliding down chute T will be hollow side up as required.

Adjustable Strip Feeding Device.—In feeding strip material, it is frequently desirable to vary the feeding stroke or adjust it accurately to meet changing conditions. The device to be described can be adapted for feeding strip material of any form or type and permits adjustment of the feeding interval or stroke to a minute degree by means of an adjusting screw. It is very efficient when moderate feeding rates are used.

The operation of the device is as follows: The plate cam N, shown in Fig. 3, is provided with a ridge around 180 degrees of the circumference which serves to raise the end of cam-lever K on which roll M is mounted. The oscillations thereby created by the cam in lever K are transmitted to the pressure lever F through the link J. Thus lever F serves to

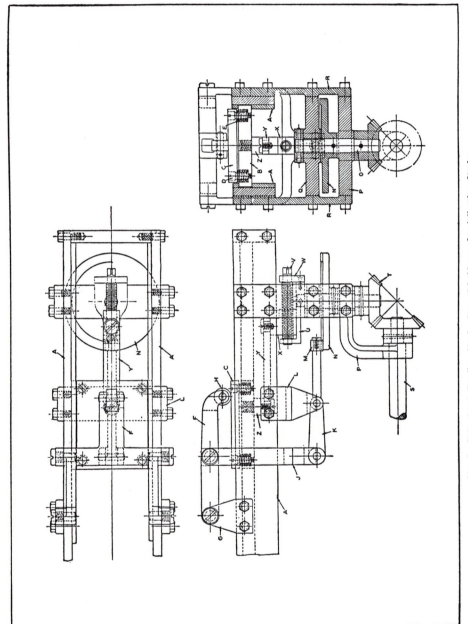

Fig. 3. Strip Stock Feeding Mechanism with Means for Adjusting Stroke.

depress and release the upper plate C which grips the stock at definite intervals, as determined by the position of cam N under the cam roller M.

The lower sliding plate B and the upper plate C, which form a unit through the connection of the screws D, are actuated by the eccentric block X through the connecting link Y. The continual rotation of block X causes the entire strip-gripping arrangement under roller H to reciprocate.

By combining the clamping motion created by cam N and the feeding movement imparted by the block X, the strip material is gripped in the jaws of the gripping assembly and given the required feeding movement. The smooth and regular operation of the device is assured by synchronizing the rotation of eccentric X with the rotation of cam N in such a manner that the gripping device will be released slightly before it has reached the end of its stroke. This will avoid the possibility of dragging the strip back slightly on the return stroke of the gripping assembly.

Strip material of any kind can be handled by the feeding device. Flat surfaces in the gripping assembly would serve well for materials that are slightly compressible. For hard materials, it may be necessary to roughen the surfaces in order to obtain a firmer hold. By altering the shape of the surfaces, it would be possible to handle materials of a form other than flat. For instance, wire can be easily controlled in a device of this type.

Referring to the construction of the device, the rails A are a part of the machine structure, and, in addition to serving as a support for the device, they have channels which guide the lower sliding plate B. The upper plate C is held in place by four fillister-head screws D, the heads serving as pilots for the upper plate. Four springs E around screws D tend to keep the upper and lower sliding plates apart. The whole assembly constitutes the strip-gripping arrangement.

The pressure lever F is supported in brackets G fastened to the rails A so that roller H rests on the surface of plate C. Link J serves to connect lever F with the cam-lever K. The cam-lever is supported in the bracket L, also fastened to the rails A, and is provided with the cam roller M which makes contact with cam N.

Cam N is fastened to the shaft O which is supported between the bearings P and Q. The bearings are fastened to the support bars R which, in turn, are attached to the rails A. The necessary power for actuating the device is obtained from the shaft S through the bevel gears T.

The stroke-adjusting arrangement consists of a screw-carrier U fastened to shaft O, the adjusting screw V, the end plate W fastened to the screw-carrier and serving to retain the adjusting screw in place, and the eccentric block X. Turning screw V causes eccentric block X to travel closer or farther away from the center of shaft O, thereby decreasing or increasing the eccentricity. The link Y serves as a connection between block X and the strip-gripping assembly through the extension block Z.

Compact Table-Feeding Mechanism.— The mechanism shown in Fig. 4 was designed to carry a chase or printing type frame P under the ram of a hydraulic press for making wax impressions from the type in the chase. The limited space available on the press prevented the use of the conventional crank movement, and made a compactly designed feeding mechanism necessary.

The table A on which the chase is loaded reciprocates between the loading position P of the chase at the left and the working position P_1 at the right. The travel of the table is guided by ways B. Table members C form a slot M which is at right angles to the center line X–X, representing the direction of rectilinear movement of the table. Block D, which is a sliding fit in this slot, is free to revolve about the stud S at one end of crank-lever F. Crankpin H is turned by crank G, which is revolved by means of the driving shaft

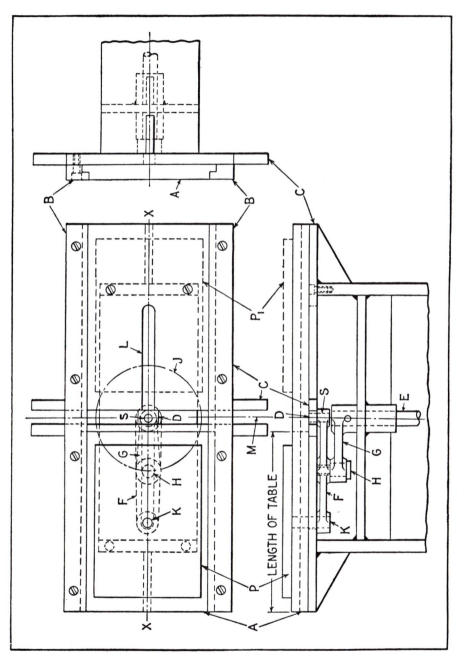

Fig. 4. Table A is Reciprocated along Center Line X-X by means of this Mechanism.

E to which it is pinned. The center line of the crankpin will follow a path shown by circle J.

The travel of one end of the crank-lever F, being fastened to block D, is confined to a reciprocating movement by the slot M. The other end of the crank-lever is fastened to table A by connecting-rod pin K. The travel of pin K, which carries the table with it, is confined by slot L formed by the table members C to a reciprocating movement. The common center line of the table and pin K follows the path indicated by center line $X–X$.

Mechanism for Operating Dial Feed and Radially Positioned Multiple Punches.—

The mechanism shown in Figs. 5 and 7 was developed for operating an indexing type dial feed and radially positioned multiple punches used for the production indenting of thin-walled tubes, such as indicated at B in the enlarged view A, Fig. 6. The tube B serves as a means of assembling or joining the wooden rod C to the cylindrical rubber piece D.

The function of the dial feed mechanism is to pick up the assembled rod C, tube B, and rubber D at E, Fig. 7 at the left, and by successive intermittent indexing movements in the direction indicated by the arrow, bring these assembled members into the position indicated by the dot-dash lines at F. While the work dwells in this position, the eight radially located cam-operated indenting punches G are advanced to produce sixteen indentations, which serve to securely fasten tube B to rod C and rubber D.

After the indenting operation, the work is indexed around toward the rear of the dial feed, where it is unloaded on a conveyor or picked up by another feeding dial. Thus one assembly is indented at each dwell period between successive indexing movements of the dial.

Referring to Figs. 5 and 7, at the left, it will be seen that the feeding dial consists primarily of two disks H, each fitted with twelve radially positioned, equally spaced slides J having U-shaped slots at their ends which pick up and carry

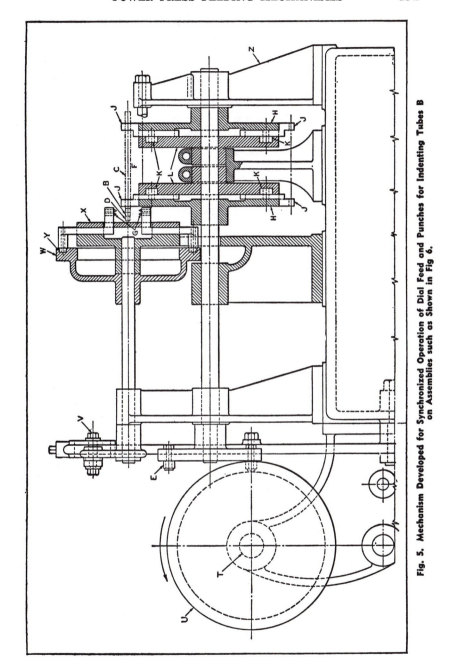

Fig. 5. Mechanism Developed for Synchronized Operation of Dial Feed and Punches for Indenting Tubes B on Assemblies such as Shown in Fig 6.

the work. Each slide J has a cam roller K which runs in a cam groove in the face of one of the two stationary cam-plates L.

The cam grooves in plates L are so laid out, as shown diagrammatically in Fig. 7, that roller K, instead of following the concentric path indicated by circle M, follows the path indicated by line N as the disk H is indexed from one position to another. This causes the slides J to carry the work along the path indicated by line P and the circles O.

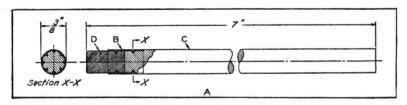

Fig. 6. Typical Assembly with Tube B Indented by Mechanism Shown in Fig. 5.

It will be noted that the work follows a path that leads away from the center of the dial or disk H as it leaves the loading position E, Fig. 7, at the left, until it reaches the position Q, after which it follows a straight horizontal path from Q to R, from which position it continues on a path that carries it back toward the center of the dial until it reaches the position indicated at S. It is necessary to have the work follow this irregular path in order to permit the tube B and rubber D, Fig. 5, to clear the indenting punches and holders as they are being indexed from Q into the indenting position at F and out again to the position R after being indented.

The intermittent indexing movements are transmitted to the feeding dial disks H from the driving shaft T, Fig. 5, through the cam U and dial driver E. The eight indenting punches are simultaneously moved inward radially to perform the indenting operation by means of the oscillating cam W operated from shaft T. The depth of the indentations can be controlled by adjusting the length of throw of

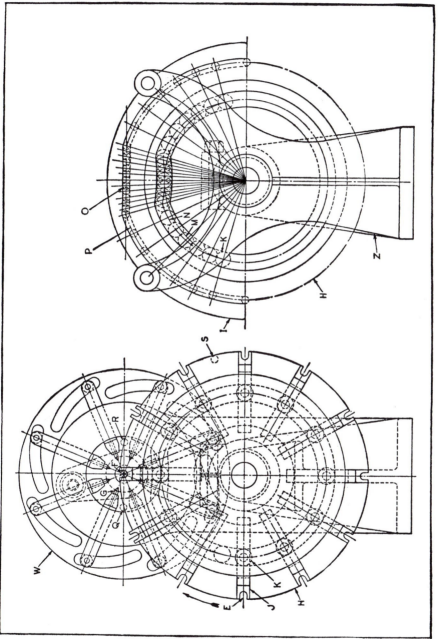

Fig. 7. (Left) End View of Indenting Punch and Dial Feed Shown in Fig. 5. (Right) Diagrammatic Layout of Cam L and Bracket Z of Fig. 5.

the cam oscillating mechanism. The oscillating movements of cam W are, of course, synchronized with the indexing movements of the work-carrying dial so that the indenting punches advance and withdraw while the work-feeding dial is stationary in one of its twelve dwell positions.

The holders of the indenting punches G are close sliding fits in the slots in the stationary head X, and have rollers Y which are running fits in their respective operating cam slots. The bracket Z, Fig. 7, at the right, supports a cover I, which keeps the work in place in the slots in slides J while it is being indexed from the loading position at E, Fig. 7, to the unloading position. The cams L, Fig. 5, are made with hubs mounted in a center pedestal equipped with a split bearing having clamping screws which provide means for individually adjusting the positions of the cams to obtain accurate alignment of dials H.

Automatic Transfer and Feeding Mechanism.—The automatic work transferring and feeding mechanism shown in Fig. 8 enables one person to operate a series of punch presses equipped to ·perform successive stamping operations. The stampings A are blanked and drawn in one press and are then blown by air pressure through a chute to the press shown in Fig. 8. The stampings are located in the die of the latter press by an automatic feeding mechanism. The feeding mechanism is shown on one press only, but other presses equipped with similar feeding devices can be added to the production line. This type of die with its lateral feeding mechanism can be adapted for handling different kinds and sizes of work. It operates at high speed and eliminates the need for a second operator in many instances.

The blanked and drawn stamping blown through the chute comes to rest against stop E, which is also shown in Fig. 9. The lateral feed-slide F now engages the stamping and pushes it forward into the spring-actuated receding jaws G which center it on the die H, Fig. 8. The slide is then withdrawn from the path of the descending punch.

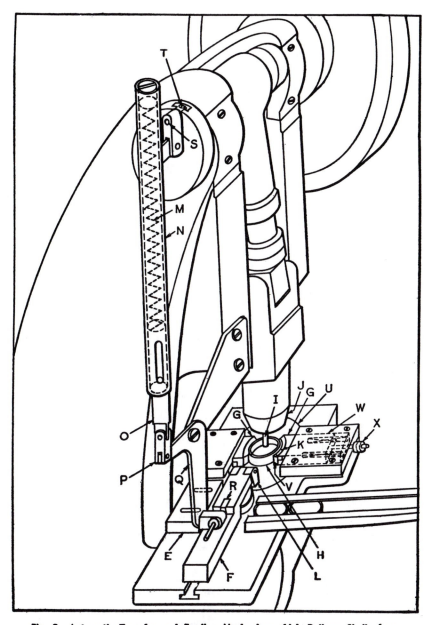

Fig. 8. Automatic Transfer and Feeding Mechanism which Delivers Shells from Another Press and Centers Them Successively on Die H.

On the down stroke of the press, a spring pin I projecting from the center of the punch comes in contact with the stamping and holds it against the die until the punch engages the work. The jaws are then pushed out of the way by contact of the tapered end J of the punch with the tapered surface K of the jaws.

On the up stroke of the ram, the stamping is pushed off the punch by pressure exerted by a spring-actuated stripper located within the punch. The stamping is held down by the spring pin I until it is gripped by the inward moving jaws. The finished stamping is pushed out of the jaws by a pawl attachment L on the front end of the feed-slide F.

The forward movement of slide F, which pushes the stamping from the jaws G, is spring-actuated, while the return movement is obtained by positive mechanical means. The yielding movement provided by the spring M on the forward stroke of the slide prevents jamming in the event that the slide is blocked in any manner, as, for instance, when a stamping is engaged by the slide before it has cleared the chute. The long spring M is contained in a two-part telescoping rod N which is connected to the slide through links O, P, Q, and R. The slide is actuated from the crankshaft of the press by crankpin S, which can be adjusted toward or away from the center of the crankshaft by screw T to vary the length of the stroke. Crankpin S is set 90 degrees ahead of the crankshaft, so that the slide moves forward during the last half of the up stroke of the press ram and during the first half of the down stroke. This setting gives the maximum time available for movement of the slide as required to permit it to clear the punch when handling large or deep work.

Changing the set-up for different operations is accomplished by simply removing the disk-shaped plate U which holds the die H and replacing it with another disk carrying the die for the new operation. The plate U, Fig. 8, is designated H in Fig. 9. Individual jaw adapters V which fit the work are attached to slides W. The jaws are opened by means of adjusting nuts X. The position of stop E is adjusted to suit the size of the work. In certain cases, the positioning attachment on the front end of the slide is also changed. The most important other modifications apply to the perforating or blanking dies, and consist of providing

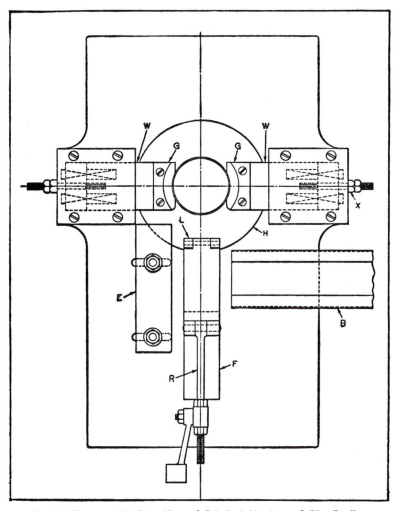

Fig. 9.   Diagrammatic Plan View of Principal Members of Die, Feeding Slide and Work Locators of Press Shown in Fig. 8.

guide pins in plate U which are extended to the punch-holder in order to maintain accurate alignment.

When two or more presses are operated in a line, their treadles are connected by linkage, so that when one press is stopped, the others also stop. The presses are run at approximately the same speed, the last ones, however, running slightly faster. With this arrangement, no damage is done in the event that a press is operated without work in its die.

A die with this feed arrangement will handle any kind of stamping work which can be fed through a chute, whether round or square, shallow or deep, or of relatively large or small size. It is only necessary to push the work into the centering jaws approximately close to the required position and it will be automatically centered. This prevents the punch from cutting into a partially entered stamping, as in the case of gravity chutes which carry the blanks directly to the die. Jamming of the die or feed rarely occurs, as previously mentioned. due to the spring action incorporated in the design.

The lateral feeding mechanism which pushes the work into the centering jaws by means of the spring-actuated slide that moves at right angles to the direction in which the stampings are fed through the chute operates rapidly and is practically trouble-free. The operating principle of the mechanism illustrated has been described as it is applied to most types of work. Certain modifications can be made to suit the requirements of special cases.

Mechanism for Ejecting Formed Part from Dovetail-Shaped Punch.—In producing the piece shown in Fig. 10, upper right, on a punch press, it is necessary to strip the work from the forming punch. Although operations of this type are not unusual, this particular operation presented some difficulty because the work was 4 inches long, and the stroke of the press only 2 1/2 inches. As it appeared inadvisable to attempt to strip the work from the punch by a

direct leverage arrangement under such conditions, the
stroke-multiplying mechanism shown was devised.

Referring to Fig. 10, lower left, the punch B is carried
in the holder A; the die, being of conventional design, is not
shown. Punch A carries the grooved bracket D in which
the ejector E slides freely in dovetailed ways. The extension
K on the forward end of the ejector passes directly under
punch B in removing work C on the up stroke.

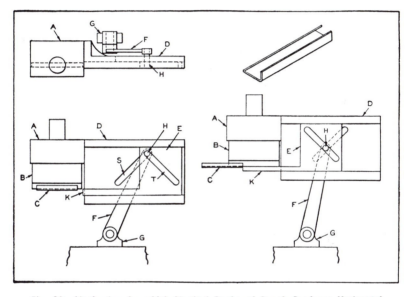

Fig. 10. Mechanism by which Vertical Stroke of Punch Produces Horizontal
Movement of Stripper Equal to Twice the Length of Punch Stroke.

A slot S is machined in bracket D at an angle of 45 degrees with the horizontal. Slide E likewise has a 45-degree
slot T, which is inclined in the reverse direction. Block G,
fastened to the bolster plate of the press, provides a bearing for the oscillating lever F, which carries the pin H at
its upper end. Pin H passes through the slots in both D and
E at the point of their intersection. In Fig. 10, lower left,
the ram of the press is shown at its lowest point, while slide
E is held at its extreme rear position.

In Fig. 10, lower right, the ram is shown as it appears after having completed more than half of its upward stroke. As the ram ascends, pin H approaches the bottom of the slot in bracket D, causing lever F to swing forward. As pin H passes through the slot in slide E, the latter member is carried forward, but because of the angularity of the slot in slide E, its movement is twice as great as that of the pin H. Thus the vertical movement of the ram is transmitted through slide E in a horizontal direction and at an increased ratio sufficient to strip work C from punch B.

Mechanism for Retarding an Automatic Feeding Device.—A two-stage forming die with an automatic feed is used to form flanges on a relay armature in two different directions. A 1-inch feed on a 2-inch stroke punch press is used for this operation. Only the last 15/32 inch of travel of the 2-inch stroke does the forming. It was necessary, therefore, to retard the automatic feed caused by the 2-inch stroke of the press. This was accomplished by the mechanism here illustrated. This retarding mechanism keeps the feeding device stationary during the forming operation.

When the punch has completed 1 17/32 inches of its travel, bellcrank H halts on the head of stop-screw K, as shown at the left in Fig. 11. However, the punch continues its travel, and completes the forming operation. As shown at the right in the illustration, when the punch moves up and the bellcrank is lifted off the stop-screw, pawl M turns ratchet N, thereby feeding a new part into the first stage of the die, moving the partially formed part from the first to the second stage of the die, and pushing a completed part out of the die.

The mechanism was assembled by fastening eyebolt A to the punch-holder with screw B. Lock-nut C and spring-housing D were threaded onto the eyebolt. A threaded bushing J was placed over one arm of the bellcrank H. The end of this arm of the bellcrank was fitted with a spring support bushing F, held to the bellcrank by cap-screw G. A pawl M

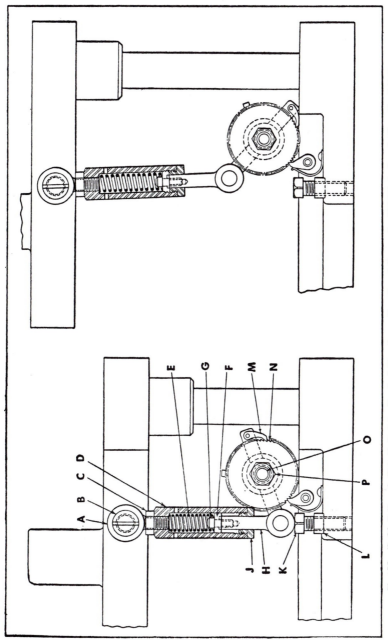

Fig. 11. (Left) Bellcrank H is Halted by Head of Stop-Screw K as Punch Descends. (Right) As Punch Rises, Pawl M Turns Ratchet N, thus Operating Automatic Feed of the Punch Press.

was pinned to the other arm of the bellcrank. Spring E was placed in the housing D, and then the bellcrank assembly was fastened to this housing by the threaded bushing J.

The arm of the bellcrank that holds the pawl was fitted over drive-shaft O of the feeding device. This arm and ratchet N were retained on the drive-shaft by means of nut P. Drive-shaft O was attached to the die-holding surface of the die set. Stop-screw K was threaded into the base of the die set. Lock-nut L is used to secure the stop-screw in the desired position after adjustments have been made.

Mechanism for Automatically Stopping Press when Stock Fails to Feed.—A certain press operation involves the feeding of two steel strips from rolls into the forming station of the machine, where pieces are cut off from each strip and assembled in automatic dies. To avoid material waste caused by failure of either strip to feed, due to a break in the material or completion of the roll, it was necessary to develop the electrically operated, automatic stop mechanism here described. With this mechanism, the positive clutch of the press is automatically tripped, thus stopping the machine, when either of the strips fails to feed. Modifications of the mechanism can be made to accommodate other types of clutches.

As shown in Fig. 12, the two steel strips X and Z, being fed in the direction indicated by the arrows, are contacted by rolls A and B, respectively. These rolls, mounted on bellcranks F and E, are held against the strip stock by springs C and D. The feeding mechanism (which is not shown) draws the thin strips over idler roll M, thus increasing the tension of the strips. The bellcranks pivot about shaft G, which is supported in the U-shaped bar H that hangs from plate J. This plate is fastened to block K, which is mounted on spindle L of the machine through which the strips pass.

The upper arms of bellcranks E and F are fitted with screws V that make contact with either of the normally open switches R and S when the feeding of the strips is inter-

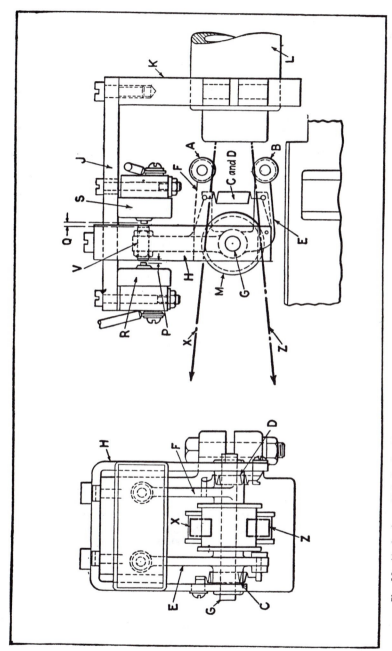

Fig. 12. Mechanism which, in Combination with that Shown in Fig. 13, Automatically Stops Press when Either Strip X or Z Fails to Feed.

rupted. This closing of the electrical circuit is accomplished through movement of bellcrank *E* or *F*, by spring *C* or *D*, depending upon which steel strip is broken, thus causing one of the screws *V* to close gap *P* or *Q*. The switches are electrically connected to solenoid *A*, Fig. 13.

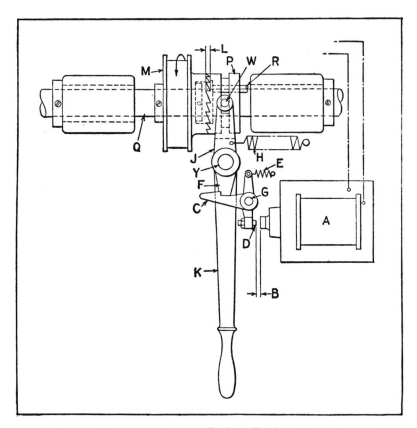

Fig. 13. When Stock Fails to Feed, Teeth on Clutch P are Automatically Disengaged from those on Hub of Pulley M, thus Stopping Press.

Referring to Fig. 13, when the electrical circuit to solenoid *A* has been completed, the contact point *D* will be moved toward the solenoid, closing gap *B*. This movement causes pawl *C* to pivot about stud *G*, thus releasing the end of pawl *F*. Spring *H*, which is connected to yoke end *J* of pawl F, then pivots the pawl around shaft Y, thus pulling the clutch teeth out of engagement at L and causing the machine to stop.

Pulley M rotates in the direction indicated by the arrow, and is held in position on shaft Q by collars. Clutch P is keyed to the shaft, and when the teeth of the clutch are in engagement with those of the pulley, key R causes the shaft to revolve. When the teeth are disengaged, the pulley is free to rotate on the stationary shaft. Pins W in yoke J fit in the groove in the clutch. The clutch can be re-engaged by lever K.

CHAPTER 17

Hoppers and Hopper Selector Mechanisms for Automatic Machines

Tool engineers and machine designers are often faced with the problem of designing mechanisms to pick up parts from hoppers for delivery to the assembly machines. By "hopper feeding" is meant the indiscriminate dumping of a load of parts into a hopper of suitable size and shape, from which the parts are picked up, in the proper position, and deposited in a track for feeding to a machine by gravity. Ordinarily, the pick-up member is so shaped that the parts cannot enter the track if they are not in the right position, and therefore are dropped back into the hopper. Occasionally, the shape of the part and the speed requirements of the machine make it necessary to pick up parts that are not all in the same position. In that case, prior to going into the assembly machine, the parts are required to pass through an auxiliary mechanism, or separator, which arranges them all in the required position.

Many types of hoppers have been designed and built with varying degrees of success. One type of hopper may work successfully for a part of a certain shape, but may prove entirely unsuitable for pieces of a different contour. A great deal of thought must be given to the selection of a hopper for any particular job. Every new problem is unique in some respect, and will necessitate variations in the type of hopper selected.

Centerboard Design of Hopper.—Fig. 1 illustrates the centerboard hopper, a highly successful type when used to pick up parts within its limitations. The hopper body may be made of cast iron or cast aluminum, or it may be of welded steel construction. Side and end section views illustrate the general construction.

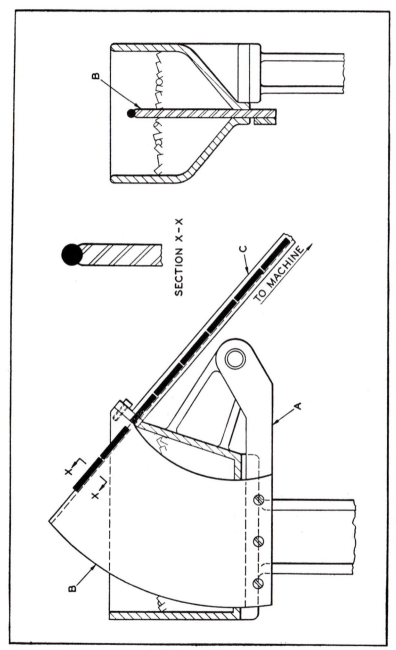

SECTION X–X

TO MACHINE

Fig. 1. Centerboard Hopper for Picking Up Rod-shaped Work or Parts of Angular Cross-section.

An arm A actuates a hardened centerboard blade B, which oscillates up and down through the mass of parts, picking up a few of them in the groove machined in its top edge. At the top of the stroke, this groove is in line with a track or tube C, and the parts slide down toward the machine. When the track is full of parts, those remaining on the centerboard fall back into the hopper. It is important, when the track is full, for the end of the last part to come flush with the end of the track. If it should project into the hopper, a jam would occur when the centerboard rose on the next stroke.

Another important design factor is to so lay out the centerboard arcs that the point of delivery, where the track joins the centerboard blade, is as high as possible. In that way, the greatest number of parts can be placed into the hopper at one time, and it will not be necessary to refill it as often. In at least one case the delivery point on several hoppers for the same machine was placed so low that the capacity of the hopper was too small for the speed at which the parts were being taken out. Careful thought to good hopper design will prevent these costly mistakes.

Section X–X shows the correct form for the top portion of the blade when used to pick up parts of round cross-section. It will be noted that the centerboard width is the same as the part diameter. However, it is machined so that it will bear on only one-quarter of the part diameter. This form has been found highly successful.

The centerboard blade should be chromium-plated. This serves two purposes. It allows the parts to slide more freely into the track, and the slippery surface thus provided prevents wedging of the parts between the blade and the bottom of the hopper on the downward stroke of the blade.

A cam is generally employed to actuate the blade, so that it will have a slow upward travel and a quick return. However, cranks have been used successfully for this purpose. They are run at speeds not greater than 40 R.P.M.

The centerboard type of hopper is recommended for all round parts having a length greater than twice the diameter. It can also be used for small disk-shaped parts by machining the groove at the top of blade to a depth of one-half the part diameter so that the parts are held as shown at *A* in Fig. 2.

An interesting application of centerboard hopper design is shown at *B* in Fig. 2. Here the blade *C* is machined at an

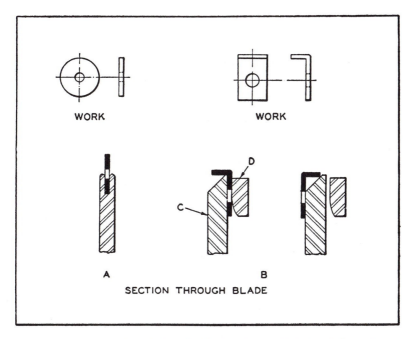

Fig. 2. Blade Designs Used for Picking Up Parts in Centerboard Hoppers.

angle, so that parts with angular projections can be picked up facing in only one direction. A stationary baffle plate *D* must be incorporated in this design to slide the parts over into the correct position at the top of the stroke. It can be seen that a part cannot be picked up in the wrong position, since it would slide off the edge of the blade, as shown at the right of Fig. 2.

Rotary Centerboard Hopper.– The rotary centerboard type of hopper can be applied to many automatic machines. By varying the blade cross-section, it can be easily adapted for parts of different shapes. This hopper has been found to be exceptionally satisfactory for channel-shaped parts of the type shown in Fig. 3.

The rotary blade A is given an intermittent motion, either by a Geneva wheel or by a pawl and ratchet arrangement.

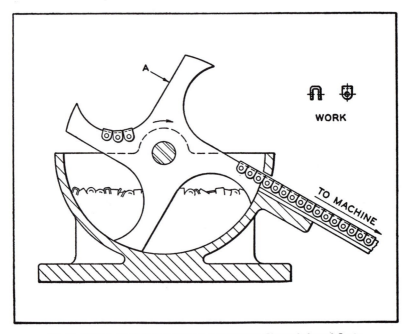

Fig. 3. Rotary Centerboard Hopper for Feeding Channel-shaped Parts to Automatic Machines.

The latter, because of its comparatively low cost, is generally used. As in the previous case, the blade should be hardened and ground, and then chromium-plated.

As with all hoppers for feeding parts, the wheel rotation should be as slow as possible, considering the feed requirements of the automatic machine which it is supplying. An ideal drive consists of a cam-operated ratchet, with a quick-return pawl and a slow blade movement. A slight dwell lobe should be incorporated in the cam at the point where the blade comes opposite the track. This gives time for the parts to slide from the hopper blade to the track. The cam should be of the harmonic motion type that allows the wheel to start slowly, accelerate gradually, and finally come to a stop. In that way, the parts that have been picked up on other arms of the rotary blade will not be knocked off by abrupt stops.

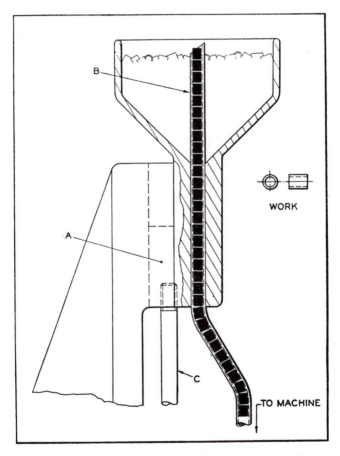

WORK

TO MACHINE

Fig. 4. Hopper Employing a Reciprocating Motion to Feed Short Parts through Stationary Tube to Machine.

Tube and Rotary Types of Hoppers.—Fig. 4 shows a good design of hopper for feeding short parts, such as illustrated at the right. The hopper body consists of a cast funnel which has a dovetailed slide *A* machined in its lower side. The feed-tube *B* is a sliding fit in a reamed hole in the hopper body as shown.

In operation, the hopper body is given an up and down movement by the actuating rod *C*. On the downward stroke, a number of parts enter the open end of the tube and slide down to the machine. The open end of the tube is machined at an angle so that, if a part falls crosswise of the opening, it will be knocked off by other parts on the upward stroke of the hopper. When the feed-tube is required to be bent, as shown, the inside diameter of the tube must be sufficiently greater than the diameter of the part so that the parts will not bind in turning the corner.

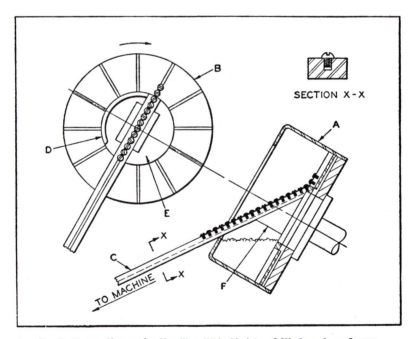

Fig. 5. Rotary Hopper for Handling Wide Variety of Work such as Screws.

The rotary type of hopper has proved practical on a wide variety of jobs. As illustrated in Fig. 5, it consists of a large round container A, mounted at an angle, with grooves machined in the baseplate B. When the baseplate rotates, some of the parts in the bottom of the hopper fall into the grooves in the baseplate and are carried up in line with the opening in track C. When the track is full, succeeding parts are carried past the track opening and fall back into the bottom of the hopper.

A baffle D is mounted on the stationary center E as shown. This prevents the parts that have been picked up from falling back into the hopper before they have passed the track opening. The track is mounted on a stationary bracket F. The set-up illustrated was designed for feeding short screws to an automatic screwdriver in such a way as to insure a continuous supply and correct positioning.

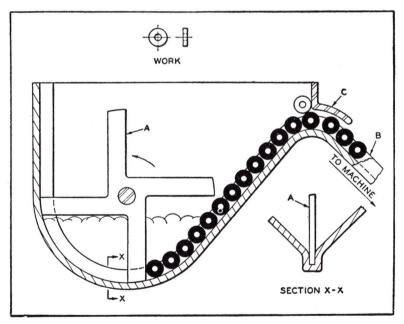

Fig. 6. Paddle-wheel Hopper for Feeding Flat Work, Varying in Shape from Square to Round.

Hopper for Feeding Flat Work of Both Round and Square Shapes.—The hopper shown in Fig. 6 is used for feeding flat work varying in shape from square to round. It is an inexpensive and very efficient design for such parts. Section X–X shows the shape of the groove at the bottom of the hopper. The parts that fall into this groove are pushed up the incline by the paddle wheel A. When the parts have been pushed up far enough, they enter the track B, down which they fall to the machine. When the track has become full, the succeeding parts ride up the baffle C as shown, and fall back into the hopper.

Barrel Hopper for Intricate Shapes.—The barrel type of hopper, while more expensive to build, is sometimes the only one that will successfully select parts of very intricate shape, including those that interlock when grouped together. It consists of a rotating hopper A, Fig. 7, which is shaped like a barrel and is open at both ends. The parts are dumped into hopper A through a loading hopper B.

On the inside of the rotating hopper are cast longitudinal fins, as shown in section X–X. Rotation of the hopper, which is mounted on bearings F and driven by the pulley G, causes the parts to flow in a steady stream down a blade C, which is set at a slight angle and connected at one end to an electric vibrator D. The other end of the blade is aligned with the track E, which leads to a machine.

The constant agitation of the parts resulting from this arrangement prevents interlocking, and some of the parts will fall on the blade in the right position. Vibration of the blade causes the correctly positioned parts to move toward the track and enter it. The loading hopper shown makes filling of the rotating hopper an easy matter.

Hopper for Feeding Rivets and Similar Shaped Parts.—The hopper shown in Fig. 8 is designed to select and feed rivets to a machine in the correct position for the riveting operation. However, it can also be used for many special parts of similar shape. The rotating portion A of the hop-

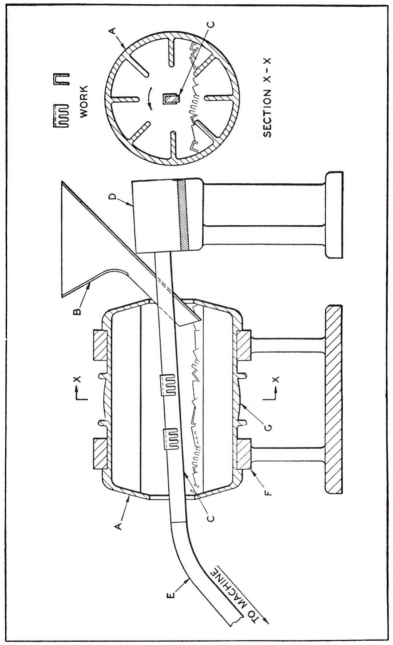

Fig. 7. Barrel Hoppers to Select and Feed Work of Intricate Shape, Including Parts that Interlock when Grouped Together.

per body contains a series of grooves as shown. Re-entrant groove *G* is machined in the stationary part *B* of the hopper body to accommodate the rivet heads. Rotation of the hopper by drive-shaft *C* allows a few rivets to enter the slots. When they are opposite the track *D*, an opening in the body allows the rivet heads to enter the track, and the rivets slide down toward the machine.

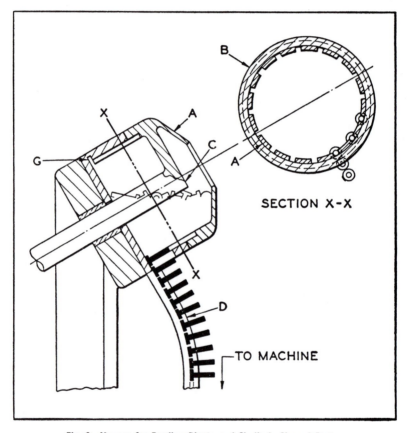

Fig. 8. Hopper for Feeding Rivets and Similarly Shaped Parts.

Tray Type Hopper for Comparatively Low Production.—

The tray type hopper is a cheaply built hopper for feeding larger parts in comparatively low production. The operator places the parts in the tray *A*, Fig. 9, and they are moved toward the track *B* by vibration. An agitator *C*, operated by a small crank, prevents jamming at the mouth of the track.

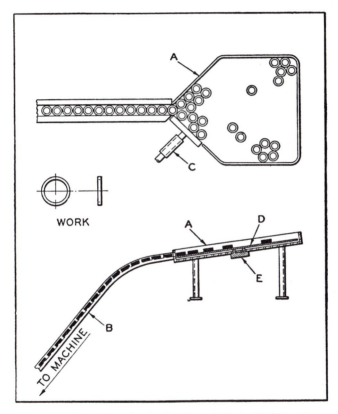

Fig. 9. Tray Hopper for Feeding Large Parts to Machines at Comparatively Low Production Rates.

The construction is simple. A vibrating plate *D*, set close to the bottom of the tray, is connected to the rods of a commercial electric vibrator *E*. The angle at which the tray must be set is determined by experiment, and is usually about 4 or 5 degrees. With the agitator operating, the parts then move readily to the mouth of the track.

Vibratory Hopper for Greater Production Require-ments.—Where greater production requirements exist, the hopper illustrated in Fig. 10 will be found applicable to a wide variety of parts. It consists of a commercial vibratory feeder *A*, suitable guiding baffles *B*, a hopper *C*, and an agitator *D*.

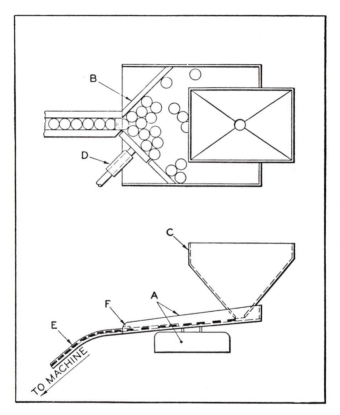

Fig. 10. Vibratory Type of Hopper for High Production Requirements.

The parts are dumped into the hopper, and the action of the vibratory feeder causes them to flow in a steady stream from its mouth. Owing to the slight incline of the pan, the parts flow toward the track *E*, being guided by the baffles.

As in the previous case, an agitator prevents jamming at the track mouth. A gate F is incorporated at that point, so that only one part can go through at a time. If a part is lying on top of another, the gate will prevent both from going through at once. The vibration will cause the lower part to enter, while the top part will fall off and enter the track in its turn.

Hopper of Magnetic Design.—An interesting hopper application is shown in Fig. 11, where Alnico magnets A are used to pick up the parts from the hopper B. The magnets

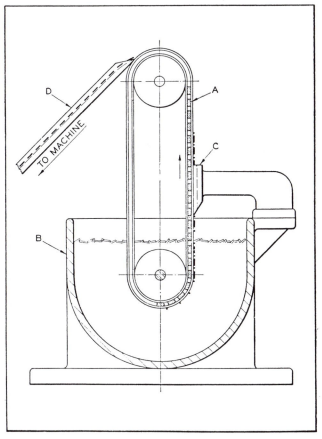

Fig. 11. Hopper which Uses Magnets to Pick Up Parts.

are incorporated in a conveyor belt which passes through the parts, some of which cling to the magnetic stations and are carried up. The stripper and side-guide arrangement C properly locates parts that were not picked up in quite the right position. In case one magnet picks up two parts, the side guide will strip the extra part, which falls back into the hopper.

At a point tangent to the top pulley, a track D strips the parts from the magnets. The track must be made of a non-magnetic material at that point. Further down, the parts pass through a demagnetizer if the magnetic properties imparted to them are objectionable.

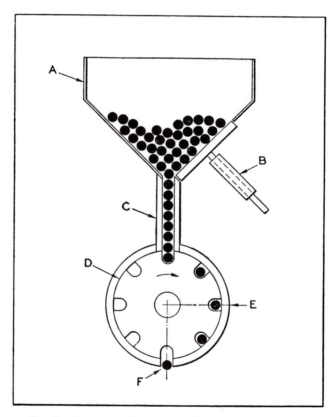

Fig. 12. Hopper for Feeding Long Rods with Agitator and Transfer Wheel which Indexes to Carry Rods to Work Station.

Simple Hopper for Feeding Long Rods.—Fig. 12 illustrates a simple hopper for feeding long rods to an automatic machine. The rods are loaded into the hopper *A*, and an agitator *B* insures a constant flow into the track *C*. At the lower end of the track, the rods enter grooves machined in a transfer wheel *D* and are carried to a work station *E* as the wheel indexes. At that point, the parts are automatically clamped and held securely while drill heads machine a hole in each end. At the ejection station *F*, the rods fall out into a box.

Spring-Actuated Pin Type Selector Mechanism.— In designing hoppers for feeding parts automatically into assembling machines, cases are often encountered where the work is required to enter the machine in a certain position for correct assembly with other components. To accomplish this, the hopper must be provided with a selector mechanism.

In the example shown in Fig. 13, one end of the workpiece is turned down to a slightly smaller diameter than the remainder of the piece, and the part is required to be fed into·the machine with the small end first. A part of this shape would ordinarily be handled by a centerboard hopper such as described earlier in this chapter. However, the centerboard blade may lift the part up in either of two positions, with the smaller end toward the machine or with the larger end facing in that direction.

To insure that the piece will enter the machine in the correct position, a selector mechanism designed as shown in the illustration may be employed. This mechanism is located at some point between the hopper and the machine. Parts fed from the hopper fall into the selector wheel of the device, which is designed so as to allow the pieces that have been picked up correctly in the hopper to pass through into the machine, while those that are in the wrong position are turned, end for end, before entering the machine.

Referring to Fig. 13, it will be seen that the selector mechanism consists of a housing A, a stationary center plug B, and an intermittently rotating ring C, driven by a ratchet wheel J. A circular groove D is machined in the center plug, which is slightly wider than the small diameter of the workpiece. A hardened pin E is pressed into the center plug, and a spring-actuated pin F is located two stations away

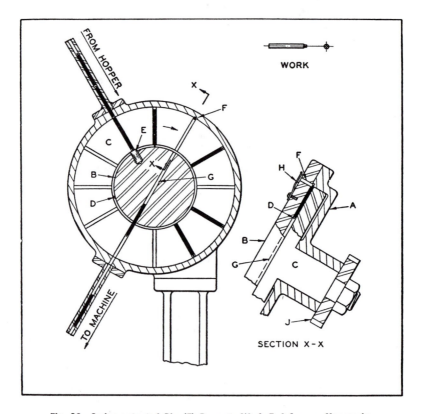

Fig. 13. Spring-actuated Pin (F) Prevents Work Fed from a Hopper in the Incorrect Position from Being Transferred to a Machine.

as shown. Pin F and the flat spring H may be seen more clearly in section $X–X$.

In operation, the parts slide down a tube from the hopper and enter a groove machined in the rotating ring C, coming to a stop against the end of the hardened pin E. When the ring indexes, the parts are carried to the idle vertical station. As they leave the pin E, the small end of those pieces that are positioned correctly will enter groove D. Pin F at the next station will not touch the parts in the groove, due to their lowered position, and thus they will be free to enter a groove G machined in the center plug, through which they slide out of the selector and into a tube leading to the machine.

An incorrectly positioned part, having its small end toward the selector housing, cannot enter groove D. Hence, when it comes opposite groove G in the center plug, the spring-actuated pin F will bear on the end and hold it from sliding down the groove. As the wheel continues to index, the part is carried around, so that when it comes opposite the tube leading to the machine it will have been turned end for end, and thus enters the machine with the small end first, as required.

In laying out this type of selector wheel, the length of the tube between the selector and the machine must be great enough to hold at least six parts—preferably more. This would insure a sufficient supply of parts to the machine in case all the pieces for a certain period of time came from the hopper in the wrong position and had to be carried around by the wheel and turned before being fed to the machine.

This type of selecting device has been found successful for a wide variety of parts. Occasionally, a great deal of ingenuity is required to design the actual method of selecting, but once this is accomplished, hopper problems can be met that would otherwise be impossible to solve.

Circular Blade Type Selector.—To handle a part formed as illustrated in Fig. 14, the rotating ring in the selector would have a circular blade A pressed into it as shown. The parts that are in the correct position as they come from the hopper rest on this blade, and when they are opposite the groove in the center plug, slide down toward the assembly machine. Those parts that are incorrectly positioned are held by the blade, as shown at section *X–X*. In this case, as in the previous example, the part is carried around to the bottom of the ring, where it is in the proper position to enter the machine. There it slides over the inclined face of the retaining blade and into the feed-tube.

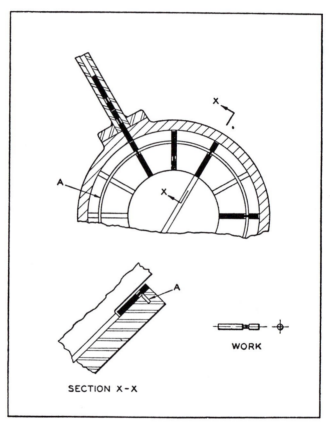

SECTION X-X

WORK

Fig. 14. Parts of Shape Shown are Uniformly Positioned by Selector that Holds Incorrectly Located Parts until They are Turned Over.

Magnetic Type Selector.– A magnetic selecting device can be used for certain parts, as shown in Fig. 15. In this case, a permanent magnet *A* is employed at the selecting station for work having a pointed end, as seen in the illustration. The magnetic attraction is not great enough to hold the parts when the point is toward the magnet, and they slide through to the machine. However, when the flat end of the part is toward the magnet, it is held and carried around by the wheel so that it enters the tube leading to the machine in the desired position.

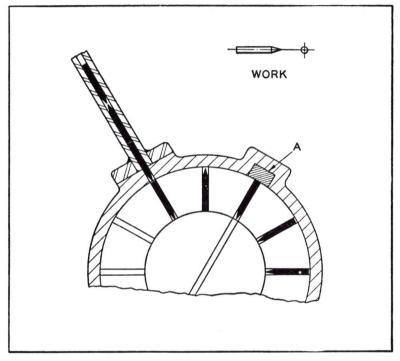

WORK

A

Fig. 15. Permanent Magnet (A) Prevents Parts that are Fed to it in the Wrong Position from Being Transferred to a Machine.

This is a cheap and efficient selecting device for certain types of parts. It is ideal for use in connection with pin-driving machines. In this case, the rotating ring must be made of some non-magnetic material, and so must the tube leading to the machine. Occasionally, it will be found necessary to provide a demagnetizing coil around the tube.

Selector for Flat Hooked Parts.—Fig. 16 illustrates a selector designed for flat, hook-shaped parts. A blade A in the cover engages the hook of incorrectly located parts and prevents them going through to the machine. When the parts are correctly positioned, the hook does not engage the

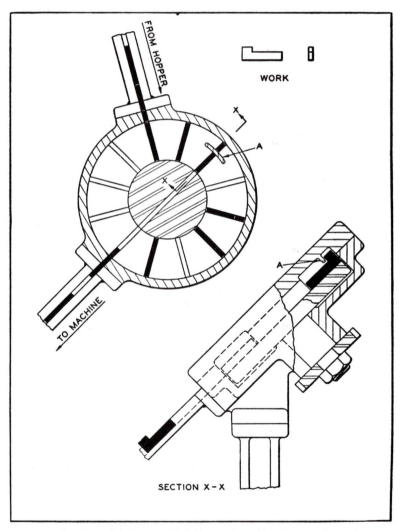

Fig. 16. Blade (A) Prevents Hook-Shaped Parts from Dropping Through to a Machine When They are Fed to the Selector in the Wrong Position.

blade, and the parts are free to go through. Section X–X shows the details of construction; it will be noted that the selector, in this case, is mounted at an angle rather than vertically. This is the most positive type of selector, and should be used whenever possible. Unfortunately, many parts do not lend themselves to such positive treatment, and spring pressure or magnetic properties must be resorted to.

Selector for Shallow Drawn Parts.–Shallow drawn parts, such as covers for various units, can easily be handled by means of the vertically mounted selector wheel shown in Fig. 17. With this arrangement, the covers enter slots in the rotating ring, and those with the open side up slide through the groove to the machine. Incorrectly placed covers are held from going through by the plugs A, and are carried around by the ring, falling into the feed-track right side up. In this case, the rotating ring must be fabricated sectionally in order to permit assembly of the plugs A.

Cup or Can Selector.–The simple device shown in Fig. 18 is used to position cans so that they will all be fed to a machine with their open ends up. In use, they slide down a tube from the hopper and hit a projecting pin A. Those that come down bottom first simply bounce from the pin and fall down the vertical tube in the same relative position. The cans that come down open end first catch on the projecting pin and flop over, so that they fall down the tube toward the machine in the desired position.

Selectors for Reversing Position of Parts.–It often happens that the easiest method of selection is to place the parts in the opposite position to that required by the sequence of operations in a particular machine. This was true in the case shown in Fig. 15, where the parts had to be fed to the machine point first, but were selected so that the opposite ends faced the machine.

In such cases, the track arrangement shown in Fig. 19 can be used to turn them end for end. This consists of a track with a break in it, arranged as shown. Two side plates

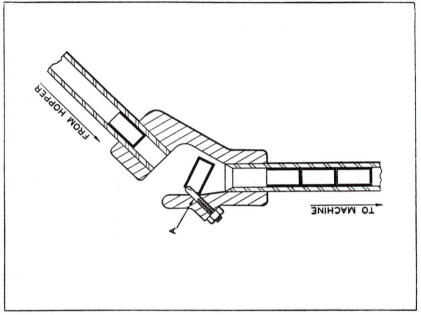

Fig. 18. Cups are Turned Over or Passed through this Selector so as to Enter a Track in a Uniform Position for Feeding to a Machine.

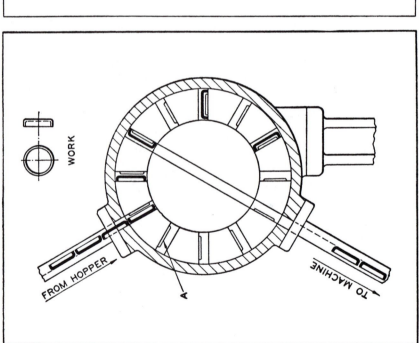

Fig. 17. Selector which Holds Incorrectly Positioned Shallow-drawn Parts by Plugs (A) and Prevents them from Being Fed to the Machine.

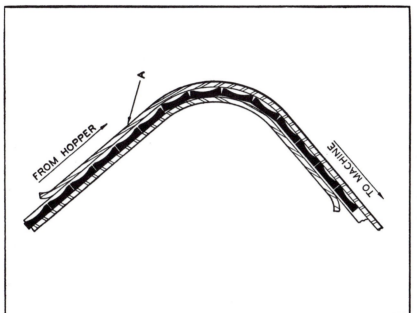

Fig. 20. Another Track Arrangement for Reversing Position of Parts Received from Hopper.

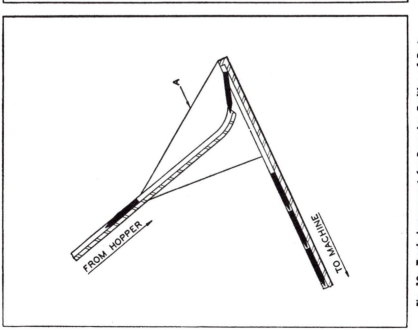

Fig. 19. Track Arrangement for Reversing Position of Parts Received from Hopper.

A prevent the parts from leaving the track. When they must be turned upside down, a simple bend in the track, as shown in Fig. 20, will accomplish this most effectively. It can be seen that in this case the cover *A* becomes the bottom of the track below the bend.

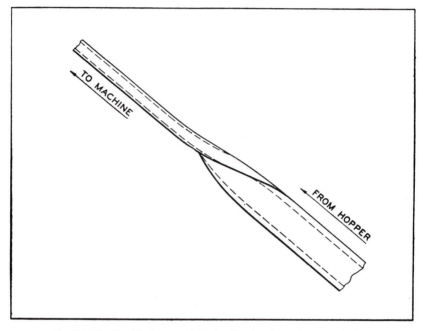

Fig. 21. Track with Quarter Twist for Turning Parts from Vertical to Horizontal Position.

Fig. 21 shows a quarter twist applied to a track. This is useful where the parts must be picked up in a vertical position by the hopper, but must be fed into the machine horizontally. Occasionally, a half twist is applied to the track to turn the parts upside down, where space requirements do not permit the use of a bend such as shown in Fig. 19.

CHAPTER 18

Miscellaneous Mechanisms

The mechanisms described in this chapter are those which were not readily classifiable in the general groups covered by the preceding chapters. They are included because of some interesting features or ingenious design.

Press Ram Mechanism that Gives Nearly Uniform Pressure During Latter Part of Stroke.— The mechanically operated presses generally employed for drawing and stamping work and for molding plastic materials produce their maximum pressures at or near the ends of their strokes. The mechanism shown diagrammatically in Fig. 1 has been designed to produce nearly a constant pressing force against the work for a considerable portion of the latter part of the ram stroke, thus giving somewhat the same characteristics as hydraulically operated presses.

The mechanism has two cranks A and B of the same length, mounted on shafts connected by two spur gears C and D of the same size. These gears, rotating in opposite directions, are so meshed that when crank A is in position 1, namely at the upper end of the stroke, the angular position of crank B is about 125 degrees in front of or ahead of crank A. The choice of this angular lead of crank B over crank A is very important in obtaining the required result, a lead angle of 125 degrees being the one best adapted for most requirements. Cranks A and B support, on their outer ends, the connecting-rods E and F, which are of the same length. The opposite ends of rods E and F are connected to the main connecting-rod G which transmits the reciprocating motion to ram H.

471

The connecting point J at which the three different members are joined follows a figure 8 path, as indicated at K. The larger part of the crank rotation—indicated by angle M—is utilized in moving the ram downward, so that the return stroke will be accomplished much more quickly.

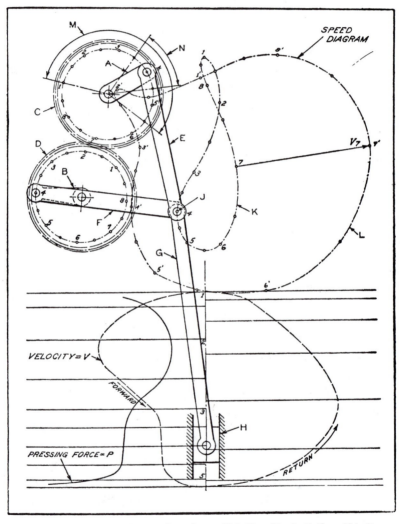

Fig. 1. Mechanism for Driving Press Ram, which Gives Nearly Uniform Velocity and Pressure During Latter Part of Down Stroke.

Besides the line K showing the path followed by connecting point J, there is also a speed diagram L which indicates the variations in its velocity. From this diagram, the speed of the ram was plotted for the down stroke, as represented by curve V. The velocity at any point on the downward stroke is indicated by the horizontal distance from the vertical reference line, which has been drawn through the center of ram H, to the given point on that part of the velocity curve which lies to the *left* of the vertical reference line. The velocity at any point on the return stroke is similarly measured, but to the given point on that part of the velocity curve which lies to the *right* of the vertical reference line as indicated in Fig. 1.

At the beginning of the stroke, the speed of the ram increases, the maximum speed being obtained at about one-third of the stroke. The speed then drops to a point where it is about one-third that of the maximum speed, and remains constant for a relatively large angular movement of the crank, as indicated by angle N. At the end of the stroke, the speed is reduced to zero.

The force diagram P, representing the force exerted by the ram, is the inverse form of the velocity diagram V. Also, in the interval during which the ram speed is practically constant, the force remains nearly constant. On the return stroke, the variation in speed is similar to that obtained with slotted crank drives.

Mechanical Equalizer for Hydraulic Press Ram.—

A mechanical equalizer designed for mounting on the ram of a Greenerd hydraulic press is shown in Fig. 2. This device is being used to distribute the pressure exerted by the ram equally between two consolidating fixtures mounted on the press platen. The application of equalized pressure to the two fixtures takes care of any variations in height during the consolidation operation.

The pressure is transmitted to two fixtures (not shown) by the pins A and B, the pressure being equalized by the arm *C* which is pivotally mounted on pin *D*. The pins *A* and *B* are shown by solid lines in their normal operating positions, with springs *E* holding the pins in contact with the equalizer arm *C*.

Assuming that the fixture in contact with pin *B* offers greater resistance to the pressure exerted by the arm than

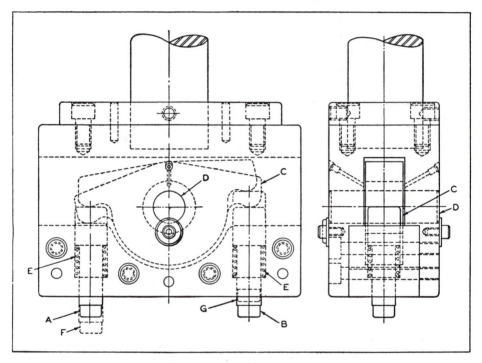

Fig. 2. Device Attached to Ram of Hydraulic Press to Obtain Equalized Application of Pressure by A and B.

does the fixture in contact with pin *A*, the pin *A* would advance while pin *B* receded, so that they would occupy relative positions such as indicated by the dotted lines at *F* and *G*.

Certain minor changes were made after the original design (illustrated) was developed, but the principle of operation remains the same.

Spring-Winding Mechanism Operated by Either Right- or Left-Hand Movement of Lever.

—A switch for controlling automobile signalling lights contains a clock movement which is started by winding up a clock spring each time the control handle is turned. Turning the handle to the right lights the green signal, indicating a right-hand turn. Turn-

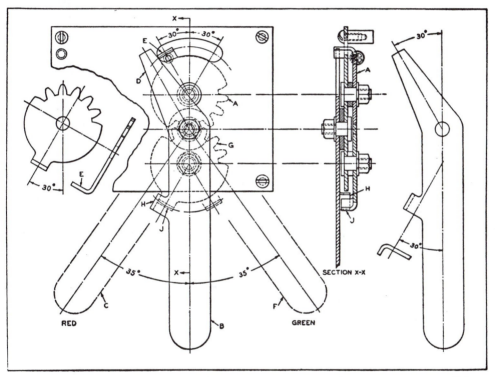

Fig. 3. Mechanism for Rotating Segment Gear in One Direction Regardless of the Direction in which the Operating Lever is Moved.

ing the handle to the left lights the red signal, indicating a left-hand turn. Turning the handle in either direction must always wind the clock spring in a clockwise direction. The problem of finding a movement which would do the winding under these conditions was solved by the mechanism shown in Fig. 3.

This mechanism serves to wind the spring regardless of which way the handle is turned. The leverage and the pressure are the same for movement in either direction. The distance the handle travels is also the same in each case.

Referring to Fig. 3, the spring to be wound is attached to segment gear A. Swinging the handle B to the position indicated by the dotted lines at C turns on the red light and rotates segment gear A in a clockwise direction through an angle of 60 degrees. This movement serves to wind the clock spring in a clockwise direction. Rotational movement is imparted to gear A by handle B through contact of the arm D with the projecting member E.

Swinging handle B to the right, into the position indicated by the dotted lines at F, serves to light the green light and also rotate the segment gear A through an angle of 60 degrees in a clockwise direction. In this case, however, rotational movement of gear A is transmitted from handle B through the segment gear G by contact of the projecting member H on handle B with the projecting member J on the segment gear G. The design of segment gear A is shown by the views to the left. Lever B is also shown in a separate view to the right.

Pivotal Joints for Special Purposes.—Pivoting joints of three different designs developed to meet special requirements are shown in Fig. 4. These joints are used to maintain proper alignment or free action of certain members. In the design shown in the upper view, the adjusting screw A pivots on the stud B which is attached to a swivel slide within the machine, a portion of which is indicated at C.

The knob at D has a curved surface at E which fits a mating surface in the lug F. As the knob is moved along the threaded portion of the screw, the end of the screw at B swings back and forth through a radial arc, the radial seat at E permitting the knob D to adapt itself to this swinging motion, which is in a horizontal plane at right angles to the center line of screw B.

In the central view of Fig. 4 a fork A used for shifting a clutch B on shaft C through the medium of the two pins D is shown. This shifting fork pivots on the screw E. A spring at F holds the shifting lever in contact with one side of the bearing, thus preventing it from vibrating.

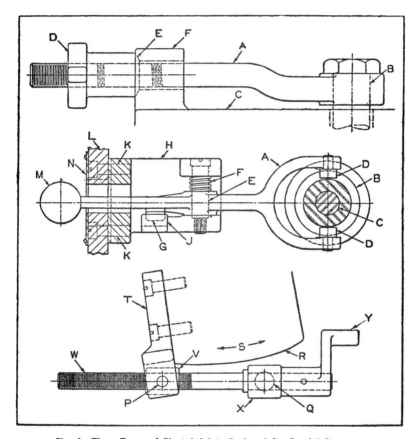

Fig. 4. Three Types of Pivotal Joints Designed for Special Purposes.

There is also a stop-lug at G which limits the sidewise movement of the lever by coming in contact with the side of lug J. This entire lever construction is attached to a bracket H, secured by screws at K to the front of the machine, a portion of which is shown at L. The shifting lever is operated by a round knob at M. A plate N is fastened over an elongated hole through which the unit is assembled.

Two pivot pins are incorporated in the arrangement shown by the lower view, which is employed for moving a swivel plate R. The pivot points are indicated at P and Q. Plate R is adjustable about a stud (not shown) with a circular oscillating movement indicated by the arrow S. On the side of plate R is mounted a block T. In a slot in the lower end of block T is carried a block V which is pivoted on two short pins P. Block V is tapped to receive screw W.

The pivot point at Q consists of a turned stem which is integral with the bearing X in which the screw W revolves when turned by the hand-lever Y, pinned to the shaft. The stem Q is free to revolve in a fixed base. Plate R, in traveling back and forth as indicated by arrow S, causes screw W to assume various angles, the swivel mountings permitting it to pivot at points P and Q.

Switch-Positioning Mechanism.—The lever arrangement shown in Fig. 5 is used on a spring winding machine which is reversed frequently by reversing the driving motor. The reversals are controlled by a double-throw drum type switch having sliding contacts. The lever mechanism serves to hold the switch in the neutral position when the mechanism is stopped to insure full disengagement of the contacts. With this arrangement, complete disengagement of the contacts is effected automatically after the operating lever has been moved to a given position.

The lever B is keyed to the shaft A, which operates the switch drum. The upper end of lever B is connected to the shifter rod G. Pin C at the lower end of rod B fits in the fork at the upper end of the lever D, which is free to oscillate on stud E. The lower end of spring F is fastened to a stationary part of the machine, while the upper end is fastened to the lower end of lever D.

The positions of the levers when the switch is in the neutral position are shown in the left-hand view of Fig. 5.

The tension of spring F, acting through lever D, serves to hold lever B upright. If the shifter rod is given slightly less than its full travel movement, the spring F will immediately return it to the neutral position.

In the right-hand view of Fig. 5, lever B is shown just before reaching its extreme left-hand position. At this

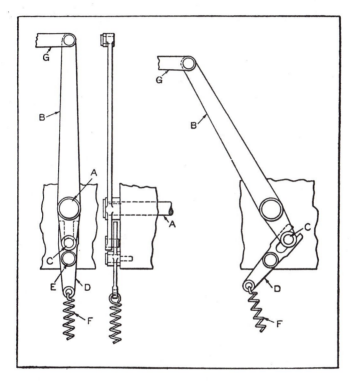

Fig. 5. (Left) Switch-operating Mechanism in Neutral Position. (Right) Switch Mechanism about to Make Contact.

point, the switch drum has not quite completed its partial revolution, and spring F, acting on pin C through lever D, still exerts sufficient pressure to return lever B to the neutral position if the movement of shifter G is not completed. However, as the movement of shifter G continues to the left, and the levers B and D more closely approach positions at right angles to each other, the pressure exerted by the fork at the end of lever D acts longitudinally on lever B and is no longer effective in returning it to the upright position. The switch contacts then remain in engagement until lever B is again returned to the position shown in the right-hand view of Fig. 5.

Follower Mechanism for Contour Milling of Grooves.— The usual method of machining a straight slot, the bottom of which changes from a parallel to a tapered surface at some point along its length is to make two passes with a milling cutter, one for each of the intersecting planes. With the proper type of contour follower, however, both surfaces can be machined in one pass, and production increased.

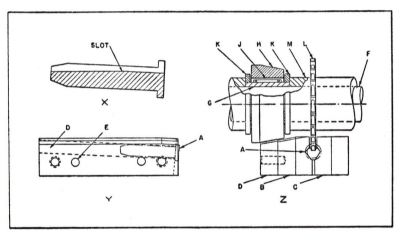

Fig. 6. Follower Mechanism Used in Milling a Contoured Slot in a Cylindrical Part as Shown at X.

The device shown at Y and Z in Fig. 6 was designed to form such a slot in a cylindrical part, as shown at X. The work A is located in V-blocks and clamped between a fixed jaw B and an adjustable jaw C, which are fitted to a 4-inch quick-acting vise. Templet D, which has the contour to be produced on the work machined on its surface, is fastened to the fixed jaw by dowel-pins E.

The follower roller H is attached to the milling machine arbor F in such a position that it rolls on the templet when the milling machine table is raised to maintain contact between the roller and templet during the feeding movement. Bushing G is made a slip fit over the arbor, and roller H is pressed on a needle bearing J, which revolves freely between two collars K. Both the templet and the roller are machined at an angle, as shown in view Z, so that the distance of the roller from the milling cutter L can be adjusted by adding or removing spacers M, thus varying the height of the cutter above the work and hence the depth of the slot.

In operation, the vise is mounted on the table of a hand milling machine. Attached to the end of the handle controlling the vertical movement of the cutter is a weight which maintains pressure on roller H, so that it is kept in contact with the templet. When the longitudinal feed of the machine is engaged, the roller follows the contour of the templet, causing the cutter to mill the slot to the same contour.

Lever and Spring Arrangement for Variably Increasing Tension on Slide.— In developing a wire-forming machine, it was necessary to provide means for gradually increasing the spring tension opposing the movement of a certain slide up to a given point in the cycle and then suddenly increasing the tension. The mechanism designed to accomplish the required variation in tension is shown in Fig. 7, the normal position of the mechanism being indicated in the upper view.

The slide A, fitted wi h roller B and cross-bar C, travels in a stationary part of the machine. Two studs D in the stationary part act as fulcrums for the levers E. Levers E carry the rods F, which pass through clearance holes in the cross-bar C. Springs H on rods F react downward against cross-bar C and upward against the rods F, thus holding levers E against the pins G, and the slide A against the pin J. In this position, springs H are only lightly compressed.

As slide A is raised, springs H are compressed by the movement of cross-bar C, the levers E being immovable, due to their contact with pins G. The tension on slide A is thus increased gradually until roller B comes in contact with the inner ends of levers E, causing the outer ends to move

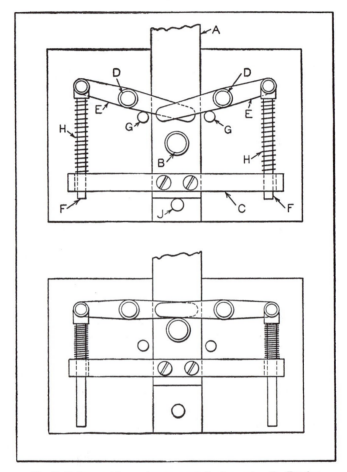

Fig. 7. Spring and Lever Arrangement for Increasing the Tension Opposing Upward Movement of Slide A.

in the reverse direction. This results in suddenly increasing the speed at which the springs are being compressed, which, in turn, causes a rapid increase in tension on the slide.

Rotating Mechanism for Creasing Flexible Material.—
The mechanism shown in Fig. 8 is designed to actuate the
jaws *A* and *B* for creasing a certain flexible material. The
jaws and their actuating mechanism comprise only the
creasing unit of a complete machine. In operation, jaw *A*
simply pivots or swings on pin *C* from the position shown
in the view at the left to that shown at the right, so that
the distance indicated at *D* is reduced to that indicated at *E*.
While this pivoting movement is taking place, jaw *B* is
pivoted inward on pin *F* a similar amount, and, in addi-

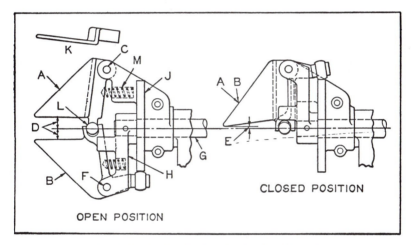

Fig. 8. Mechanism for Creasing Flexible Material.

tion, is rotated approximately one-half revolution. As a result of these two movements, jaw *B* is brought into position
beside jaw *A*, as shown in the view to the right, in which
the outlines of jaws *A* and *B* coincide.

The jaws are actuated as described by the shaft G, pinned
to arm *H*. As shaft *G* rotates, the follower roll on an arm
projecting from jaw *B* rides up on the lobe of the face cam
J, causing the jaw to pivot or swing inward. A fork on jaw
B, which fits over a ball *L* machined on jaw *A*, transmits
the required pivoting motion to the latter jaw. Helical

springs M maintain a constant opening pressure on the
jaws. The jaws are offset, as indicated by view K, to permit
them to be closed, as indicated in the view to the right.

Mechanism Designed to Crimp Ends of Heavy Paper
Cylinder.—The lever and toggle action mechanism shown in
Fig. 10 is designed to crimp the upper end of a heavy paper
cylinder B, Fig. 9, for a length C preparatory to folding and
flattening it over the end of an inside core or spool of twine
A. After the paper has been folded over the twine, it is
secured in place by cementing a circular label over the end.

To enable the end of the paper cylinder to be satisfactorily
folded and flattened, the mechanism was designed to produce twelve crimps, equally spaced about its circumference,
as shown in the plan view of Fig. 9. This required twelve
individual toggle units, arranged in a circle around a central operating unit. Two of the toggle units are shown in
the closed position in Fig. 10.

The spool X, Fig. 10, is lightly clamped in position, with
the paper cylinder extending a distance Z beyond its upper
end. The clamping means employed consists of a V-clamp
A which slides on two tightening rods B, only one of which
is shown. The tightening rods, with the spool located between them, pass through the frame of the mechanism and
are fitted with clamping nuts. While the work is being
clamped in place, the lever-shaped jaws C and D move upward, clearing the spool and paper.

The entire unit then moves down to the lower position,
but with the jaws C and D held open. These jaws are next
closed around the end of the paper cylinder, crimping it in
twelve places, as required. After this has been accomplished,
it is a simple matter to fold over the end of the paper and
flatten it down into place to receive the label

Referring to the construction of the mechanism, housing
E is a drum-shaped part to which is attached a series of
brackets F. In the illustration, the drum is cut through the
center to show two of the twelve units C and D. Each of

these units is pivoted on its own stud Q in the drum. The twelve angular extended ends G of levers D enter a groove in yoke H that is free to slide over the central shaft J. The central shaft has a pad K attached to the lower end which holds the spool down and acts as a control center.

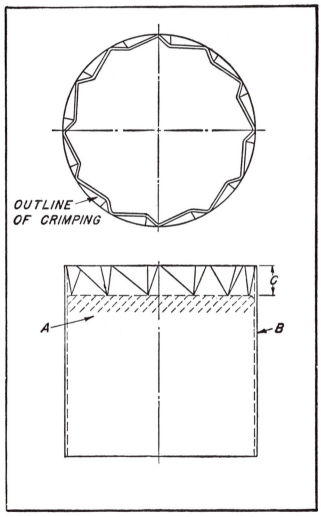

Fig. 9. Twelve Equally Spaced Crimps are Produced in Heavy Paper Cylinder by Mechanism Shown in Fig. 10.

The lever-shaped jaws C have straight extended ends L to which are attached, by means of connecting links M, a series of twelve right-angle levers N, all of which engage the groove in yoke P. Yoke P operates jaws C, bringing them into the open or closed position, while yoke H operates jaws D; thus the simultaneous action of the two series of

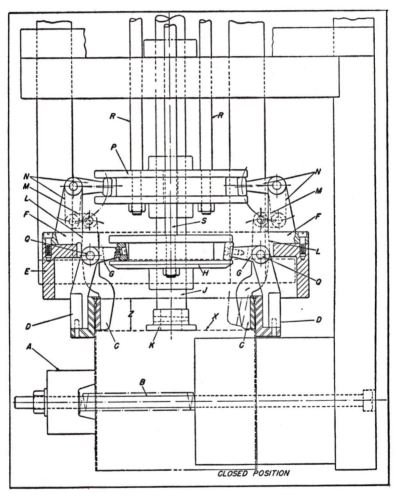

CLOSED POSITION

Fig. 10. Mechanism Designed to Crimp Ends of Heavy Paper Cylinders as Shown in Fig. 9.

jaws serves to crimp the circumference of the heavy paper. Two rods R attached to an operating mechanism within the machine serve to slide the yoke P up and down on the central shaft J. In a similar manner, two rods S operate the yoke H.

With the spool and paper cylinder clamped in place, the central shaft J moves down until pad K touches the spool. As this is done, the entire unit carried on drum E moves down into position. Jaws C are then closed by an upward movement of yoke P, after which jaws D are closed as yoke H slides upward along the shaft from the open to the closed position.

Selective Timing Mechanism for Actuating a Control Lever.—The timing mechanism shown in Figs. 11 and 12 has been used successfully on one of the textile machines manufactured by the James Hunter Machine Co., North Adams, Mass. As shown in the illustrations, it differs from the conventional type of timing devices. It covers a wide range of timing requirements and can be set for such operating intervals or periods as 1, 2, 3, 4, and 5 minutes, or 8, 16, 24, 32, and 40 minutes. An important feature of this mechanism is its extreme simplicity, the selection of the various intervals or time periods being instantly accomplished by merely turning a knob. This eliminates the necessity for locating or relocating various fingers or cam lobes about a disk, as in conventional timing devices.

The specific purpose of the mechanism illustrated is to impart one forward movement to lever V, Fig. 11, for a predetermined number of revolutions of shaft B. This forward movement of lever V can be used to release a clutch, make an electrical contact, or perform any duty necessary for starting other mechanisms or machines at the selected time intervals.

Crank A on the drive-shaft B transfers a reciprocating movement to the bellcrank C through the connecting-rod D. Bellcrank C is free to turn on stud E, and through its pawl

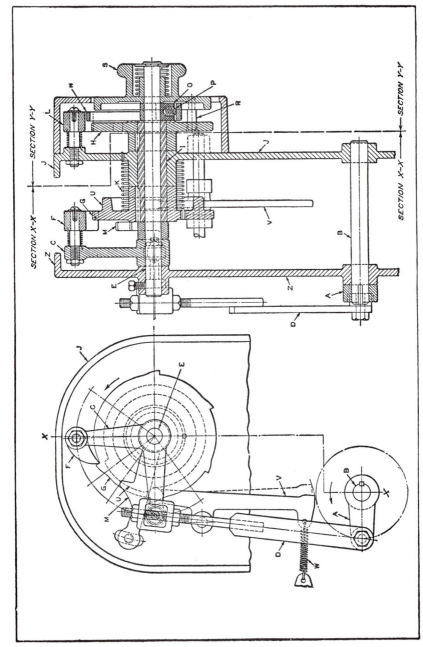

Fig. 11. By Setting Control Knob S to Any Number from 1 to 5, Lever V will be Moved Forward after Any Number of Revolutions of Shaft B from One to Five.

F, rotates the ratchet wheel G. The ratchet wheel H rotates in unison with ratchet wheel G, as both members are keyed to sleeve I, which is free to turn on stud E and extends through the stationary frame J. A helical torsion spring K, mounted between ratchet G and frame J, tends to rotate the two ratchets in an opposite direction to that imparted by bellcrank C and pawl F. Ratchets G and H are divided into six equal parts, five of which have teeth that are engaged by pawls F and L, respectively. Referring to Fig. 12, these teeth are marked No. 1_1, No. 2_1, No. 3_1, No. 4_1, and No. 5_1.

Referring again to Fig. 11, pawl F is released from ratchet G at the end of its stroke by cam M, but the ratchets are normally prevented from rotating under the action of spring K by pawl L mounted on frame J. Lever N, the function of which is to lift pawl L from ratchet H, is prevented from rotating too freely on stud E by the friction block O and spring P, and is carried forward under pawl L by the cam lobe Q, shown in Fig. 12, on ratchet H, and backward by pin R in control knob S. Pin R can be placed in any one of the holes in ratchet H numbered from 1 to 5. The pointer T on ratchet H, shown in Fig. 12, and numbers 1, 2, 3, 4, and 5 on the control knob S are used in making the required setting. The cam U is extended on the side of ratchet G and contacts with the lever V, held against it by the spring W, as shown in Fig. 11.

Shaft B rotates constantly when the mechanism is in operation, and through crank A, connecting-rod D, bellcrank C, and pawl F advances ratchets G and H one tooth for each revolution. Cam U is an integral part of ratchet G and acts upon lever V only when ratchets G and H are in the position shown. The position of ratchets G and H at the start of the timing cycle determines the number of revolutions of shaft B for each forward movement of lever V.

This is accomplished as follows: Assume that pin R is placed in hole No. 3, as indicated in Fig. 12. While cam U is acting upon lever V, the lobe Q on ratchet H carries lever

N under pawl L, lifting it clear of ratchet H. When pawl F is released from ratchet G by cam M, ratchets G and H turn backward through the action of spring K, until pin R comes in contact with lever N and carries it backward in time to allow pawl L to drop and engage tooth No. 3_1. Thus, during the third revolution of shaft B, the ratchets and cam U again come into position to move lever V forward.

Should pin R be placed in hole No. 4, lever N would be carried back from under pawl L by pin R one tooth later

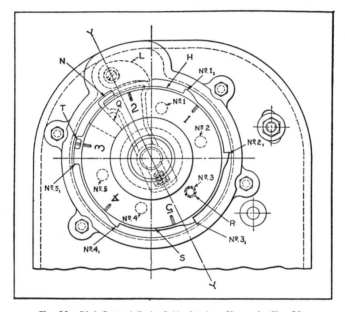

Fig. 12. Dial Control End of Mechanism Shown in Fig. 11.

when the ratchets were returned by spring K, and pawl L would engage tooth No. 4_1. Lever V would then be moved forward during the fourth revolution of shaft B. Should pawl L fail to engage any of the teeth, spring K would be prevented from being unwound by a safety stop, which is just in front of the end of pawl L on the vertical center line, Fig. 12. This stop prevents the pointer T from making a complete revolution.

Pin R is so arranged that it cannot be removed from in front of lever N. Therefore, if hole No. 2 is selected, lever N will be carried backward during the setting, and after imparting the forward movement to lever V, the return of the ratchets will be stopped on tooth No. 2_1. The forward movement of lever V will then take place during the second revolution of shaft B.

This mechanism can be designed for a different number of timing periods by dividing a cycle of the ratchets into one more division than the number of timing periods desired and proportioning the stroke of the connecting-rod accordingly.

Amplifying Mechanism for Precision Measuring Instrument.—A movement-amplifying mechanism developed to transmit movement from the contact or measuring point to the indicating pointer of precision measuring instruments has, as its most important part, a metal strip of rectangular cross-section which is twisted into a helix, as shown at A and B, Fig. 13. This twisted part, of unusual design, is employed in instruments for taking precision measurements of length, weight, pressure, electrical energy, etc., which require an amplifying unit that will operate with a minimum of frictional and energy loss and without back pressure.

The mechanism described and illustrated is protected by patents of Aktiebolaget C. E. Johansson, of Eskilstuna, Sweden. It has been employed in extensometers, electrocardiographs, micro-monometers, variometers, and surface finish testing instruments.

The metal strip A, Fig. 13, is twisted into the required helical form by fastening each end rigidly and winding from the center. The winding operation is continued until the metal has been formed sufficiently to retain the helical shape permanently. When the strip twisted in this manner is held at each end and stretched, the center of the strip will rotate about an axis which is the center of the cross-section of the strip. Actually, one end of the twisted strip is held in a fixed position, while the other end is attached to a lever or crank connected with the measuring point of the instrument, as shown in Fig. 14. The indicating pointer P is secured to the center of the twisted strip. With this arrangement, the indicating pointer will be moved over a graduated scale when the measuring point at the lower end of member A is moved.

In the case of the twisted solid strip A, Fig. 13, the metal in the center of the section is compressed in winding, and

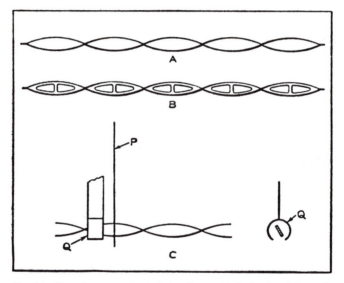

Fig. 13. Two Types of Metal Strip, A and B, Indicating Pointer P, and Split Tube Q Used in Amplifying Mechanism Shown in Fig. 14.

elongated when the strip is stretched. To correct this condition, a series of perforations may be cut out of the central portion of the strip, giving it the form at B. Such a strip requires less energy to operate and also gives a greater rotative movement with a given tension on the strip than the one shown at A. The relation between the cross-section of the strip, elongation, pitch of winding, and the stretching force required to produce rotation has been determined by trying different combinations of cross-section dimensions, pitch of twist, and size and number of perforations.

Within a certain range, the rotation of a strip about its center is practically directly proportional to the elongation. On one type of experimental strip, this portion of the curve covers a range of about 60 degrees. The rotation of the strip within this range is approximately 18 degrees for an elongation of 0.00039 inch. Tests show that a force of one gram produces a rotation of 5 to 7 degrees.

Another strip which requires a much lower operating force and produces a much higher amplification gives such a high rotative or amplifying effect that it does not need to be perforated if used within a range of 145 degrees rotation. By "operating force" is meant the force required to hold the pointer in the starting or zero position. The latter strip is 0.0042 by 0.0002 inch in cross-section, 1.5748 inches long, and has a twist of 2160 degrees.

By varying the dimensions of the cross-section, length, and pitch of the twist in the strip, it is possible to produce many different amplification ratios. The strips mentioned are only examples, and do not show the full possibilities of their use in amplifying mechanisms. The twisted strips, when properly mounted in an instrument, are surprisingly strong. The elongating force or tension required to produce rotation of the strip about its axis can be reduced to a minute fraction of the amount normally required by balancing the normal or initial tension with a permanent magnet.

The "Mikrokator" amplifying and indicating mechanism shown in Fig. 14 is fitted with a strip B like the one shown at B, Fig. 13. Spindle A, Fig. 14, which carries the measuring point at its lower end, is forced downward against stop C by a coil spring. To provide a frictionless support for the spindle at the lower end, it is fastened to a metal diaphragm D. This diaphragm is cut out, as shown by the plan view E, so as to provide maximum flexibility and not interfere with the free movement of the spindle.

The upper end of the spindle is fastened directly to horizontal spring F and the horizontal member of spring "knee" G. One end of the twisted strip is fastened directly to the vertical member of spring knee G. The other end is fastened

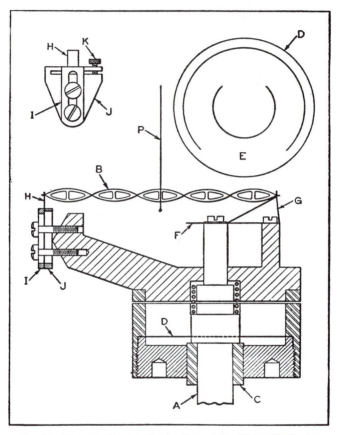

Fig. 14. Diagrams Showing Construction of Amplifying Mechanism of Precision Gage.

to the adjustable spring support H. An upward movement of the spindle will cause the vertical member of spring knee G to move to the right. This movement of the spring knee results in an elongation of the twisted strip, and causes pointer P, fastened to the center of twisted strip B, to rotate across the scale of the instrument. Varying the height of the vertical member of the spring knee changes the ratio of amplification between the spindle and the pointer.

The adjustable spring support H is used to adjust the pointer position and movement to suit the scale. For economical production, it is more practical to produce identical scales than an individual scale for each instrument. To permit the use of identical scales, adjustable spring support H is provided for adjusting each mechanism until pointer movement corresponds exactly to scale graduations.

This adjustment is accomplished by an upward or downward movement of the plates I and J. These plates have elongated holes through which are passed the screws that clamp the plates to the frame. If the plates are moved up, the portion of spring support H that projects above the plates is reduced, and thus the spring support is stiffened; this causes greater elongation of the strip and greater movement of the pointer for a given movement of the spindle. If the plates are moved down, the portion of spring support H that projects above the plates is increased, and thus the spring support becomes more flexible, and, as it bends more easily, results in less elongation of the strip and a smaller movement of the pointer for a given movement of the spindle.

After the pointer has been adjusted so that it corresponds approximately to the scale graduations, the final adjustment is made by means of screw K. Adjusting plate J is slotted at the top, as shown, so that by turning screw K in a right-hand direction, the tongue at the top of the plate is moved upward, thus shortening the distance the spring projects above the supporting plate. By turning the screw in a left-hand direction, the tongue at the top of the plate is lowered, thus increasing the amount the spring projects.

This adjustment of the tongue has the same effect on the elongation of the strip and pointer movement as the adjustment of the plates I and J; but as the adjustment is made by means of a screw, minute adjustments of the pointer movement are possible. These movements can be adjusted until the actual spindle movement, as checked with gage-blocks, is made to correspond exactly with the indicated movement of the pointer on the scale. The adjusting plates can be changed to increase or decrease the initial tension.

As the only damping effect on the pointer movement is furnished by the resistance of the air, it is very important that the weight and the inertia be reduced to a minimum. The pointer, which is mounted in the center of the strip, is made of tapered glass tubing. The tubing at the large end is approximately 0.0024 inch in diameter, and, at the small or outer end, 0.0012 inch in diameter. As a pointer of such small diameter would be very difficult to see, it is provided with a small circular disk just below the tip. This glass tube is so flexible that it can be bent as easily as a hair without danger of breakage, and will return to its original shape after bending.

If the pointer is allowed to swing freely from the extreme right or plus 0.003 inch reading back to zero, it takes about three-fourths second for it to move this distance and come to an absolute stop. On some production measuring applications, where a damping interval of three-fourths second is too great, a quicker damping effect is obtained by having the strip rotate in a drop of oil. The oil is held in a short length of split tubing which encircles the strip close to the point at which the pointer is fastened to the strip, as shown at Q, view C, Fig. 13. This makes it possible to obtain almost instant damping of the pointer movement.

The highest amplification on a standard instrument of this type is 3000 to 1. On this instrument, a movement of 0.0001 inch of the measuring tip causes a movement of 0.300 inch of the pointer. On the corresponding instrument, with a scale graduated in metric units, a movement of 0.001 millimeter of the measuring tip causes the pointer to move 3 millimeters.

The highest amplification provided on a special instrument had a ratio of 27,600 to 1. In this case, a movement of 0.001 millimeter of the measuring tip produces a pointer movement of 27.6 millimeters. The scale has graduations for each 0.00002 millimeter. The width of these graduations is 0.55 millimeter. On a corresponding instrument graduated in the English system a movement of 0.0001 inch of the measuring tip produces a pointer movement of 2.76 inches. Scale graduations are 0.028 inch wide for each 0.000001 inch.

Mechanism for Obtaining Uniform Adjustment of Guide Rollers.– A number of strands of wire are fed into a machine for producing a woven wire product by passing the wires over grooved rollers. The rollers are spaced to meet certain specifications. At times, it is necessary to change the positions of the rollers in order to increase the "spread" of the wires. The adjustment must be accomplished while the machine is in operation, and the spacing between the wires must be uniform throughout the total spread. The two views in Fig. 15 show the design of a guide mechanism that fulfills these requirements. When applied to the machine, the mechanism is located in a vertical position instead of in the horizontal position shown in Fig. 15.

A stationary part A of the machine has a dovetail groove in it to receive a series of blocks B, several of which are shown. These blocks are free to slide in part A, with the exception of the block at the extreme right-hand end, which is pinned in position. Each of the blocks B has a roller C which supports a strand of wire, and all the blocks carry an externally threaded bushing D, the head flange of which fits into a recess in the adjacent block B. Shaft E passes through the assembly of blocks and bushings, and is splined, so that any rotative motion given it through the crank-handle F is transmitted to bushings D through a key in the bore of each bushing. The screws that fasten the keys in the bores of bushings D are shown in the head flanges of the bushings.

The lower view in Fig. 15 shows the mechanism with the rollers set in a position of minimum spread. It will be noted that each block B is in contact with the succeeding block. As shaft E is rotated by handle F, the housings D are rotated with it, causing each block B to be moved away from its adjacent block by an amount equal to the lead of the thread on bushing D multiplied by the number of rotations given shaft E.

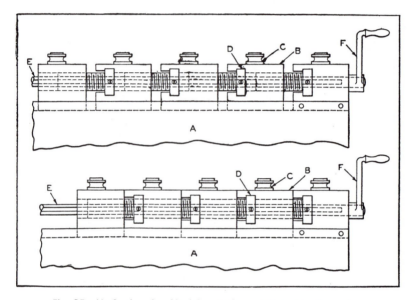

Fig. 15. Mechanism for Obtaining Uniform Adjustment of Spacing between Rollers C.

The upper view shows the mechanism as it appears after handle F has been given two turns. As each of the bushings D is rotated, it withdraws from the block into which it is threaded, causing that block and all those to the left of it to be moved the same amount. The increase in the spread between the two outside rollers represents the accumulative effect of the axial movement of bushings D; and as each of these bushings is rotated the same amount, the increase in the center distance between any two rollers is the same.

Link Mechanism for Operating Combination Furnace Door and Work Plate.—A door for a modern high-temperature furnace must be so designed that it can be opened quickly; it must be a tight fit in the closed position and take up a minimum amount of space. A door designed to meet these requirements is used on muffle type furnaces operating at temperatures up to 1000 degrees C. (1832 degrees F.). The door of the furnace is suspended on a multiple-

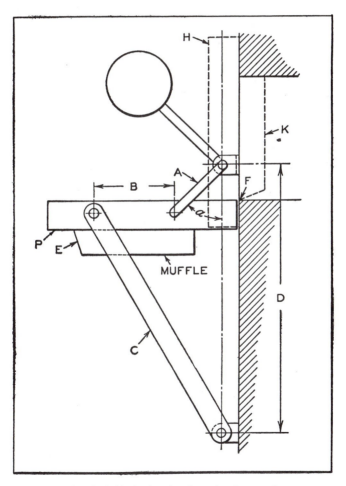

MUFFLE

Fig. 16. Link Mechanism for Operating Furnace Door.

link arrangement, so that when it is opened it rotates through an angle of 90 degrees, which brings its upper side into a horizontal position for supporting the parts requiring heat-treatment.

The door plate shown at P in Fig. 16 is actuated by a link A and is guided by a longer link or arm C. When the door is closed by rotating link A in a clockwise direction, plate P will be in the vertical position indicated by dotted lines at H, with the muffle in the position shown by the dotted lines at K. After rotating arm A about 135 degrees counter-clockwise, the door reaches the open horizontal position shown by the full lines. On the first part of the closing movement, the end of the door that is nearest the furnace rises without withdrawing from the furnace far enough to leave an opening through which small parts might fall.

The front edge of the muffle must be inclined as shown at E in order to permit it to clear the edge of the door opening at F. The door is counterbalanced by a weight attached to arm A, which serves also as a crank for operating the door. The length or dimension B between the pivoting points on plate P and the length of link C can be calculated if the length of link A and dimension D, as well as the angle a are given, using the formulas:

$$B = \frac{AD \; (1 + \cos a)}{A \; (1 + \sin a) + D}$$

and

$$C = A + D - B$$

Example—$A = 3$ inches; $D = 12$ inches; $a = 45$ degrees; $\sin a = 0.707$; and $\cos a = 0.707$.

Substituting the numerical values in the preceding formulas, we have,

$$B = \frac{3 \times 12 \; (1 + 0.707)}{3 \; (1 + 0.707) + 12} = 3.59 \text{ inches}$$

and

$$C = 3 + 12 - 3.59 = 11.41 \text{ inches}$$

The link mechanism described can also be used to advantage for other purposes, such as supporting tables where space is limited, as it permits the tables to be easily folded upward against the walls.

Stripper Mechanism for Wire-Forming.—In the operation of a wire-forming machine, one end of wire W—shown in the accompanying diagram, Fig. 17—is twisted around a pin, as indicated in the second view from the top, and then stripped off the pin in preparation for a subsequent operation. The twisted wire must be held in contact with die-plate K, as shown in the third view from the top, for a short time after pin F is withdrawn. Stripping fingers S must then be raised out of the way after the work has been stripped. The stripping operation is performed by frictionally operated levers in the manner shown by the diagrams.

The diagram at the top of Fig. 17 shows the levers immediately before the stripping operation is started. Shaft A, supported by bearing B, is given an intermittent oscillating motion by a cam (not shown). Shaft A is keyed to lever C, which carries pin F around which wire W is twisted. Lever E swivels on pin H, carried on lever C, and is shaped on the free end to form stripping fingers S. Lever D, which is made of bronze, is split (as shown) to permit it to be clamped around the finished hub extension of bearing B, the frictional resistance being adjusted by tension spring T. The hole in the outer end of lever D is slightly elongated and engages pin I, which actuates lever E.

The assembly is shown in its lower resting position in the top view of Fig. 17. Stripping fingers S of lever E are raised against the under side of lever C to allow space for the entrance of the initial forming die, which, at this point, has completed its work and withdrawn. As lever C is raised to withdraw pin F, lever D, being frictionally attached to bearing B, is not raised immediately; but, as pin H rises with lever C, lever E is caused to swivel on fulcrum pin I:

As lever C swings upward, the outer end of lever E is

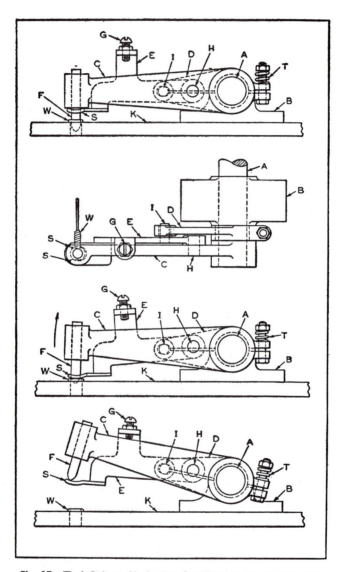

Fig. 17. (Top) Stripper Mechanism for Wire-forming Machine with Wire W Twisted around Mandrel F. (Center) Stripper Fingers about to Strip Wire W from Pin F. (Bottom) Mechanism with Pin F and Stripper Fingers S Raised to their Highest Points.

caused to swing downward, as shown in the third view from the top of Fig. 17. Since fingers S at the outer end of lever E press the work downward against die-plate K, the movement of lever E is restricted, and continued movement of lever C in the direction indicated by the arrow causes lever D to slip on the hub of bearing B. The frictional resistance of lever D on the hub of bearing B thus reacts as a downward pressure on the outer end of lever E, as indicated in this same view, and in the plan view shown in the second view from the top of Fig. 17.

As the movement of lever C continues, its position relative to lever E continues to change until screw G on lever E comes in contact with the upper edge of lever C. From this point, the entire assembly continues its movement as a unit, as shown in the bottom view of Fig. 17, until lever C reaches its extreme upper position, where it will permit the entrance of a forming die. As the motion of lever C is reversed, the frictional resistance of lever D on the hub of bearing B reacts on lever E in the reverse direction, so that the outer end of lever E is immediately raised until it comes in contact with the under side of lever C, when the entire assembly moves as a single unit until it reaches the position illustrated in the top diagram of Fig. 17, ready for the next stripping operation to be performed.

Differential Screw Design.—The use of a differential screw for very fine adjustment has well known advantages. However, these often seem to be outweighed by the difficulty of obtaining the required perfection in two co-axial threads differing minutely from each other in pitch. It is possible, however, to make up differential screws with effective pitches of only a few thousandths inch from screws of no extraordinary quality, and these can be used for very delicate adjustments of quite heavy members. Such a differential screw is shown in Fig. 18.

As shown in this illustration, a movable part of width M is to move a distance of plus or minus d relative to the fixed member, which has a width F. Let the pitch of the coarser threaded section of the screw which passes through the fixed member be designated P_t, and the pitch of the finer threaded section of the screw which passes through the movable member as P_m.

If the coarse-threaded part of the screw is at the extreme left in the fixed member, and the movable member is at the extreme right on the fine-pitch thread, then:

1. For each turn of the screw, the coarse-threaded part of the screw will move to the right a distance of P_t and the movable member will move to the left on the fine-pitch thread a distance of P_m.

2. The movable member will thus have a net movement to the right with relation to the fixed member that is equal to $P_t - P_m$.

3. Now if the movable member is to have a total net movement to the right of $2d$, then $\dfrac{2d}{P_t - P_m}$ turns of the screw will be required.

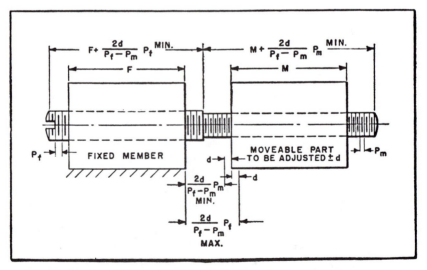

Fig. 18. Diagram of Differential Screw Showing Formulas for Calculating Minimum Lengths of Coarse- and Fine-threaded Sections.

4. For that number of turns of the screw, the coarse-pitch portion will travel through the fixed member a distance equal to $\dfrac{2d}{P_t - P_m} \times P_t$; hence, the minimum length of the coarse-pitch section must be $F + \dfrac{2d}{P_t - P_m} \times P_t$ as is indicated in Fig. 18.

5. Similarly, the movable member will travel on the fine-pitch section a distance equal to $\dfrac{2d}{P_t - P_m} \times P_m$, and the minimum length of that section will be $M + \dfrac{2d}{P_t - P_m} \times P_m$.

6. If both threaded sections are of minimum length, the maximum distance between the fixed and movable members will be equal to $\dfrac{2d}{P_t - P_m} \times P_t$, and the minimum distance will be equal to $\dfrac{2d}{P_t - P_m} \times P_m$.

Example—If the fixed and movable members are each 1 inch wide and a total range of adjustment of 1/8 inch in either direction is required, what will be the minimum length of the coarser thread, if it has 20 threads per inch, and of the finer thread, if it has 32 threads per inch? What will be the pitch of an equivalent thread that will provide the same fineness of adjustment? What will be the minimum and maximum distances between the fixed and movable members?

$$\textit{Solution—} F = 1;\ M = 1;\ d = \frac{1}{16} = 0.0625;$$

$$P_t = \frac{1}{20} = 0.05;\ P_m = \frac{1}{32} = 0.03125$$

Minimum length of coarse thread =

$$1 + \frac{2 \times 0.0625}{0.05 - 0.03125} \times 0.05$$

$$= 1 + \frac{0.125 \times 0.05}{0.01875} = 1.333 \text{ inches}$$

Minimum length of finer thread =

$$1 + \frac{2 \times 0.0625}{0.05 - 0.03125} \times 0.03125$$

$$= 1 + \frac{0.125 \times 0.03125}{0.01875} = 1.208 \text{ inches}$$

Pitch of equivalent thread =
$$0.05 - 0.03125 = 0.01875 \text{ inch}$$
Minimum distance between members = 0.208 inch
Maximum distance between members = 0.333 inch

Differential Screw Micrometer Mechanism.—A differential screw mechanism developed by the National Physical Laboratory of England to obtain the magnification of mi-

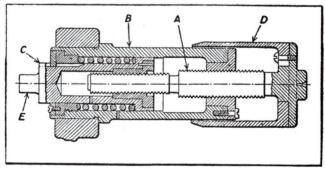

Fig. 19.   Cross-sectional View of Differential Screw Mechanism
of Micrometer.

crometer readings is shown in Fig. 19. In this design, two differential screws of relatively coarse pitch are employed to increase the accuracy of the micrometer reading. The larger screw A has 20 threads per inch, and the smaller one 25 threads per inch. Both threads are right-hand.

Screw A is engaged by a fixed nut at the right-hand end of barrel B, while the finer thread screw passes through a nut secured in sliding plunger C, the exposed end E of which forms one of the measuring anvils. A spring maintains contact between the micrometer screw and the two internal threads. The net movement of the plunger is $1/20 - 1/25 = 0.01$ inch for a complete rotation of the screw. The edge of thimble D, attached to the screw, is graduated in 100 divisions, each representing 0.0001 inch, the distance separating adjacent lines being about 0.04 inch.

The magnification with this arrangement is 400, compared with about 60 for an ordinary micrometer having 40 threads per inch and a thimble of about 1/2 inch diameter. A travel of 1 inch of the main screw moves the plunger only 0.2 inch. The total range of the instrument is, therefore, considerably reduced.

Pump with Flexible Rubber-Tube Action.—A pump designed to isolate the liquid or gas being pumped from the pump mechanism itself is a development of the Downingtown Mfg. Co., Downingtown, Pa. The operation of this pump, which is known as the Downingtown-Huber Squeegee type pump, is based upon the alternate squeezing and releasing of a rubber tube by a rocking compressor ring.

As shown in the diagrams, Fig. 20, the pump consists of six main parts. To the drive-shaft in the center is keyed an off-center rotor. This, in turn, can be keyed in any one of three positions to an adjustable eccentric. (In both diagrams the rotor is shown keyed in the central position.) The shaft, off-center rotor, and adjustable eccentric rotate as a unit in the compressor ring, and impart a rocking action to it.

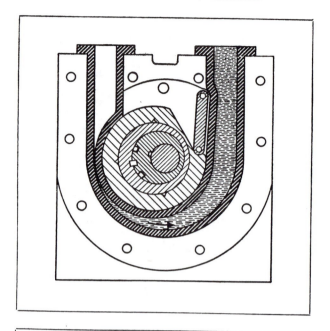

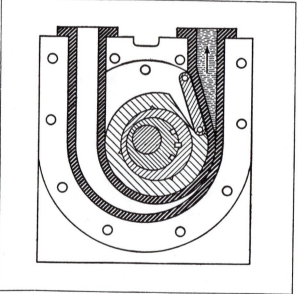

Fig. 20. Diagrams Showing Two Stages of Operation of Squeegee
Type Pump in which Rotating Eccentric Member Forces Liquid
in Rubber Tube from Inlet to Outlet.

This rocking action pushes the compressor ring out radially against the tube and continues progressively along the curved portion of the tube until it approaches the discharge end, when the tube is released and compression begins again at the intake end. Two stages of this action are shown in Fig. 20. Compression of the tube in this manner advances the liquid or gas being pumped toward the discharge side, while expansion of the tube back to its normal diameter produces a vacuum drawing more liquid or gas in from the intake side. A tube guide plate attached to the discharge side of the compressor ring prevents expansion of the tube beyond its normal diameter at this point where the tube is not surrounded by the pump housing or the compressor ring.

Thus, the liquid or gas being pumped is totally enclosed within the tube while passing through the pump. The tube itself can be made of pure gum rubber and various acid and oil resisting synthetic materials; in addition, it can be lined with synthetic materials for corrosion resistance or to prevent contamination. Solutions containing solids, whether abrasive or otherwise, cause little wear, as the inside of the tube is a smooth, continuous surface. The compressor ring merely rocks against the tube without any rubbing action, and hence causes practically no wear. A new tube is readily installed by simply fitting it inside the housing which forms its support and backing.

This pump is available in a fractional gallons per minute size which has a capacity range of from 0 to 6 gallons per hour, and develops a lift of 25 feet and a discharge pressure of 25 pounds per square inch. It weighs 3 pounds in a bronze housing, or 1 pound in a plastic housing. In the larger sizes, the capacities range from one-half gallon per minute up to 50 gallons per minute in the single-stage type, and up to 100 gallons per minute in the double-stage type.

Mechanism for Straightening Fine Wire.—Fig. 21 shows a comparatively simple machine designed for straightening coiled nickel - silver wire 1/64 inch in diameter. Straight pieces of this wire were required in lengths of 3, 6, and 12 inches. As it is difficult to straighten such short lengths, pieces 2 feet or more in length are straightened on the machine and then cut off to the required lengths. Wire in lengths as short as 2 feet, however, is generally drawn through the straightening machine twice. The end of the wire which leaves the machine last at the first pass is fed in first at the second pass. In some cases, more than two passes of the wire through the machine are necessary.

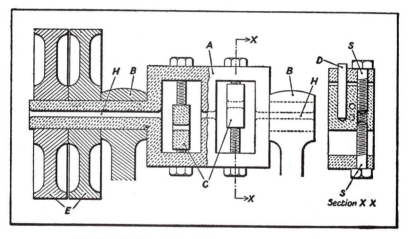

Fig. 21. Machine Designed for Straightening Fine Wire.

The machine is not complicated to make or to use. Similar designs are employed in the hosiery machine building trade for straightening the wires on which jacks are pivoted. Frame *A* of the machine is free to revolve in bearings *B*. It is shown driven by belt through the fast and loose pulleys *E*, although any other convenient form of drive can be provided. A suitable speed is about 100 R.P.M. A hole *H*, slightly larger in diameter than the wire to be straightened, is drilled through the center of the frame. Extreme accuracy in the diameter of this hole is not required, 1/8 inch being suitable for wires up to 3/32 inch in diameter.

Supported in slots in the frame A are the blocks C. These blocks are adjustable lengthwise in the slots by means of the screws S, while the dowels D prevent the blocks from turning. A hole of the same size as that drilled through the center of the frame is drilled in each of the blocks. These holes are so positioned that they can, by adjustment of the blocks, be aligned with the center hole in the frame, or they can be offset by an amount depending on the diameter of wire to be straightened.

In operation, the wire is fed into the center hole on one side of the machine until about 1/2 inch projects at the other side. This projecting piece is then gripped in a hand vise. The blocks are adjusted to positions that will cause the wire to be bent as it is pulled through the machine. The amount of this bending can best be found by trial, but it should be just sufficient to remove any kinks from the wire. The power is now applied, causing the frame to revolve, and the wire is pulled through the machine.

To prevent the wire from being scored, the ends of the holes in both the bending blocks and the frame should be provided with fairly large radii.

Mechanism for Taking Up Slack in Sprocket Chain.— A mechanism employing a weight-and-lever arrangement for taking up slack in a sprocket chain and for keeping the chain tight while it is in operation is shown in Fig. 22. On the non-movable base A is mounted a bracket B carrying a short shaft C, which supports a hinged lever P. Slidably mounted on the base is a unit D which carries the adjustable plate E. This plate supports the idler sprocket F, which is free to revolve on stud G. The function of the idler sprocket is to continuously exert sufficient pull on the sprocket chain H to keep it tight.

The cast-iron weight K, which exerts a pull in the direction indicated by arrow L, applies a pulling action to the connecting links M and N in the directions indicated by the arrows. This action transmits a compound movement to the hinged lever P, pulling back on rod Q and exerting sufficient pull on slide D to keep the chain H tight. It will be noted that the tendency of the weight to shorten the distance R as lever P pivots about the short shaft C causes the upper end of lever P to be forced to the left against the adjusting nut S on the thread at the outer end of rod Q.

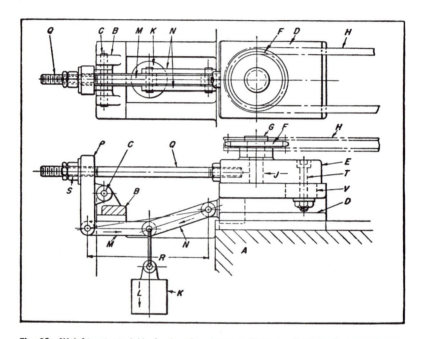

Fig. 22. Weight-actuated Mechanism Developed to Take up Slack in Sprocket Chain.

Plate E is adjustable on slide D and permits periodical adjustment to take up any major slack which may develop in the chain. Four screws of the type shown at T hold the plate in position, while the slide D is provided with an elongated slot V which permits adjustment of plate E. Adjusting nut S on rod Q permits setting the mechanism so that weight K will be in its most effective position for holding the chain under the tension required to take up the slack in the sprocket chain.

Mechanism for Operating Automatic Hoisting and Stacking Tongs.—The problem of providing more floor space to meet the requirements of companies engaged in armament work was solved in some cases by placing manufactured parts and partial assemblies into labeled barrels, boxes, or drums, and then piling up these packed containers to a height of several tiers. While this was found a satisfactory method of obtaining extra floor space, it required the use of a crane and good material-handling equipment. The quick-action automatic tongs, shown in Fig. 23, with operating mechanism designed as illustrated, were constructed especially for use in performing this work.

Referring to Fig. 23, center bar A and latches B are flame-cut from rough structural steel 3/4 inch thick. The center bar is purposely made heavy, so that it will descend by gravity when released. The four bars H are welded to the sides of the pivot bars, and the two lower cross-bars C are welded across the lower ends of bars H, as shown in the upper right-hand corner. The toothed gripping jaws D, four in number, are designed to obtain an equalizing grip on the container to be lifted. This feature permits the jaws to clamp and lift either round barrels, or containers or boxes having flat sides.

When the tongs are open, latches B are hooked over the two arc-shaped sections F that form part of center bar A. The tongs are then placed around the object to be lifted. A slight pull of the hand-chain G opens the latches and frees bars H and jaws D, so that they instantly approach each other and firmly grip any object to be handled, such as barrel K. The load is then hoisted by the crane hook and placed in the desired location.

When the crane hook is lowered, the center bar descends by gravity, as provided for by the vertical clearance slot S cut through its center. When it has descended far enough, latch hooks B are caused to snap together over the arc-shaped sections F through the pulling action of spring T

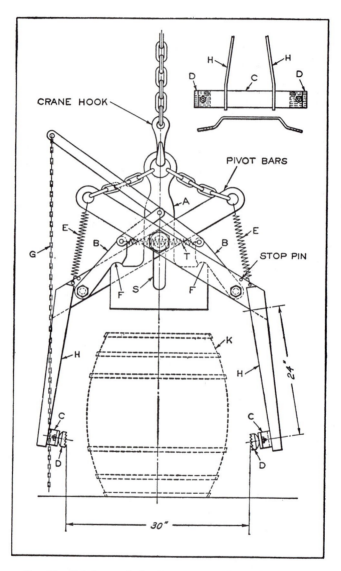

Fig. 23. Hoisting and Stacking Tongs with Mechanism for Gripping Barrel or Box when Control Chain G is Pulled and which Releases Load when Crane Hook is Lowered.

attached to their ends. When the crane hook is again raised, the latches cause the jaws to open so that the tongs are ready to be lowered for lifting the next container.

In removing packages from the tops of the stacked tiers, the operation of the device is simply reversed. The two light-tension springs E serve to prevent the jaws from collapsing or coming entirely together.

In building this equipment, no precision fits are necessary. The center block has a clearance of about 1/8 inch between the pivot bars. A separator bushing through the long vertical slot and around the pivot bolt has a similar clearance. These tongs grip securely anything that can be placed between them. Although the distance between the open jaws D is indicated as 30 inches in the illustration, this opening can be adjusted by providing several holes for the latch fulcrum bolts at different positions.

Friction Clutch for Grinding Machines Operated at High Speeds.—In Fig. 24 is shown a friction clutch that can be nicely balanced to permit operation at high speeds without perceptible vibration. The various parts of the clutch, shown in Fig. 25, are balanced separately and as a unit. The body A, Fig. 24, is made of semi-steel and finished all over. It has a bushing K and runs free when not engaged with the expanding ring B. The expanding ring is also made of semi-steel and is so shaped that it will balance itself. The raised portion at J shows the method of distributing the metal to obtain the required balance. At C is a hardened steel cone that slides along the shaft to operate the finger D. It is rounded at the front end M which acts as a cam when in contact with the point of the set-screw I. The set-screw I is screwed into a tapped hole in finger D and is locked in position by a check-nut. The round point of this screw is hardened to prevent wear.

The clutch finger D is a steel casting, shaped to assist in balancing the unit. It is pivoted about the screw F, which is threaded into and fastened to the expanding ring B. The

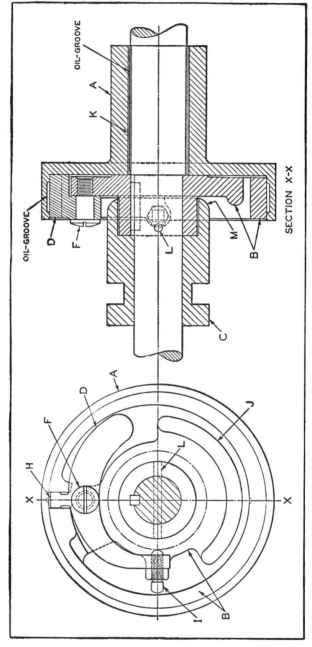

Fig. 24. Friction Clutch Designed for Use at High Speeds.

lever arm H is a part of finger D and serves to expand ring B. Ring B is keyed to the shaft and is held in position by a pin L.

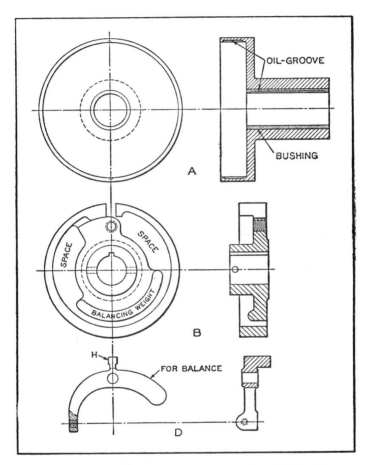

Fig. 25. Detail of Parts Comprising Clutch Shown in Fig. 24.

To engage the clutch, the operator simply slides the cone C toward the expanding finger D, bringing the cam face M into contact with the set-screw I and moving it outward, thus causing finger D to pivot about screw F. The lever arm H fitted in the slot in ring B causes the ring to expand

and grip the inner friction surface of the clutch body *A*. The expansion of ring *B* causes the entire unit to turn as one member. Cone *C* is counterbored to fit over the hub of member *B*, so as to enable the clutch cone to operate as near the center of the clutch ring as possible. All rotating parts of the clutch are lubricated with oil.

Mechanism for Controlling Cutter-Head Slide of Cam-Generating Device.—An automatic cam-generating device with a templet and follower-pin mechanism for controlling the cutter-head slide is shown in Fig. 26. One of the chief advantages of the mechanism here illustrated is the comparative ease with which the flat-plate master profile or contour templet *A* can be developed, laid out, and machined. Much work is saved as compared with developing the layout on a master cylindrical cam blank and machining the cam groove directly on the cylindrical blank.

Use of the flat-plate templet is especially advantageous in the case of slow-moving, uniform-motion cams in which the developed profile diagram of the various rise-and-fall surfaces consists of simple inclined straight lines. Development templets made from steel plate or flat ground stock are employed, as shown at *A* in the illustration, for the mechanism that controls the slide on which the cutter *B* is mounted for generating cam groove *C* in cylindrical blank *D*.

The cam blank *D* in which groove *C* is to be cut is mounted on an arbor between centers with a driving pin *E* arranged to rotate the work when the spindle *F* is rotated by turning worm-shaft *G* either with a hand-crank or a motor equipped with a reduction-gear drive. The worm on shaft *G*, meshing with the worm-wheel on shaft *F*, rotates pinion *H*, as well as the work. Pinion *H*, meshing with rack *J* on slide *K*, causes slide *K* to be traversed at right angles to the axis of shaft *F* simultaneously with the rotation of cam blank *D*.

As slide *K* moves transversely, the cam profile of templet *A* transmits longitudinal movement to slide *Q* through follower-pin *L*. The compound cutter-slides *M* and *N*, mounted

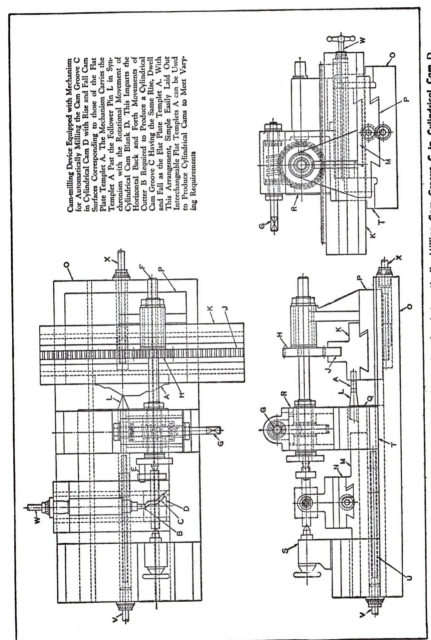

Cam-milling Device Equipped with Mechanism for Automatically Milling the Cam Groove C in Cylindrical Cam D with Rise and Fall Cam Surfaces Corresponding to those of the Flat Plate Templet A. The Mechanism Carries the Templet A Past the Follower Pin L in Synchronism with the Rotational Movement of Cylindrical Cam Blank D. This Imparts the Horizontal Back and Forth Movements of Cutter B Required to Produce a Cylindrical Cam Groove C Having the Same Rise, Dwell and Fall as the Flat Plate Templet A. With This Arrangement, Simple Easily Laid Out Interchangeable Flat Templets A can be Used to Produce Cylindrical Cams to Meet Varying Requirements

Fig. 26. Cam-milling Device Equipped with Mechanism for Automatically Milling Cam Groove C In Cylindrical Cam D Using Flat Templet A.

on slide Q, thus traverse the cutter B longitudinally to the left or right along the axis of the cam blank D in accordance with the profile or contour of the cam templet A. This simultaneous rotation of the cylindrical cam blank D and traversing movement of the rotating end-milling cutter B results in milling the cam groove C with the required rise-and-fall contours predetermined by the contour or profile of cam templet A. During the cam milling operation the follower-pin L is kept in contact with templet A by means of a cable and weight (not shown), which is attached to the end of the slide M.

Referring to the structural details of the device, base O is provided with machined ways throughout its entire length and carries the two longitudinal slides P and Q. Slide P is cast integral with the bracket bearing which supports work-spindle F, and it also carries the cross-slide K with rack J and templet A.

Slide Q, supporting follower-pin L, carries the compound slide M. Slide N is mounted on slide M and carries the spindle of cutter B. Headstock R and tailstock S are mounted on bridges T and U, respectively. Gear H, which is cast integral with a long sleeve bearing and is keyed to the work-spindle F, drives rack J secured to cross-slide K, thus imparting the traverse movement to templet A.

Compound slide M is provided with a long feed-screw V, which enables the cutter B to be set in any desired position along the surface of the work. Cross-slide N, carrying the spindle head, is provided with cross-feed screw W which enables cutter B to be adjusted radially to any desired distance from the center of the work. Screw W is also used for feeding the cutter into the work. The setting of slide P to accommodate templets of various sizes is accomplished by means of screw X.

Mechanism for Adjusting Throw or Radial Position of Block on Rotating Arm.— The mechanism here described was designed to permit a very fine, continuous adjustment of block *A*, Fig. 27, left-hand diagram, along radial arm *B* while the arm is rotating about axis *X–X*. The arrangement of the mechanism, as adopted, is shown diagrammatically in the center diagram of Fig. 27, while an alternative arrangement (considered, but not adopted) is shown in the right-hand diagram of Fig. 27.

Screw *C* is used to move the block, and is, in turn, driven by shaft *D* through bevel gears. Shaft *D* carries one wheel of differential *E*, the other wheel of which is driven by a

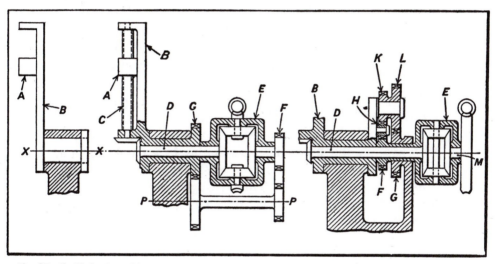

Fig. 27. (Left) Section of Mechanism that is Required to Have the Throw of Block A Adjusted while Arm B is Rotating. (Center) Sectional View of Mechanism Designed for Adjusting Throw of Block A. (Right) Sectional View of Alternative Throw-adjusting Arrangement.

train of gears from wheel *G*, which is fixed to arm *B*. The intermediate shaft rotates about a fixed axis *P–P*.

The ratio of the gear train beginning with driver *G* and ending with driven gear *F* is unity, and because of the reverse idler, shown at the right-hand end of the gear train, gears *F* and *G* rotate in opposite directions. Thus, when the housing of the differential is stationary, shaft *D* rotates in the same direction and at the same speed as arm *B*, and the position of block *A* on the arm is fixed. If, however, the differential housing is rotated, then its motion is added to, or subtracted from, that of shaft D, which, consequently, rotates relative to arm B and thus moves block A inward or outward. The block is, therefore, controlled by the worm and wheel drive to the differential housing. The drive to the mechanism and to arm B is through gear G.

In the alternative arrangement, shown in the right-hand diagram of Fig. 27, the ratio of the epicyclic train F to G must be 1 to 2, in order that the differential housing may be rotated in the same direction as, but at half the speed of, arm B. This requires the number of teeth in G, multiplied by the number of teeth in K, divided by the product of the number of teeth in L, and the number of teeth in F to be equivalent to the ratio 1 to 2, and also necessitates the use of intermediate gear H.

When shaft M is stationary, the rotation of the differential housing E causes shaft D to rotate in the same direction and at double the speed, so that it rotates in synchronism with arm B, and block A thus remains in a fixed position. When shaft M is rotated, the motion is transmitted to screw C and the block is moved. This arrangement, however, was not adopted because the drive to the arm was more difficult to arrange.

INDEX

PAGE

Adjustment mechanism for radial position of block on rotating arm _____520

Amplifying mechanism for precision measuring instruments__491

Angular movement, crank and link mechanisms for increasing _____ 251, 254

    link mechanism for increasing_____254

Automatic cam generating device, mechanism for controlling cutter head slide of_____518

Automatic feed, for a drill press_____360

    for centerless grinder _____379

    for placing drawn shell in punching die_____420

    mechanism for retarding _____440

Automatic hoisting and stacking tongs, mechanism for operating _____513

Automatic shifting, mechanism for_____370

Automatic stop for roll driving mechanism_____100

Automatic stroke length variation, mechanism for_____222

Automatic stroke reversing mechanism_____145

Automatic transfer and feeding mechanism_____434

Automatic variable lift cam mechanism_____242

Automatic work-locating mechanism for milling machine 132, 135

Automatic work-reversing and transferring mechanism_____415

Automatic wrapping machine, mechanism for operating slides alternately _____417

Back-gear shifter, automatic_____325

Ball and socket mechanism for operating sleeve valve_____292

Ball bearings, for rotation and axial reciprocation_____297

Barrel, hoisting and stacking tongs for_____513

Barrel hopper for intricate shapes_____454

Bearings, ball, for rotation and axial reciprocation_____297

Belt drive, quick-change two-speed_____311

Box-nailing machine, feeding mechanism for_____403

Brake and circuit breaker for milling machine spindle_____130

Brake, clutch equipment for quick-acting_____102

Cable-winding machine, reversing mechanism for_____148

Cam, designed to operate on alternate revolutions_____ 11

    for changing position of reciprocating motion_____173

PAGE

Cam, for intermittent mechanism... 31

for intermittent motion.. 33

for longer stroke without larger operating space.................... 1

for net-making machine, compound,.. 7

for reciprocating motion 1

for shaft speed control.. 16

for variable reciprocating movement..................................... 214

for variable stroke mechanism.. 234

high-lift, low-pressure-angle 10

multiple, and lever mechanism... 5

Cam actuated intermittent worm drive mechanism............... 29

Cam and crank for feeding mechanism................................... 276

Cam controlled variable speed drive...................................... 316

Cam drive, variable-lift mechanism for.................................. 226

Cam generating device, mechanism for controlling cutter head
slide 518

Cam mechanism, for automatic variable lift........................ 242

for indexing work-table and feeding drill........................... 17

for intermittent rotary motion.. 40

for tracing complex path.. 7

to provide reciprocating motion with locked rest periods........186

with variable quick-drop adjustment..................................... 4

Cam-operated toggle and lever mechanism............................. 14

Cam speeds, mechanism for changing...................................... 327

Carriage feeding mechanism adjustable from 1 to 50 microns
per revolution 351

Carriage locking and releasing mechanism......................... 122

Centerboard hopper, design of.. 446

rotary 450

Centerless grinder, automatic magazine feed for............... 379

Centerless oscillating motion, mechanism for............ 246, 250

Chain-driven table, reversing mechanism for...................... 137

Chain-making machine, dial transfer mechanism for.......... 410

Chart recording pen, reciprocating mechanism for............. 172

Chobert riveting machine, operating mechanism............... 385

Clamping mechanism and release for tailstock center........ 127

Clutch, for quick-acting brake.. 102

friction, for grinding machines... 516

reversing, positive type 158

PAGE

Clutch mechanism for overload release 93
Collet operated mechanism and spindle control for
   screw machines 365
Compound cam mechanism 7
Cone pulley with high-ratio epicyclic reduction gearing 305
Constant-speed pull-roll for winding metal strips 329
Contour milling of grooves, follower mechanism for 480
Control, external, for mechanism within rotating member 295
Control lever, selective timing mechanism for actuating 487
Crank and cam for feeding mechanism 276
Crank and link, to increase angular movement of shaft 251, 254
   to reduce oscillating motion 256
Crank motion, with rest period for reciprocating slide 206, 207
   space linkage for transmitting oscillating 263
Crankpin, adjustable, for variable-stroke motion 229, 231
Crank to provide irregular reciprocating motion 224
Creasing flexible material, rotating mechanism for 483
Crimping mechanism for ends of heavy paper cylinder 484
Cutting speed, mechanism for maintaining constant 333

Dial feed and radially positioned multiple punch mechanism 430
Dial transfer mechanism for chain-making machine 410
Differential drive, single chain 177
Differential screw design 503
Differential screw micrometer mechanism 506
Disk saw, straight line reciprocating mechanism for 163
Drill, feeding and work-table indexing, cam and gear for 17
   pneumatic, reciprocating and rotary motion for 285
Drilling machine, special, indexing mechanism for 363
Drill press, automatic feeding mechanism for 360
Drive, belt, two-speed quick-change 311
   for intermittently rotated and locked shafts 52
   intermittent friction 54
   intermitted worm 29
   light, safety release mechanisms for 104
   single chain differential 177
   unit with overload slip mechanism 97

Ejector for formed parts, mechanism for 438

PAGE

Feed, mechanism adjustable from 1 to 50 microns
    per revolution _____351
  mechanism for obtaining coarse and fine_____350
  mechanism for retarding automatic_____440
  variable, arrangement for automatic wheel dressing device_358
Feeding and transfer mechanism, automatic_____434
Feeding mechanism, adjustable from 1 to 50 microns
    per revolution _____351
  adjustable, for strip material_____425
  automatic, for centerless grinding_____379
  automatic, for drill press_____360
  automatic, for placing drawn shell in punching die_____420
  automatic, mechanism for retarding_____440
  for box-nailing machine _____403
  for coarse and fine feed_____350
  for filling container and applying covers_____387
  for retarding automatic_____440
  for screw milling machine_____357
  for thin plates, suction cup equipped_____405
  for wooden pegs _____393
  intermittent_____401
  intermittent, to operate two slides from one cam_____395
  quick-acting intermittent _____ 56
  rapid-motion short-stroke, for wire_____399
  table _____428
Feeding washer-shaped blanks to die, mechanism for_____422
Feed-screw dials, variable stroke mechanism for graduating_222
Feed screw, reversing mechanism for_____150
Feed trip mechanism for lathe overloads_____ 98
Film projector indexing mechanism_____ 38
Flexible material, rotating mechanism for creasing_____483
Follower mechanism for contour milling of grooves_____480
Friction clutch for grinding machines_____516
Friction drive, for irregular intermittent motion_____ 54
  for stepless speed variation_____313
  mechanism for rewinding roll_____332
Furnace door and work plate, link mechanism for operating_499

PAGE

Gear and cam mechanism, for indexing work-table and
    feeding drill _____ 17
    for intermittent rotary motion_____ 40
Gear and link mechanism for synchronized motion_____201
Gear and rack, for intermittent rotary motion_____ 58
    for uniform reciprocating motion_____175
Gear and star-wheel indexing device_____ 44
Gear drive, for motion reversal and dwell_____141
    intermittent _____ 23
Gear for indexing mechanism, modified helical_____ 27
Gearing, high ratio epicyclic reduction_____305
    high speed intermittent_____ 45
Geneva mechanism, for indexing at uniform rotational speed__ 80
    for indexing small film projector_____ 38
    for intermittent reversible rotation_____ 83
    for precise intermittent indexing_____ 77
Graduating feed-screw dials, variable stroke mechanism for___222
Grinding machine, high speed friction clutch for_____516
    variable feed arrangement for automatic wheel dressing____358
Guide rollers, mechanism to obtain uniform adjustment of____ 497

Hoisting and stacking tongs, mechanism for operating_____513
Hopper, barrel type for intricate shapes_____454
    centerboard design of _____446
    circular blade type selector for_____463
    cup or can selector for_____467
    for feeding long rods_____461
    for feeding rivets and similar shaped parts_____454
    for feeding small cylindrical parts_____383
    magnetic type _____459
    magnetic type selector for_____464
    paddle wheel type _____454
    rotary centerboard _____450
    selector for flat hooked parts_____466
    selector for reversing position of parts_____467
    selector for shallow drawn parts_____467
    spring actuated pin-type selector mechanism for_____461
    tray type for comparatively low production_____456

PAGE

Hopper, tube and rotary types of _____452
    vibratory type _____458
Hour-glass cam for intermittent mechanism_____ 31
Hydraulic drive, precision speed control mechanism for_____345

Indexing mechanism, for film projector_____ 38
    Geneva motion for uniform rotational speed_____ 80
    Geneva wheel for intermittent_____ 77
    on special drilling machine_____363
    self-locking, lever operated _____ 73
    simple gear and star-wheel device for_____ 44
    with modified helical gear_____ 27
Indexing work-table and feeding drill, cam for_____ 17
Intermittent drive, spur gear mechanism for_____ 23
Intermittent feeding mechanism _____401
    quick-acting _____ 56
    to operate two slides_____395
Intermittent gear mechanism, for high speed operation_____ 45
    for rotation with rest period_____ 20
    for uniform reciprocating motion_____175
Intermittent indexing, Geneva mechanism for precise_____ 77
    Geneva mechanism for uniform speed_____ 80
Intermittent motion, cam transmitted_____ 11
    friction drive for irregular_____ 54
    from a uniformly reciprocating slide_____ 33
    from uniform motion to reciprocating_____ 35
    mechanism for reciprocating _____244
    ratchet mechanism _____63, 65
    timing interval changing _____ 60
Intermittent rotary motion, adjustable_____ 25
    from hour-glass or cylindrical cam_____ 31
    gear and cam mechanism for_____ 40
    Geneva wheel mechanism for reversible_____ 83
    irregular _____ 47
    rack and gear assembly for_____ 58
    smooth stopping mechanism for_____ 49
    with locking device for shafts_____ 52
    worm drive mechanism, cam actuated _____ 29

PAGE

Jaw vise, mechanism for operating floating................................125
Joints, pivotal, for special purposes................................476

Lever and link motion to operate slide, locking................................109
Lever and spring mechanism to variably increase tension
    on slide 481
Lever device for machine overload relief................................ 95
Lever, mechanism for tripping rotating................................ 89
Lever mechanism, for adjusting arc-shaped levers................................129
    multiple cam 5
    to clamp bolts and operate locking pin................................110
    to operate ratchet with special coil adjustment................................ 67
Linear and rotary motion, foot-operated mechanism for................................293
Linkage mechanism, for operating slide and plunger................................166
    for increasing angular movement................................254
Locating work in milling machines, mechanism for............132, 135
Locking and releasing mechanism for traveling carriage................................122
Locking lever, for two positions................................109
Locking mechanism, for locking pin and clamping bolts in
    one movement 110
    non-reversing rack-and-pinion 120
    positive, for rack-and-pinion motion................................114
    rack-and-pinion for vise 115

Magazine feed, automatic, for centerless grinder................................379
Magnetic hopper, design of................................459
Measuring instruments, amplifying mechanism for................................491
Micrometer, differential screw mechanism for................................506
Milling machine, work-locating mechanism for............132, 135
    brake and circuit breaker for spindle of................................130
    controlled infeed mechanism for screw milling................................357
    follower mechanism to contour mill grooves for................................480
    mechanism for coarse and fine feed................................350
    self-locking indexing mechanism for................................ 73
    shaping attachment, oscillating mechanism for................................277
Movement amplifying mechanism for measuring instruments..491
Multiple punches and dial feed, mechanism for operating........430

PAGE

Net-making machine, cam for _____ 7

Oscillating cam, adjustable _____279
Oscillating crank motion, transmitting by space linkage
    mechanism _____263
Oscillating mechanism for milling machine shaping
    attachment _____277
Oscillating motion, centerless _____ 246, 250
    crank and link mechanisms for increasing_____251, 254
    for feeding mechanism, crank and cam for_____276
    link mechanism for increasing_____254
    mechanism for applying to driven shafts_____273
    mechanism for doubling _____257
    mechanism for imparting to adjustable arm_____269
    mechanism for reducing _____256
    mechanism for variable intermittent_____244
    transmitted by universal joints_____260
    transmitted from vertical to horizontal plane_____265
    simple mechanism for producing_____271
    vertical, mechanism to convert from horizontal_____263
    with interrupting control, mechanism for_____275
Overload feed-trip mechanism for lathes_____ 98
Overload release, clutch mechanism for_____ 93
Overload relief device for machine protection_____ 95
Overload slip mechanism, drive unit with_____ 97

Paper, mechanism designed to crimp ends of_____484
Pivotal joints for special purposes_____476
Planing convex surfaces, mechanism for_____250
Positioning mechanism, for block on rotating arm_____520
    for switches _____478
Press, mechanism for automatic stopping when stock fails
    to feed _____442
    variable stroke toggle lever mechanism for_____234
Press ram, hydraulic, mechanical equalizer for_____473
Press ram mechanism, for uniform pressure during stroke___471
Press stopping mechanism to operate when stock fails to feed__442
Pressure applied to a rod, mechanism to regulate_____240
Pressure pad, cam mechanism for operating_____ 14

PAGE

Pull roll for winding metal strips, constant speed_____329
Pump, flexible rubber tube action for_____507
Punching die, automatic feed for_____420
Punch press, mechanism for ejecting formed parts from_____438

Quick-drop adjustment cam, variable_____ 4

Rack-and-pinion, self-locking non-reversing mechanism_____120
    non-reversible motion _____117
    positive lock for _____114
    with locking motion _____115
Ratchet and pawl mechanism, for conveyor bands or belts_____ 67
Ratchet mechanism, double action reversing_____ 75
    intermittent motion _____ 63
    noiseless _____ 69
    reversing remotely controlled _____ 68
    to convert reciprocating to rotary motion_____ 71
    to prevent reverse rotation of shaft_____ 69
    with idle period _____ 65
Reciprocating and rotary motion for pneumatic drill_____285
Reciprocating and rotary motion, mechanism for combined____290
Reciprocating mechanism, double, with displaced operating
    positions _____210
    for chart recording pen_____ 172
    straight line for disk saw_____163
    with variable point of reversal_____152
Reciprocating motion, adjustable, cam operated lever for_____214
    automatic_____145
    ball bearings for _____297
    conversion of rotary to constant velocity_____183
    for two slides, crank mechanism for_____198, 203
    for wire-forming machine_____171
    gear and link mechanism for_____201
    intermittent, from uniform motion_____ 35
    irregular, crank-operated _____224
    link mechanism for _____201
    linkage mechanisms for _____166
    long stroke for guiding wire on reel_____169

PAGE

Reciprocating motion, mechanism and lock for............173
  mechanism to obtain from rotary...........177, 182, 183, 188, 192
  rack and intermittent gear for..............................175
  straight line 164, 165
  swivel joint mechanism for................................195
  to continuous rotary motion, ratchet for converting............ 71
  variable 219, 236
  variable, cam mechanism for...............................242
  variable intermittent, mechanism for......................244
  variable stroke, cam mechanism for.................226, 234
  variable stroke, mechanism for..................229, 231
  with positively locked rest periods...............186, 192
Reciprocating shaft with partial rotation, mechanism for........290
Reciprocating slide, crank motion with rest period for....206, 207
  hand control for 373
  mechanism for varying speed of............................215
Recording pen, reciprocating mechanism for................172
Retarding mechanism for automatic feeding device...............440
Reversing, automatic stroke 145
Reversing and transferring work, mechanism for...............415
Reversing clutch, positive type..............................158
Reversing device, with variable reversal points..............152
Reversing driving shaft, mechanism for producing
    speed change by...301
Reversing mechanism, for cable winding machine...............148
  for carriage traversing screw.............................143
  for chain-driven table 137
  for feed screw 150
  ratchet with remote control............................... 68
  with dwell at each reversal...............................141
Reversing ratchet movement, double-action................... 75
Reversing transfer mechanism 408
Rivets, hopper for feeding....................................454
  tubular, mechanism for inserting and heading...............385
Roll driving mechanism, automatic stop for if material
    breaks 100
Rollers, guide, mechanism to obtain uniform adjustment of....497
Roll rewinding mechanism, friction drive.................... 332

PAGE

Rotary and linear motion, foot-operated mechanism for............293
Rotary and reciprocating motion, combined, mechanism for......290
Rotary and tube types of hoppers..452
Rotary motion, adjustable intermittent..................................... 25
    intermittent, rack and gear for................................... 58
    irregular intermittent 47
    irregular, mechanism for obtaining..................................337
    mechanism for reversing 143
    mechanism to convert to reciprocating......177, 182, 183, 188, 192
    ratchet to convert reciprocating motion to..................... 71
    two-gear speed reduction mechanism for uniform.............304
Rotary or oscillating motion, mechanism for applying to
    shafts 273
Rotating and traversing movement, mechanism for producing..282
Rotating mechanism, for creasing flexible material................483
    intermittent, for smooth operation................................. 49
Rotating member, external control for mechanism within........295

Safety catch and stop, for controlling rotation of shaft............377
Safety mechanisms for light drives on special machines............104
Screw design, differential..503
Screw machines, spindle control and collet operating mecha-
    nism for 365
Screw micrometer mechanism, differential............................506
Screw milling machine, controlled infeed mechanism for..........357
Selective timing mechanism for actuating a control lever........487
Selector, for cups or cans, hopper....................................467
    for flat hooked parts, hopper......................................466
    for hoppers to reverse position of parts..........................467
    for shallow drawn parts, hopper..................................467
    magnetic type for hoppers..464
    mechanism, design of for hoppers................................461
    mechanism for hoppers, circular blade type....................463
Shaft indexing gear.. 27
Shaft, intermittently rotated and locked............................ 52
    mechanism for advancing gear driven............................ 335
    mechanism for applying rotary or oscillating motions to......273
    mechanism for starting, stopping, changing speed, and re-
    versing 318

PAGE

Shaft, mechanism for varying speed and direction of rotation...320

    rotating reversing mechanism ...143

    rotation, stop with safety clutch for controlling...377

    speed control, cam for...16

    two-way stop for angular movement of...88

Sheet transfer mechanism...413

Shifter, automatic back-gear...325

Shifting, automatic, mechanism for...370

Slide and plunger, linkage mechanisms for...166

Slide, mechanism for varying speed of...215

Slides, mechanism for alternately operating two...417

    mechanism to operate...198, 203

Sliding block mechanisms...188

Speed changing mechanism, cam and worm...320

    starting, stopping, and reversing...318

    with reversing drive shaft...301

Speed control, automatic variable...333

    mechanism for maintaining constant cutting speed...333

Speed reducer, cone pulley and epicyclic reduction gear for...305

    mechanism for ...327

    two-gear mechanism for...304

    wabble gear mechanism for...323

    stepless, friction drive for...313

Spindle brake and circuit breaker, for milling machine...130

Spindle control and collet operated mechanism for screw machines ...365

Spring winding mechanism, lever operated...475

Sprocket chain, mechanism for taking up slack in...511

Stop, two-way, for angular movement of shaft...88

    with safety catch for controlling rotation of shaft...377

Straightening mechanism for fine wire...509

Straight line reciprocating mechanism for disk saw...163

Straight line reciprocating motion, mechanism for...164, 165

Straw-baling press, double toggle lever mechanism for...272

    oscillating motion for...271

Strip feeding device, adjustable...425

Stripper mechanism for wire forming...501

Stroke length, mechanism for controlling...240

PAGE

Suction cup and mechanism for picking up and feeding thin
  plates 405
Switch positioning mechanism 478
Swivel joint mechanism to operate slide..............................195
Synchronous motion, mechanism for....................................336

Table feeding, mechanism for..428
Tailstock center, releasing and clamping mechanism for........127
Taking-up mechanism for slack in sprocket chain..................511
Tension on slide, mechanism to variably increase..................481
Timing interval, intermittent motion for changing.............. 60
Timing mechanism, selective, for actuating a control lever......487
Toggle and lever mechanism, cam operated............................ 14
  for operating baling presses..272
Toggle lever mechanism, variable stroke for presses..............234
Toggle lever press, variable stroke mechanism for..................227
Transfer and feeding mechanism, automatic..........................434
Transfer mechanism, dial, for chain-making machine..............410
  for sheets 413
  reversing 408
Transmission, variable speed..314
Traversing and rotating movement, mechanism for producing..282
Tray type hopper for comparatively low production..................456
Tripping mechanism for lathe feed overloads.......................... 98
  for rotating lever.. 89
  rotating shaft operated..86, 87
Tube and rotary types of hoppers..452
Twisting-wire, mechanism to vary pitch of twist....................341

Universal joints to transmit oscillating motion......................260

Valve gear, mechanism for pneumatic drill............................285
Valve, sleeve type, ball and socket mechanism for..................292
Variable quick drop adjustment cam...................................... 4
Variable speed drive, cam controlled....................................316
Variable speed, mechanism to control precision of................345
Variable speed transmission..314
Vibratory hopper 458
Vise with locking motion, rack-and-pinion operated..............115

PAGE

Wabble gear mechanism for speed reducer_____323

Washer-shaped blanks, mechanism for feeding concave_____422

Wheel dressing device, automatic, variable feed arrangement
    for _____358

Winding springs, lever mechanism for_____475

Wire crimping machine, hand control for reciprocating slides 373

Wire feeding, mechanism for corrugated_____395

Wire feeding mechanism for rapid-motion short stroke_____399

Wire-forming machine, lever and spring arrangement to vari-
    ably increase tension on slide_____481

    mechanism for controlling slack in_____335

    reciprocating motion for_____171

    stripper mechanism for_____501

Wire stitching carriage, locking and releasing mechanism_____122

Wire, straightening mechanism for fine_____509

Wire twisting, mechanism to vary pitch of twist_____341

Worm drive mechanism, intermittent_____ 29

Wrapping machine, mechanism for operating slides alter-
    nately _____417